The Elements of Bacteriological Technique

by J. W. H. Eyre

Copyright © 7/14/2015
Jefferson Publication

ISBN-13: 978-1515074519

Printed in the United States of America

All rights reserved. No part of this book may be reprinted or reproduced or utilized in any form or by any electronic, mechanical, or other means, now known or hereafter invented, including photocopying and recording, or in any form of storage or retrieval system, without prior permission in writing from the publisher.'

Contents

BACTERIOLOGICAL TECHNIQUE. 4
I. LABORATORY REGULATIONS. 4
    Guy's Hospital. 4
II. GLASS APPARATUS IN COMMON USE. 5
III. METHODS OF STERILISATION. 18
    STERILISING AGENTS. 19
IV. THE MICROSCOPE. 32
    FOOTNOTES: 44
V. MICROSCOPICAL EXAMINATION OF BACTERIA AND OTHER MICRO-FUNGI. 44
    APPARATUS AND REAGENTS USED IN ORDINARY MICROSCOPICAL EXAMINATION. ... 44
VI. STAINING METHODS. 53
VII. METHODS OF DEMONSTRATING BACTERIA IN TISSUES. 64
VIII. CLASSIFICATION OF FUNGI. 71
IX. SCHIZOMYCETES. 74
X. NUTRIENT MEDIA. 81
    FOOTNOTES: 90
XI. CULTURE MEDIA. 90
    ORDINARY OR STOCK MEDIA. 90
    FOOTNOTES: 98
XII. SPECIAL MEDIA. 98
XIII. INCUBATORS. 115
    FOOTNOTES: 118
XIV. METHODS OF CULTIVATION. 118
    FOOTNOTES: 133
XV. METHODS OF ISOLATION. 133
XVI. METHODS OF IDENTIFICATION AND STUDY. 137
    FOOTNOTES: 170
XVII. EXPERIMENTAL INOCULATION OF ANIMALS. 171
    FOOTNOTES: 189
XVIII. THE STUDY OF EXPERIMENTAL INFECTIONS DURING LIFE. 189
XIX. POST-MORTEM EXAMINATIONS OF EXPERIMENTAL ANIMALS. 199
XX. THE STUDY OF THE PATHOGENIC BACTERIA. 205
    FOOTNOTES: 207
XXI. BACTERIOLOGICAL ANALYSES. 207
APPENDIX. 242
    METRIC AND IMPERIAL SYSTEMS OF WEIGHTS AND MEASURES. 242

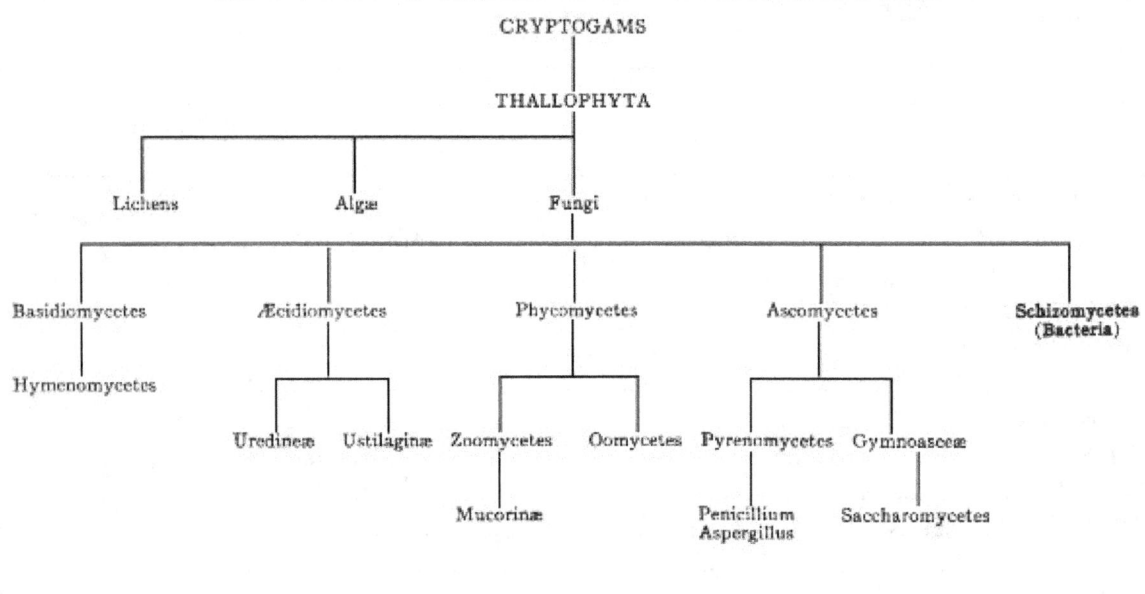

BACTERIOLOGICAL TECHNIQUE.

I. LABORATORY REGULATIONS.

The following regulations are laid down for observance in the Bacteriological Laboratories under the direction of the author. Similar regulations should be enforced in all laboratories where pathogenic bacteria are studied.

Guy's Hospital.

BACTERIOLOGICAL DEPARTMENT.
HANDLING OF INFECTIVE MATERIALS.
  The following Regulations have been drawn up in the interest of those working in the Laboratory as well as the public at large, and will be strictly enforced.
  Their object is to avoid the dangers of infection which may arise from neglect of necessary precautions or from carelessness.
  Everyone must note that by neglecting the general rules laid down he not only runs grave risk himself, but is a danger to others.
REGULATIONS.
  1. Each worker must wear a gown or overall, provided at his own expense, which must be kept in the Laboratory.
  2. The hands must be disinfected with lysol 2 per cent. solution, carbolic acid 5 per cent. solution, or corrosive sublimate 1 per mille solution, after dealing with infectious material, and **before using towels**.
  3. On no account must Laboratory towels or dusters be used for wiping up infectious material, and if such towels or dusters do become soiled, they must be immediately sterilised by boiling.
  4. Special pails containing disinfectant are provided to receive any waste material, and nothing must be thrown on the floor.
  5. All instruments must be flamed, boiled, or otherwise disinfected immediately after use.
  6. Labels must be moistened with water, and not by the mouth.

7. All disused cover-glasses, slides, and pipettes after use in handling infectious material, etc., must be placed in 2 per cent. lysol solution. A vessel is supplied on each bench for this purpose.

8. All plate and tube cultures of pathogenic organisms when done with, must be placed for immediate disinfection in the boxes provided for the purpose.

9. No fluids are to be discharged into sinks or drains unless previously disinfected.

10. Animals are to be dissected only after being nailed out on the wooden boards, and their skin thoroughly washed with disinfectant solution.

11. Immediately after the post-mortem examination is completed each cadaver must be placed in the zinc animal-box—*without removing the carcase from the post-mortem board*—and the cover of the box replaced, ready for carriage to the destructor.

12. Dead animals, when done with, are cremated in the destructor, and the laboratory attendant must be notified when the bodies are ready for cremation.

13. None of the workers in the laboratory are allowed to enter the animal houses unless accompanied by the special attendant in charge, who must scrupulously observe the same directions regarding personal disinfection as the workers in the laboratories.

14. No cultures are to be taken out of the laboratory without the permission of the head of the Department.

15. All accidents, such as spilling infected material, cutting or pricking the fingers, must be at once reported to the bacteriologist in charge.

II. GLASS APPARATUS IN COMMON USE.

The equipment of the bacteriological laboratory, so far as the glass apparatus is concerned, differs but little from that of a chemical laboratory, and the cleanliness of the apparatus is equally important. The glassware comprised in the following list, in addition to being clean, must be stored in a sterile or germ-free condition.

Test-tubes.—It is convenient to keep several sizes of test-tubes in stock, to meet special requirements, viz.:

1. 18×1.5 cm., to contain media for ordinary tube cultivations.
2. 18×1.3 cm., to contain media used for pouring plate cultivations, and also for holding sterile "swabs."
3. 18×2 cm., to contain wedges of potato, beetroot, or other vegetable media.
4. 13×1.5 cm., to contain inspissated blood-serum.

The tubes should be made from the best German potash glass, "blue-lined," stout and heavy, with the edge of the mouth of the tube *slightly* turned over, but not to such an extent as to form a definite rim. (Cost about $1.50, or 6 shillings per gross.) Such tubes are expensive it is true, but they are sufficiently stout to resist rough handling, do not usually break if accidentally allowed to drop (a point of some moment when dealing with cultures of pathogenic bacteria), can be cleaned, sterilised, and used over and over again, and by their length of life fully justify their initial expense.

A point be noted is that the manufacturers rarely turn out such tubes as these absolutely uniform in calibre, and a batch of 18 by 1.5 cm. tubes usually contains such extreme sizes as 18 by 2 cm. and 18 by 1.3 cm. Consequently, if a set of standard tubes is kept for comparison or callipers are used each new supply of so-called 18 by 1.5 cm. tubes may be easily sorted out into these three sizes, and so simplify ordering.

5. 5×0.7 cm., for use in the inverted position inside the tubes containing carbohydrate media, as gas-collecting tubes.

These tubes, "unrimmed," may be of common thin glass as less than two per cent. are fit for use a second time.

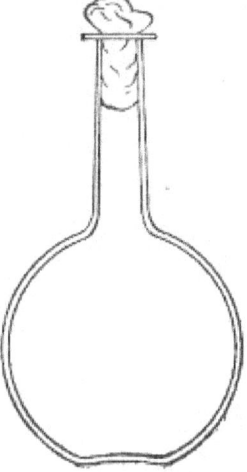

Fig. 1.—Bohemian flask.

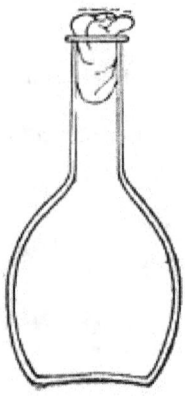

Fig. 2.—Pear-shaped flask.

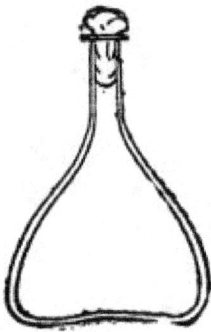

Fig. 3.—Erlenmeyer flask (narrow neck).

Bohemian Flasks (Fig. 1).—These are the ordinary flasks of the chemical laboratory. A good variety, ranging in capacity from 250 to 3000 c.c., should be kept on hand. A modified form, known as the "pear-shaped" (Fig. 2), is preferable for the smaller sizes—*i. e.*, 250 and 500 c.c.

Erlenmeyer's Flasks (Fig. 3).—Erlenmeyer's flasks of 75, 100, and 250 c.c. capacity are extremely useful. For use as culture flasks care should be taken to select only such as have a narrow neck of about 2 cm. in length.

Kolle's Culture Flasks (Fig. 4).—These thin, flat flasks (to contain agar or gelatine, which is allowed to solidify in a layer on one side) are extremely useful on account of the large nutrient surface available for growth. A surface cultivation in one of these will yield as much growth as ten or twelve "oblique" tube cultures. The wide mouth, however, is a disadvantage, and for many purposes thin, flat culture bottles known as **Roux's bottles** (Fig. 5) are to be preferred.

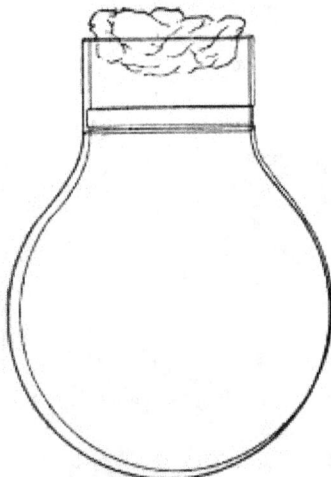

Fig. 4.—Kolle's culture flask.

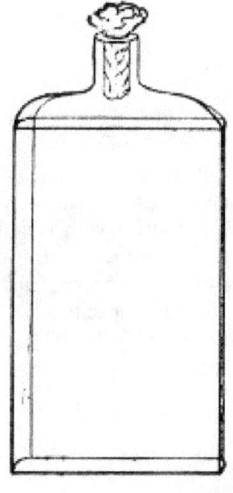

Fig. 5.—Roux's culture bottle.

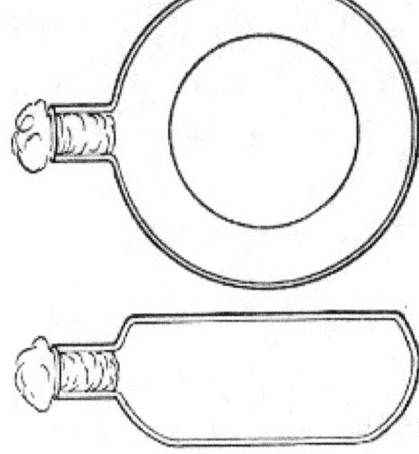

Fig. 6.—Guy's culture bottle.

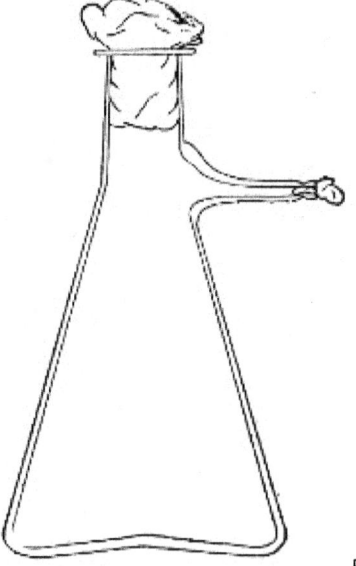

Fig. 7.—Filter flask.

An even more convenient pattern is that used in the author's laboratory (Fig. 6), as owing to the greater depth of medium which it is possible to obtain in these flasks an exceedingly luxuriant growth is possible; the narrow neck reduces the chance of accidental contamination to a minimum and the general shape permits the flasks to be stacked one upon the other.

Filter Flasks or Kitasato's Serum Flasks (Fig. 7).—Various sizes, from 250 to 2000 c.c. capacity. These must be of stout glass, to resist the pressure to which they are subjected, but at the same time must be thoroughly well annealed, in order to withstand the temperature necessary for sterilisation.

All flasks should be either of Jena glass or the almost equally well-known Resistance or R glass, the extra initial expense being justified by the comparative immunity of the glass from breakage.

Petri's Dishes or "Plates" (Fig. 8, *a*).—These have now completely replaced the rectangular sheets of glass introduced by Koch for the plate method of cultivation. Each "plate" consists of a pair of circular discs of glass with sharply upturned edges, thus forming shallow dishes, one of slightly greater diameter than the other, and so, when inverted, forming a cover or cap for the smaller. Plates having an outside diameter of 10 cm. and a height of 1.5 cm. are the most generally useful. A batch of eighteen such plates is sterilised and stored in a cylindrical copper box (30 cm. high by 12 cm. diameter) provided with a "pull-off" lid. Inside each box is a copper stirrup with a circular bottom, upon which the plates rest, and by means of which each can be raised in turn to the mouth of the box (Fig. 9) for removal.

Capsules (Fig. 8, *b* and *c*).—These are Petri's dishes of smaller diameter but greater depth than those termed plates. Two sizes will be found especially useful—viz., 4 cm. diameter by 2 cm. high, capacity about 14 c.c.; and 5 cm. diameter by 2 cm. high, capacity about 25 c.c. These are stored in copper cylinders of similar construction to those used for plates, but measuring 20 by 6 cm. and 20 by 7 cm., respectively.

Graduated Pipettes.—Several varieties of these are required, viz.:
1. Pipettes of 1 c.c. capacity graduated in 0.1 c.c.
2. Pipettes of 1 c.c. capacity graduated in 0.01 c.c. (Fig. 10, *a*).

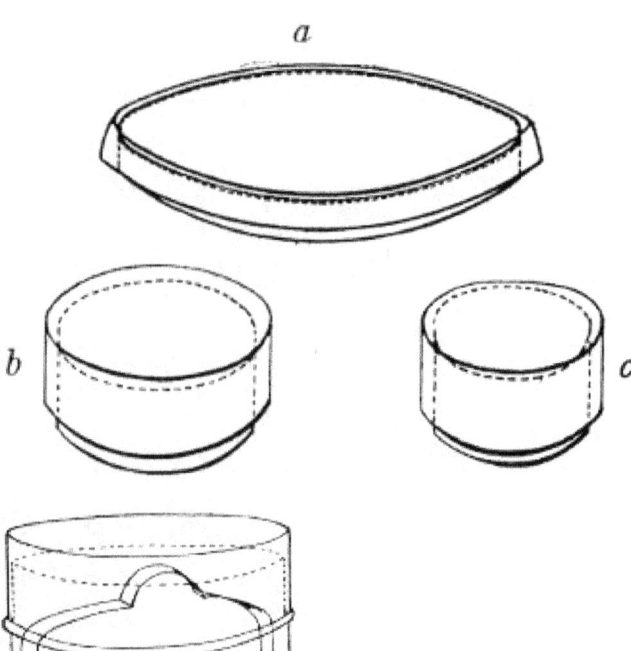

Fig. 8.—Petri dish (a), and capsules (b, c).

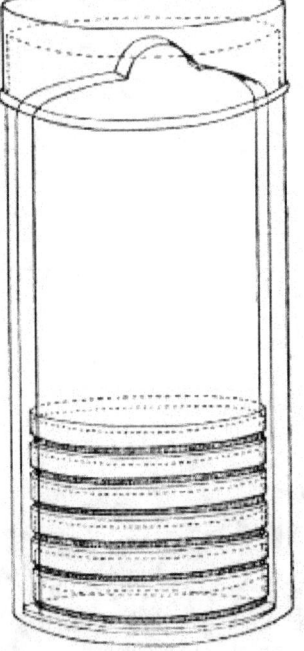

Fig. 9.—Plate box with stirrup.

3. Pipettes of 10 c.c. capacity graduated in 0.1 c.c. (Fig. 10, *b*).

These should be about 30 cm. in length (1 and 2 of fairly narrow bore), graduated to the extreme point, and having at least a 10 cm. length of clear space between the first graduation and the upper end; the open mouth should be plugged with cotton-wool. Each variety should be sterilised and stored in a separate cylindrical copper case some 36 by 6 cm., with "pull-off" lid, upon which is stamped, in plain figures, the capacity of the contained pipettes.

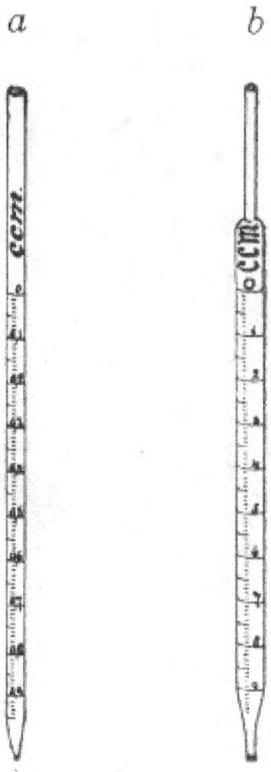

Fig. 10.—Measuring pipettes, a and b.

The laboratory should also be provided with a complete set of "Standard" graduated pipettes, each pipette in the set being stamped and authenticated by a certificate from one of the recognised Physical Measurement Laboratories, such as Charlottenburg. These instruments are expensive and should be reserved solely for standardising the pipettes in ordinary use, and for calibrating small pipettes manufactured in the laboratory. Such a set should comprise, at least, pipettes delivering 10 c.c., 5 c.c., 2.5 c.c., 2 c.c., 1 c.c., 0.5 c.c., 0.25 c.c., 0.2 c.c., 0.1 c.c., 0.05 c.c., and 0.01 c.c., respectively.

In the immediately following sections are described small pieces of glass apparatus which should be prepared in the laboratory from glass tubing of various sizes. In their preparation three articles are essential; first a three-square hard-steel file or preferably a glass-worker's knife of hard Thuringian steel for cutting glass tubes etc.; next a blowpipe flame, for although much can be done with the ordinary Bunsen burner, a blowpipe flame makes for rapid work; and lastly a bat's-wing burner.

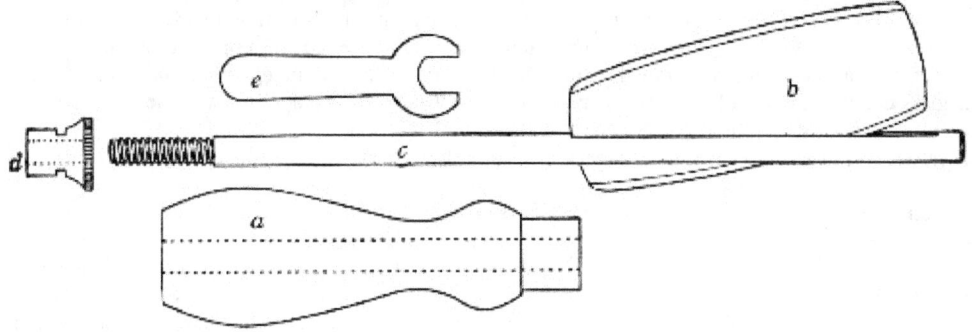

Fig. 11.—Glass-cutting knife. a. handle. b. double edged blade. c. shaft. d. locking nut. e. spanner for nut.

1. The glass-cutting knife. This article is sold in two forms, a bench knife (Fig. 11) and a pocket knife. The former is provided with a blade some 8 cm. in length and having two cutting edges. The cutting edge when examined in a strong light is seen to be composed of small closely set teeth, similar to those in a saw. The knife should be kept sharp by frequent stroppings on a sandstone hone. The pocket form, about 6-cm. long over all, consists of a small spring blade with one cutting edge mounted in scales like an ordinary pocket knife.

2. For real convenience of work the blowpipe should be mounted on a special table connected up with cylindrical bellows operated by a pedal. That figured (Fig. 12) is made by mounting a teak top 60 cm. square upon the uprights of an enclosed double-action concertina bellows (Enfer's) and provided with a Fletcher's Universal gas blowpipe.

3. An ordinary bat's-wing gas-burner mounted at the far corner of the table top is invaluable in the preparation of tubular apparatus with sharp curves, and for coating newly-made glass apparatus with a layer of soot to prevent too rapid cooling, and its usually associated result—cracking.

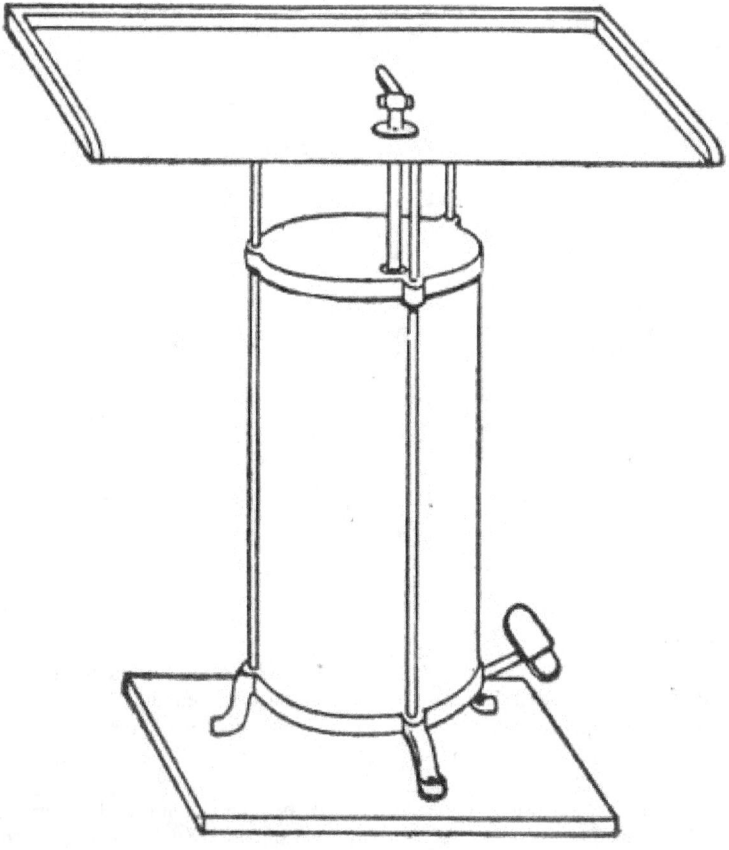

Fig. 12.—Glass blower's table with Enfer's foot bellows.

6. **Sedimentation tubes 5×0.5 cm.**, for sedimentation reactions, etc., and for containing small quantities of fluid to be centrifugalised in the hæmatocrit. These are made by taking 14-cm. lengths of stout glass tubing of the requisite diameter and heating the centre in the Bunsen or blowpipe flame. When the central portion is quite soft draw the ends quickly apart and then round off the pointed ends of the two test-tubes thus formed. With the glass-cutting knife cut off whatever may be necessary from the open ends to make the tubes the required length.

A rectangular block of "plasticine" (modelling clay) into which the conical ends can be thrust makes a very convenient stand for these small tubes.

Capillary Pipettes or Pasteur's Pipettes (Fig. 13 *a*).—These little instruments are invaluable, and a goodly supply should be kept on hand. They are prepared from soft-glass tubing of various-sized calibre (the most generally useful size being 8 mm. diameter) in the following manner: Hold a 10 cm. length of glass tube by each end, and whilst rotating it heat the central portion in the Bunsen flame or the blowpipe blast-flame until the glass is red hot and soft. Now remove it from the flame and steadily pull the ends apart, so drawing the heated portion out into a roomy capillary tube; break the capillary portion at its centre, seal the broken ends in the flame, and round off the edges of the open end of each pipette. A loose plug of cotton-wool in the open mouth completes the capillary pipette. After a number have been prepared, they are sterilised and stored in batches, either in metal cases similar to those used for the graduated pipettes or in large-sized test-tubes—sealed ends downward and plugged ends toward the mouth of the case.

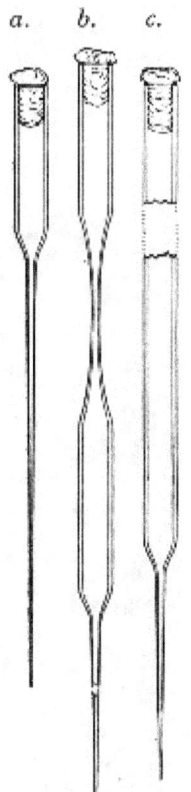

Fig. 13.—Capillary pipettes. a, b, c.

The filling and emptying of the capillary pipette is most satisfactorily accomplished by slipping a small rubber teat (similar to that on a baby's feeding bottle but *not perforated*) on the upper end, after cutting or snapping off the sealed point of the capillary portion. If pressure is now exerted upon the elastic bulb by a finger and thumb whilst the capillary end is below the surface of the fluid to be taken up, some of the contained air will be driven out, and subsequent relaxation of that pressure (resulting in the formation of a partial vacuum) will cause the fluid to ascend the capillary tube. Subsequent compression of the bulb will naturally result in the complete expulsion of the fluid from the pipette (Fig. 14).

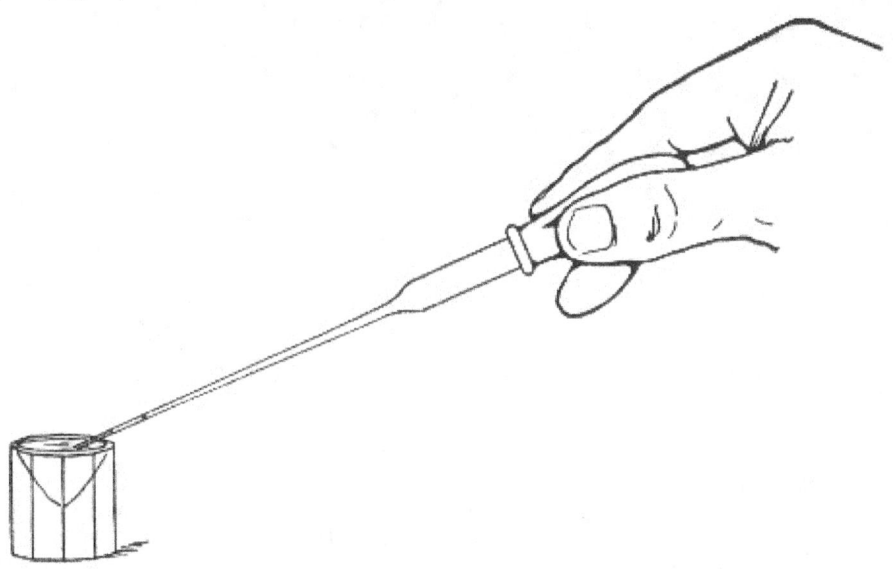

Fig. 14.—Filling the capillary teat-pipette.

A modification of this pipette, in which a constriction or short length of capillary tube is introduced just below the plugged mouth (Fig. 13, *b*), will also be found extremely useful in the collection and storage of morbid exudations.

A third form, where the capillary portion is about 4 or 5 cm. long and only forms a small fraction of the entire length of the pipette (Fig. 13, *c*), will also be found useful.

"Blood" Pipettes (Fig 15).—Special pipettes for the collection of fairly large quantities of blood (as suggested by Pakes) should also be prepared. These are made from *soft* glass tubing of 1 cm. bore, in a similar manner to the Pasteur pipettes, except that the point of the blowpipe flame must be used in order to obtain the sharp shoulder at either end of the central bulb. The terminal tubes must retain a diameter of at least 1 mm., in order to avoid capillary action during the collection of the fluid.

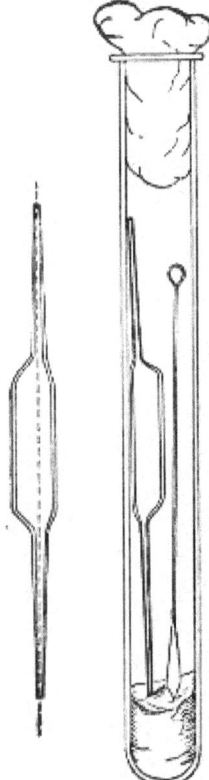

Fig. 15.—Blood pipettes and hair-lip pin in a test-tube.

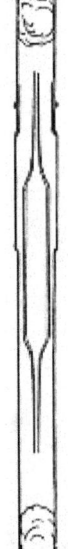

Fig. 16.—Blood-pipette in metal thermometer case.

For sterilisation and storage each pipette is placed inside a test-tube, resting on a wad of cotton-wool, and the tube plugged in the ordinary manner. As these tubes are used almost exclusively for blood work, it is usual to place a lance-headed hare-lip pin or a No. 9 flat Hagedorn needle inside the tube so that the entire outfit may be sterilised at one time.

For the collection of small quantities of blood for agglutination reactions and the like, many prefer a short straight piece of narrow glass tubing drawn out at either extremity to almost capillary dimensions. Such pipettes, about 8 cm. in length over all, are most conveniently sterilized in ordinary metal thermometer cases (Fig. 16).

Graduated Capillary Pipettes (Fig. 17).—These should also be made in the laboratory—from manometer tubing—of simple, convenient shape, and graduated by the aid of "standard" pipettes (in hundredths) to contain such quantities as 10, 50, and 90 c. mm., and carefully marked with a writing diamond. These, previously sterilised in large test-tubes, will be found extremely useful in preparing accurate percentage solutions, when only minute quantities of fluid are available.

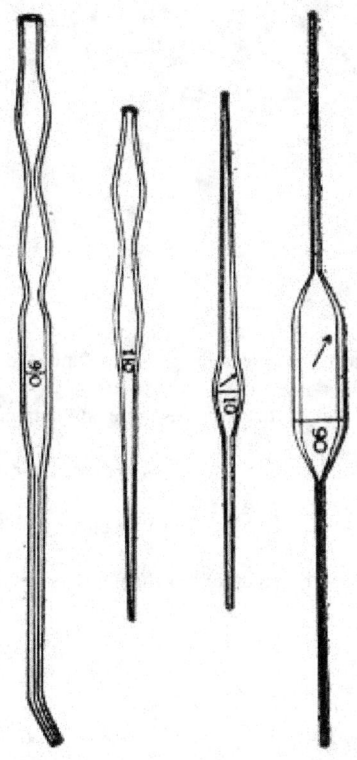

Fig. 17.—Capillary graduated pipettes.

Automatic ("Throttle") Pipettes.—These ingenious pipettes, introduced by Wright, can easily be calibrated in the laboratory and are exceedingly useful for graduating small pipettes, for measuring small quantities of fluids, in preparing dilutions of serum for agglutination reactions, etc. They are usually made from the Capillary Pasteur pipettes (Fig. 13, *a*). The following description of the manufacture of a 5 c. mm. pipette will serve to show how the small automatic pipettes are calibrated.

1. Select a pipette the capillary portion of which is fairly roomy in bore and possesses regular even walls, and remove the cotton-wool plug from the open end.

2. Heat the capillary portion near the free extremity in the by-pass flame of the bunsen burner and draw it out into a very fine hair-like tube and break this across. This hair-like extremity will permit the passage of air but is too fine for metallic mercury to pass.

3. From a standard graduated pipette deliver 5 c. mm. clean mercury into the upper wide portion of the pipette.

4. Adjust a rubber teat to the pipette and by pressure on the bulb gradually drive the mercury in an unbroken column down the capillary tube until it is stopped by the filiform extremity.

5. Cut off the capillary tube exactly at the upper level of the column of mercury, invert it and allow the mercury to run out.

6. Snap off the remainder of the capillary tube from the broad upper portion of the pipette which is now destined to form the covering tube or air chamber, or what we may term the "barrel." This barrel now has the lower end in the form of a truncated cone, the upper end being cut square. Remove the teat.

7. Introduce the capillary tube into this barrel with the filiform extremity uppermost, and the square cut end projecting about 0.5 cm. beyond the tapering end of the barrel.

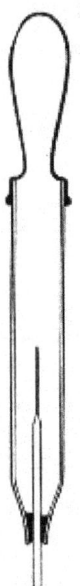

Fig. 18.—Throttle pipette—small capacity.

8. Drop a small pellet of sealing wax into the barrel by the side of the capillary tube and then warm the tube at the gas flame until the wax becomes softened and makes an air-tight joint between the capillary tube and the end of the barrel.

9. Fit a rubber teat to the open end of the barrel, and so complete a pipette which can be depended upon to always aspirate and deliver exactly 5 c. mm. of fluid.

Slight modification of this procedure is necessary in making tubes to measure larger volumes than say 75 c. mm. Thus to make a throttle pipette to measure 100 c. mm.:

1. Take a short length of quill tubing and draw out one end into a roomy capillary stem, and again draw out the extremity into a fine hair point, thus forming a small Pasteur pipette with a hair-like capillary extremity.

2. With a standard pipette fill 100 c. mm. into the neck of this pipette, and make a scratch with a writing diamond at the upper level (a) of the mercury meniscus (Fig. 19, A).

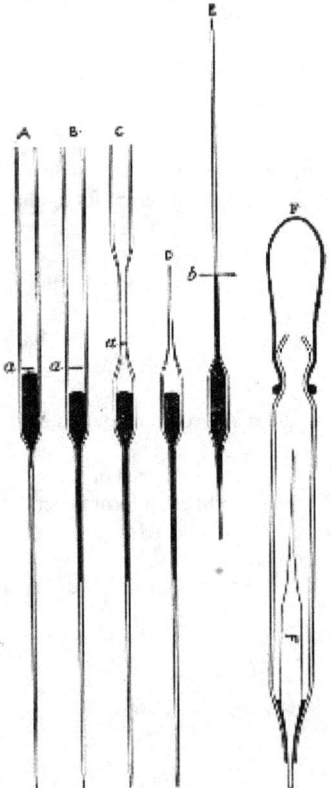

Fig. 19.—Making throttle pipettes—large capacity

Now force the mercury down into the capillary stem as far as it will go, so as to leave the upper part of the tube in the region of the diamond scratch empty (Fig. 19, B).

3. Heat the tube in the region of the diamond scratch in the blowpipe flame, and removing the tube from the flame draw it out so that the diamond scratch now occupies a position somewhere near the centre of this new capillary portion (Fig. 19, C).

4. Heat the tube in this position in the peep flame of the Bunsen burner, and draw it out into a hair-like extremity. Snap off the glass tube, leaving about 5 mm. of hair-like extremity attached to the upper capillary portion (Fig. 19, D). Allow the glass to cool.

5. Lift up the bulb by the long capillary stem and allow the mercury to return to its original position—an operation which will be facilitated by snapping off the hair-like extremity from the long piece of capillary tubing.

6. Mark on the capillary stem with a grease pencil the position of the end of the column of mercury (Fig. 19, E.)

7. Warm the capillary tubing at this spot in the peep flame of the Bunsen burner, and draw it out very slightly so that when cut at this position a pointed extremity will be obtained.

8. With a glass-cutting knife cut the capillary tube through at the point "b," and allow the mercury to run out.

9. Now apply a thick layer of sealing wax to the neck of the bulb.

10. Take a piece of 5 mm. bore glass tubing and draw it out as if making an ordinary Pasteur pipette.

11. Break the capillary portion off so as to leave a covering tube similar to that already used for the smaller graduated pipettes. Into this covering tube drop the graduated bulb and draw the capillary stem down through the conical extremity until further progress is stopped by the layer of sealing wax.

12. Warm the pipette in the gas flame so as to melt the sealing wax and make an air-tight joint.

13. Fit an india-rubber teat over the open end of the covering tube, and the automatic pipette is ready for use (Fig. 19, F).

Sedimentation Pipettes (Fig. 20).—These are prepared from 10 cm. lengths of narrow glass tubing by sealing one extremity, blowing a small bulb at the centre, and plugging the open end with cotton-wool; after sterilisation the open end is provided with a short piece of rubber tubing and a glass mouthpiece. When it is necessary to observe sedimentation reactions in very small quantities of fluid, these tubes will be found much more convenient than the 5 by 0.5 cm. test-tubes previously mentioned.

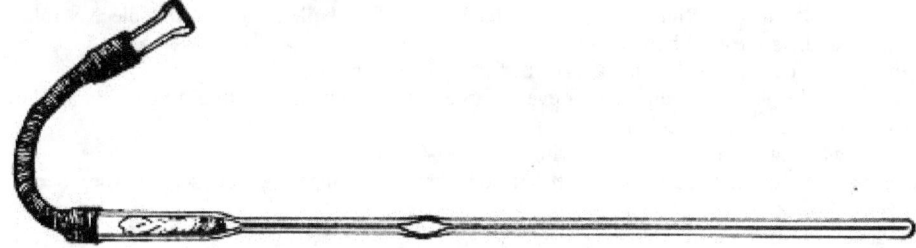

Fig. 20.—Sedimentation pipette.

Pasteur pipettes fitted with india-rubber teats will also be found useful for sedimentation tests when dealing with minute quantities of serum, etc.

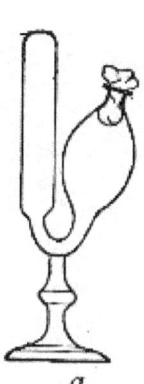 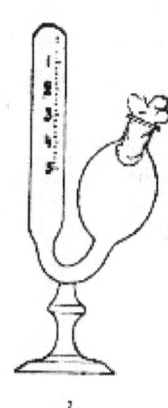 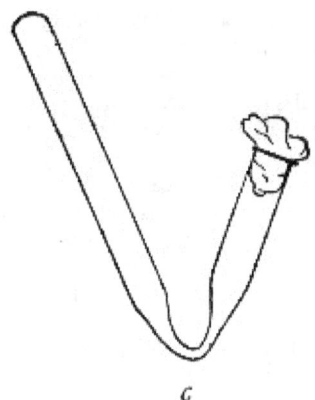

a   *b*   *c*   Fig. 21.—Fermentation tubes.

Fermentation Tubes (Fig. 21).—These are used for the collection and analysis of the gases liberated from the media during the growth of some varieties of bacteria and may be either plain (*a*) or graduated (*b*). A simple form (Fig. 21, *c*) may be made from 14 cm. lengths of soft glass tubing of 1.5 cm. diameter. The Bunsen flame is applied to a spot some 5 cm. from one end of such a piece of tubing and the tube slightly drawn out to form a constriction, the constricted part is bent in the bat's-wing flame, to an acute angle, and the open extremity of the long arm sealed off in the blowpipe flame. The open end of the short arm is rounded off and then plugged with cotton-wool, and the tube is ready for sterilisation.

CLEANING OF GLASS APPARATUS.

All glassware used in the bacteriological laboratory must be thoroughly cleaned before use, and this rule applies as forcibly to new as to old apparatus, although the methods employed may vary slightly.

To Clean New Test-tubes.—

1. Place the tubes in a bucket or other convenient receptacle, fill with water and add a handful of "Sapon" or other soap powder. See that the tubes are full and submerged.

2. Fix the bucket over a large Bunsen flame and boil for thirty minutes—or boil in the autoclave for a similar period.

3. Cleanse the interior of the tubes with the aid of test-tube brushes, and rinse thoroughly in cold water.

4. Invert the tubes and allow them to drain completely.

5. Dry the tubes and polish the glass inside and out with a soft cloth, such as selvyt.
New flasks, plates, and capsules must be cleaned in a similar manner.
To Clean New Graduated Pipettes.—
1. Place the pipettes in a convenient receptacle, filled with water to which soap powder has been added.
2. Boil the water vigorously for twenty minutes over a Bunsen flame.
3. Rinse the pipettes in running water and drain.
4. Run distilled water through the pipettes and drain.
5. Run rectified spirits through the pipette and drain as completely as possible.
6. Place the pipettes in the hot-air oven (*vide* page 31), close the door, open the ventilating slide, and run the temperature slowly up to about 80° C. Turn off the gas and allow the oven to cool.

Or 6a. Attach each pipette in turn to the rubber tube of the foot bellows, or blowpipe air-blast, and blow air through the pipette until the interior is dry.

Glassware that has already been used is regarded as *infected*, and is treated in a slightly different manner.
Infected Test-tubes.—
1. Pack the tubes in the wire basket of the autoclave (having previously removed the cotton-wool plugs, caps, etc.), in the vertical position, and before replacing the basket see that there is a sufficiency of water in the bottom of the boiler. Now attach a piece of rubber tubing to the nearest water tap, and by means of this fill each tube with water.
2. Disinfect completely by exposing the tubes, etc., to a temperature of 120° C. for twenty minutes (*vide* page 37).

(If an autoclave is not available, the tubes must be placed in a digester, or even a large pan or pail with a tightly fitting cover, and boiled vigorously for some thirty to forty-five minutes to ensure disinfection.)

3. Whilst still hot, empty each tube in turn and roughly clean its interior with a stiff test-tube brush.
4. Place the tubes in a bucket or other convenient receptacle, fill with water and add a handful of Sapon or other soap powder. See that the tubes are full and submerged.
5. Fix the bucket over a large Bunsen flame and boil for thirty minutes.
6. Cleanse the interior of the tubes with the aid of test-tube brushes, and rinse thoroughly in cold water.
7. Drain off the water and immerse tubes in a large jar containing water acidulated with 2 to 5 per cent. hydrochloric acid. Allow them to remain there for about fifteen minutes.
8. Remove from the acid jar, drain, rinse thoroughly in running water, then with distilled water.
9. Invert the tubes and allow them to drain completely.

Dry the tubes and polish the glass inside and out with a soft cloth, such as selvyt.
Infected flasks, plates, and capsules must be treated in a similar manner.

Flasks which have been used only in the preparation of media must be cleaned immediately they are finished with. Fill each flask with water to which some soap powder and a few crystals of potassium permanganate have been added, and let boil over the naked flame. The interior of the flask can then usually be perfectly cleaned with the aid of a flask brush, but in some cases water acidulated with 5 per cent. nitric acid, or a large wad of wet cotton-wool previously rolled in silver sand, must be shaken around the interior of the flask, after which rinse thoroughly with clean water, dry, and polish.

Infected Pipettes.—
1. Plunge infected pipettes immediately after use into tall glass cylinders containing a 2 per cent. solution of lysol, and allow them to remain therein for some days.
2. Remove from the jar and drain. Boil in water to which a little soap has been added, for thirty minutes.
3. Rinse thoroughly in cold water.
4. Immerse in 5 per cent. nitric acid for an hour or two.
5. Rinse again in running water to remove all traces of acid.
6. Complete the cleaning as described under "new pipettes."

When dealing with graduated capillary pipettes employed for blood or serum work (whether new or infected), much time is consumed in the various steps from 5 onward, and the cleansing process can be materially hastened if the following device is adopted.

Fit up a large-sized Kitasato's filter flask to a Sprengel's suction pump or a Geryk air pump (see page 43). To the side tubulure of the filter flask attach a 20 cm. length of rubber pressure tubing having a calibre sufficiently large to admit the ends of the pipettes.

Next fill a small beaker with distilled water. Attach the first pipette to the free end of the rubber tubing, place the pipette point downward in the beaker of water and start the pump (Fig. 22).

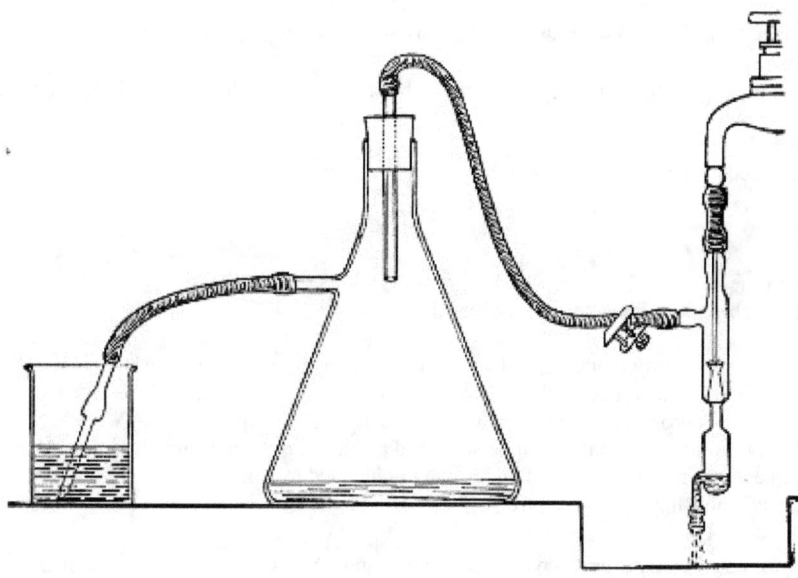

Fig. 22.—Cleaning blood pipettes.

When all the water has been aspirated through the pipette into the filter flask, fill the beaker with rectified spirit and when this is exhausted refill with ether. Detach the pipette and dry in the hot-air oven.

Slides and cover-slips (Fig. 23), when first purchased, have "greasy" surfaces, upon which water gathers in minute drops and effectually prevents the spreading of thin, even films.

Microscopical Slides.—The slides in general use are those known as "three by one" slips (measuring 3 inches by 1 inch, or 76 by 26 mm.), and should be of good white crown glass, with ground edges.

New slides should be allowed to remain in alcohol acidulated with 5 per cent. hydrochloric acid for some hours, rinsed in running water, roughly drained on a towel, dried, and finally polished with a selvyt cloth.

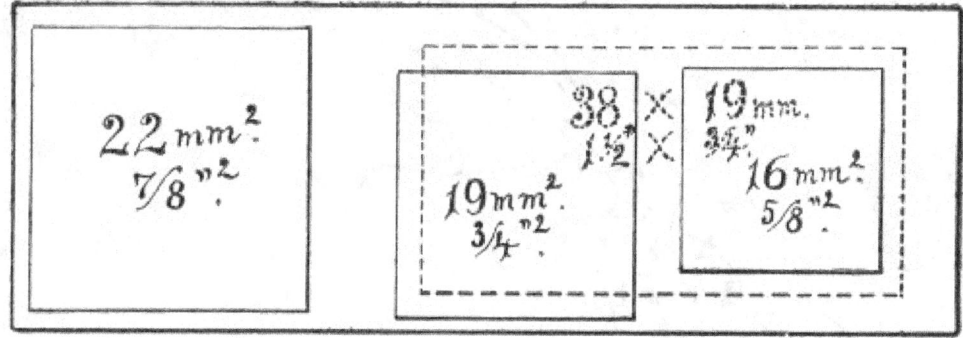

Fig. 23.—Slides and cover-slips, actual size.

If only a few slides are required for immediate use a good plan is to rub the surface with jeweler's emery paper (Hubert's 00). A piece of hard wood 76×26×26 mm. with a piece of this emery paper gummed tightly around it is an exceedingly useful article on the microscope bench.

Cover-slips.—The most useful sizes are the 19 mm. squares for ordinary cover-glass film preparations, and 38 by 19 mm. rectangles for blood films and serial sections; both varieties must be of "No. 1" thickness, which varies between 0.15 and 0.22 mm., that they may be available for use with the high-power immersion lenses.

Cover-slips should be cleaned in the following manner:

1. Drop the cover-slips one by one into an enamelled iron pot or tall glass beaker, containing a 10 per cent. solution of chromic acid.
2. Heat over a Bunsen flame and allow the acid to boil gently for twenty minutes.

Note.—A few pieces of pipe-clay or pumice may be placed in the beaker to prevent the "spurting" of the chromic acid.

3. Turn the cover-slips out into a flat glass dish and wash in running water under the tap until all trace of yellow colour has disappeared. During the washing keep the cover-slips in motion by imparting a rotatory movement to the dish.
4. Wash in distilled water in a similar manner.
5. Wash in rectified spirit.
6. Transfer the cover-slips, by means of a pair of clean forceps, previously heated in the Bunsen flame to destroy any trace of grease, to a small beaker of absolute alcohol.

Drain off the alcohol and transfer the cover-slips, by means of the forceps, to a wide-mouthed glass pot, containing absolute alcohol, in which they are to be stored, and stopper tightly.

Note.—After once being placed in the chromic acid, the cover-slips must on no account be touched by the fingers.

Used Slides and Cover-slips.—Used slides with the mounted cover-slip preparations, and cover-slips used for hanging-drop mounts, should, when discarded, be thrown into a pot containing a 2 per cent. solution of lysol.

After immersion therein for a week or so, even the cover-slips mounted with Canada balsam can be readily detached from their slides.

Slides.—
1. Wash the slides thoroughly in running water.
2. Boil the slides in water to which "sapon" has been added, for half an hour.
3. Rinse thoroughly in cold water.
4. Dry and polish with a dry cloth.

Cover-slips.—
1. Wash the cover-slips thoroughly in running water.
2. Boil the cover-slips in 10 per cent. solution of chromic acid, as for new cover-slips.
3. Wash thoroughly in running water.
4. Pick out those cover-slips which show much adherent dirty matter, and rub them between thumb and forefinger under the water tap. The dirt usually rubs off easily, as it has become friable from contact with the chromic acid.
5. Return all the cover-slips to the beaker, fill in *fresh* chromic acid solution, and treat as new cover-slips.

Note.—*Test-tubes, plates, capsules*, etc., which, from long use, have become scratched and hazy, or which cannot be cleaned in any other way, may be dealt with by immersing them in an enamelled iron bath, containing water acidulated to 1 per cent. with hydrofluoric acid, for ten minutes, rinsing thoroughly in water, drying, and polishing.

PLUGGING TEST-TUBES AND FLASKS.

Before sterilisation all test-tubes and flasks must be carefully plugged with cotton-wool, and for this purpose best absorbent cotton-wool (preferably that put up in cylindrical one-pound packets and interleaved with tissue paper—known as surgeons' wool) should be employed.
1. For a test-tube or a small flask, tear a strip of cotton-wool some 10 cm. long by 2 cm. wide from the roll.
2. Turn in the ends neatly and roll the strip of wool lightly between the thumb and fingers of both hands to form a long cylinder.
3. Double this at the centre and introduce the now rounded end into the open mouth of the tube or flask.
4. Now, whilst supporting the wool between the thumb and fingers of the right hand, rotate the test-tube between those of the left, and gradually screw the plug of wool into its mouth for a distance of about 2.5 cm., leaving about the same length of wool projecting.

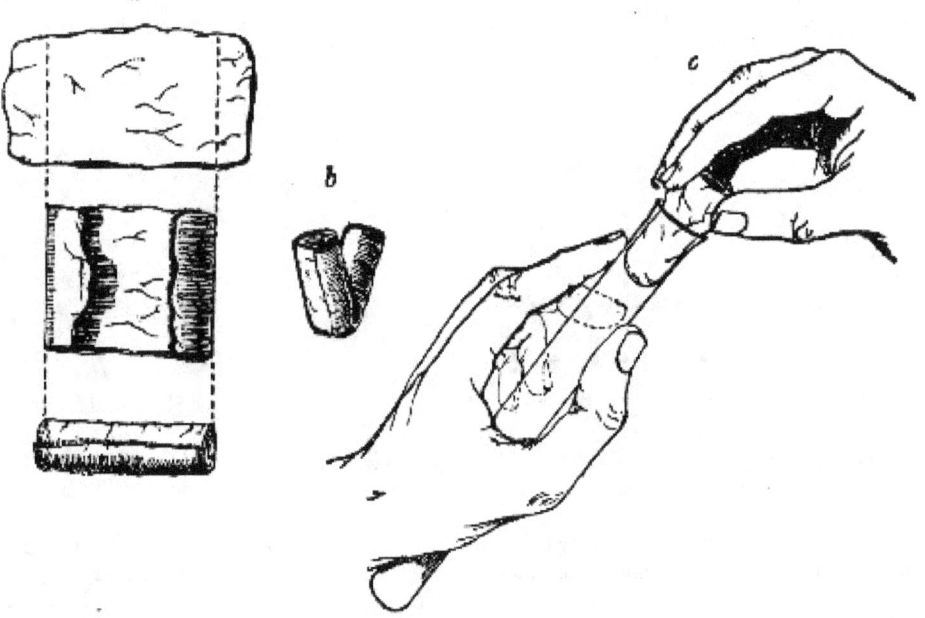

Fig 24..—Plugging test-tubes: a, cylinder of wool being rolled; b, cylinder of wool being doubled; c, cylinder of wool being inserted in tube.

The plug must be firm and fit the tube or flask fairly tightly, sufficiently tightly in fact to bear the weight of the glass plus the amount of medium the vessel is intended to contain, but not so tightly as to prevent it from being easily removed by a screwing motion when grasped between the fourth, or third and fourth, fingers, and the palm of the hand.

For a large flask a similar but larger strip of wool must be taken; the method of making and inserting the plug is identical.

III. METHODS OF STERILISATION.

STERILISING AGENTS.

Sterilisation—*i. e.*, the removal or the destruction of germ life—may be effected by the use of various agents. As applied to the practical requirements of the bacteriological laboratory, many of these agents, such as electricity, sunlight, etc., are of little value, others are limited in their applications; others again are so well suited to particular purposes that their use is almost entirely restricted to such.

The sterilising agents in common use are:

Chemical Reagents.—*Disinfectants* (for the disinfection of glass and metal apparatus and of morbid tissues).

Physical Agents. Heat.—(*a*) *Dry Heat:*
1. Naked flame (for the sterilisation of platinum needles, etc.).
2. Muffle furnace (for the sterilisation of filter candles, and for the destruction of morbid tissues).
3. Hot air (for the sterilisation of all glassware and of metal apparatus).

(*b*) *Moist Heat:*
1. Water at 56° C. (for the sterilisation of certain albuminous fluids).
2. Water at 100° C. (for the sterilisation of surgical instruments, rubber tubing, and stoppers, etc.).
3. Streaming steam at 100° C. (for the sterilisation of media).
4. Superheated steam at 115° C. or 120° C. (for the disinfection of contaminated articles and the destruction of old cultivations of bacteria).

Filtration.—
1. Cotton-wool filters (for the sterilisation of air and gases).
2. Porcelain filters (for the sterilisation of various liquids).

METHODS OF APPLICATION.

Chemical Reagents, such as belong to the class known as antiseptics (*i. e.*, substances which inhibit the growth of, but do not destroy, bacterial life), are obviously useless. Disinfectants or germicides (*i. e.*, substances which destroy bacterial life), on the other hand, are of value in the disinfection of morbid material, and also of various pieces of apparatus, such as pipettes, pending their cleansing and complete sterilisation by other processes. To this class (in order of general utility) belong:

Lysol, 2 per cent. solution; Perchloride of mercury, 0.1 per cent. solution; Carbolic acid, 5 per cent. solution; Absolute alcohol; Ether; Chloroform; Camphor; Thymol; Toluol; Volatile oils, such as oil of mustard, oil of garlic.

Formaldehyde is a powerful germicide, but its penetrating vapor restricts its use. These disinfectants are but little used in the final sterilisation of apparatus, chiefly on account of the difficulty of effecting their complete removal, for the presence of even traces of these chemicals is sufficient to so inhibit or alter the growth of bacteria as to vitiate subsequent experiments conducted by the aid of apparatus sterilised in this manner.

Note.—Tubes, flasks, filter flasks, pipettes, glass tubing, etc., may be rapidly sterilised, in case of emergency, by washing, in turn, with distilled water, perchloride of mercury solution, alcohol, and ether, draining, and finally gently heating over a gas flame to completely drive off the ether vapor. Chloroform or other volatile disinfectants may be added to various fluids in order to effect the destruction of contained bacteria, and when this has been done, may be completely driven off from the fluid by the application of gentle heat.

Dry Heat.—The *naked flame* of the Bunsen burner is invariably used for sterilising the platinum needles (which are heated to redness) and may be employed for sterilising the points of forceps, or other small instruments, cover-glasses, pipettes, etc., a very short exposure to this heat being sufficient.

Ether Flame.—In an emergency small instruments, needles, etc., may be sterilised by dipping them in ether and after removal lighting the adherent fluid and allowing it to burn off the surface of the instruments. Repeat the process twice. It may then be safely assumed that the apparatus so treated is sterile.

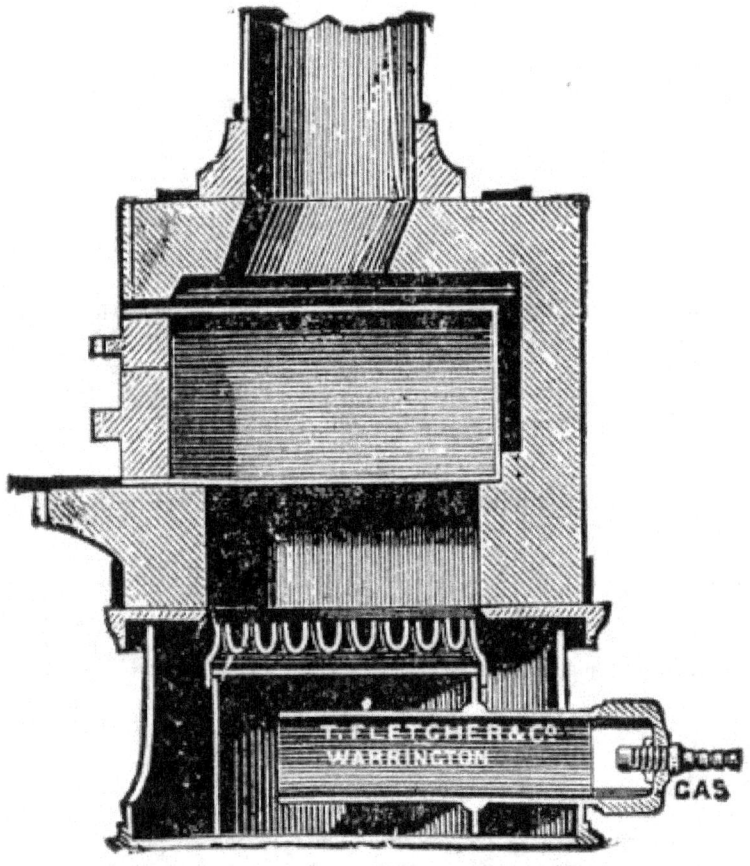

Fig. 25.—Muffle furnace.

Muffle Furnace (Fig. 25).—Although this form of heat is chiefly used for the destruction of the dead bodies of small infected animals, morbid tissues, etc., it is also employed for the sterilisation of porcelain filter candles (*vide* p. 42).

Filter candles are disinfected immediately after use by boiling in a beaker of water for some fifteen or twenty minutes. This treatment, however, leaves the dead bodies of the bacteria upon the surface and blocking the interstices of the filter.

To destroy the organic matter and prepare the filter candle for further use proceed as follows:

1. Roll each bougie up in a piece of asbestos cloth, secure the ends of the cloth with a few turns of copper wire, and place inside the muffle (a small muffle 76×88×163 mm. will hold perhaps four small filter candles).

2. Light the gas and raise the contents of the muffle to a white heat; maintain this temperature for five minutes.

3. Extinguish the gas, and when the muffle has become quite cold remove the filter candles, and store them (without removing the asbestos wrappings) in sterile metal boxes.

Note.—The too rapid cooling of the candles, such as takes place if they are removed from the muffle before it has cooled down to the room temperature, may give rise to microscopic cracks and flaws which will effectually destroy their efficiency.

Hot Air.—Hot air at 150° C. destroys all bacteria, spores, etc:, in about thirty minutes; a momentary exposure to a temperature of 175° to 180° C. will effect the same result and offers the more convenient method of sterilisation. This method is only applicable to glass and metallic substances, and the small bulk of cotton-wool comprised in the test-tube plugs, etc. Large masses of fabric are not effectually sterilised by dry heat—short of charring—as its power of penetration is not great.

Sterilisation by hot air is effected in the hot-air oven (Fig. 18). This is a rectangular, double-walled metal box, mounted on a stand and heated from below by a large Bunsen burner. The interior of the oven is provided with loose shelves upon which the articles to be sterilised are arranged, either singly or packed in square wire baskets or crates, kept specially for this purpose. One of the sides is hinged to form a door. The central portion of the metal bottom, on which the Bunsen flame would play, is cut away, and replaced by firebrick plates, which slide in metal grooves and are easily replaced when broken or worn out. The top of the oven is provided with a perforated ventilator slide and two tubulures, the one for the reception of a centigrade thermometer graduated to 200° or 250°C., the other for a thermo-regulator. An ordinary mercurial thermo-regulator may be used but it is preferable to employ a regulating capsule of the Hearson type (see p. 219) with a spring arm adjusted to the lever so that when the boiling-point of the capsule (*e. g.*, 175°C.) is reached the gas supply is absolutely cut off and the jet cannot again be lighted until the spring-arm has been readjusted by hand. The thermo-regulator is by no means a necessity, and may be replaced by a large bore thermometer with a sliding platinum point, connected with an electric bell, which can be easily adjusted to ring at any given temperature. Even if the steriliser is provided with the capsule regulator above described the contact thermometer should also be fitted.

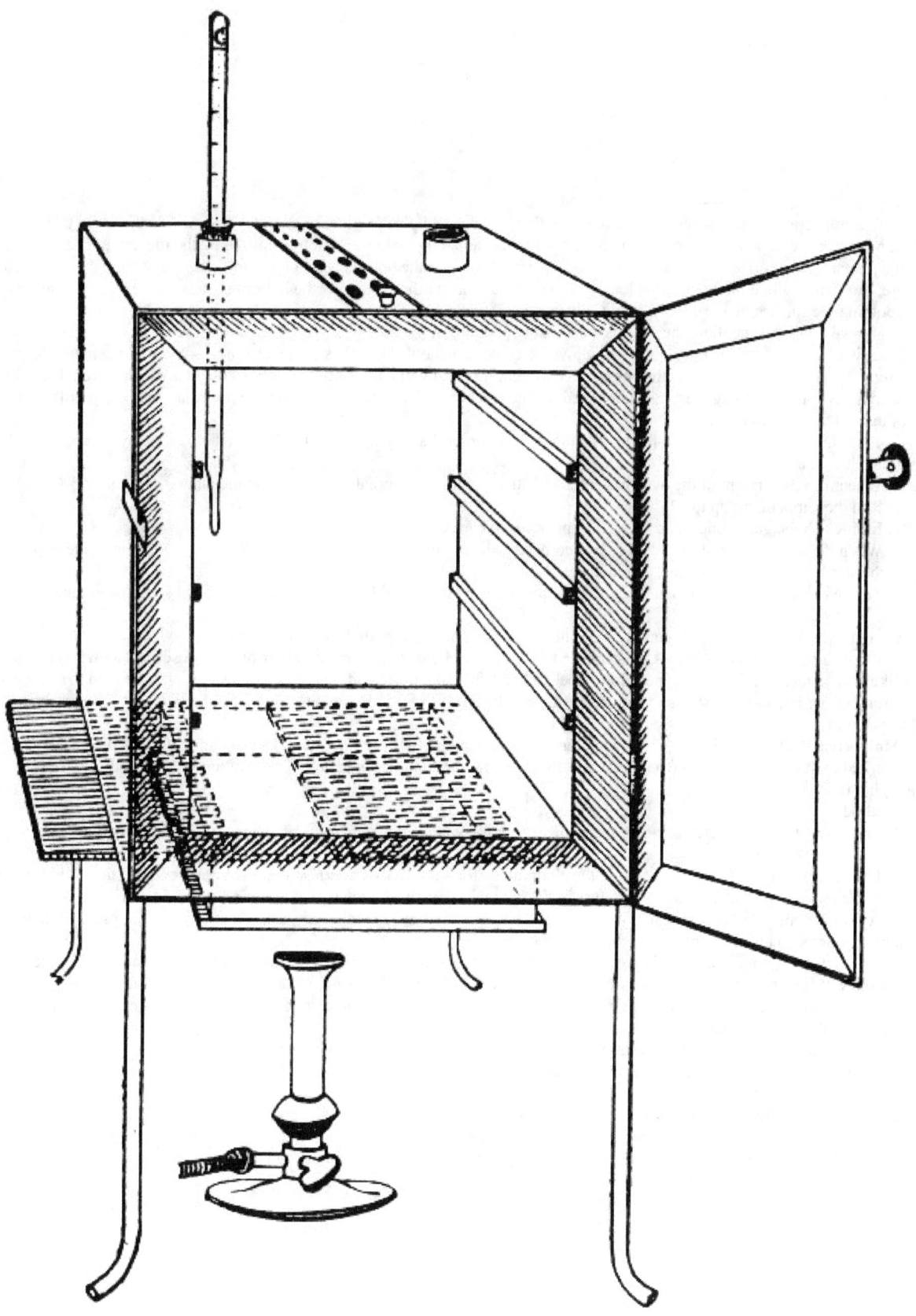

Fig. 26.—Hot-air oven.

To Use the Hot-air Oven.—

1. Place the crates of test-tubes, metal cases containing plates and pipettes, loose apparatus, etc., inside the oven, taking particular care that none of the cotton-wool plugs are in contact with the walls, otherwise the heat transmitted by the metal will char or even flame them.

To prepare a wire crate for the reception of test-tubes, etc., cover the bottom with a layer of thick asbestos cloth; or take some asbestos fibre, moisten it with a little water and knead it into a paste; plaster the paste over the bottom of the crate, working it into the meshes and smoothing the surface by means of a pestle. When several crates have been thus treated, place them inside the hot-air oven, close the door, open the ventilating slide, light the gas, and run the temperature of the interior up to about 160° C. After an interval of ten minutes extinguish the gas, open the oven door, and allow the contents to cool. The asbestos now forms a smooth, dry, spongy layer over the bottom, which will last many months before needing renewal, and will considerably diminish the loss of tubes from breakage.

Copper cylinders and large test-tubes intended for the reception of pipettes are prepared in a similar manner, in order to protect the points of these articles from injury.

2. Close the oven door, and open the ventilating slide, in order that any moisture left in the tubes, etc., may escape; light the gas below; set the electric alarm to ring at 100°C.

3. When the temperature of the oven has reached 100°C., close the ventilating slide; reset the alarm to ring at 175°C.

4. Run the temperature up to 175°C.

5. Extinguish the gas at once, and allow the apparatus to cool.

6. When the temperature of the interior, as recorded by the thermometer, has fallen to 60°C.—*but not before*—the door may be opened and the sterile articles removed and stored away.

Note.—Neglect of this precautionary cooling of the oven to 60° C. will result in numerous cracked and broken tubes.

On removal from the oven, the cotton-wool plugs will probably be slightly brown in colour.

Metal instruments, such as knives, scissors, and forceps, may be sterilised in the hot-air oven as described above, but exposure to 175° C. is likely to seriously affect the temper of the steel and certainly blunts the cutting edges. If, however, it is desired to sterilise surgical instruments by hot air, they should be packed in a metal box, or boxes, and heated to 130° C. and retained at that temperature for about thirty minutes.

Moist Heat.—*Water at 56° C.*—This temperature, if maintained for thirty minutes, is sufficient to destroy the vegetative forms of bacteria, but has practically no effect on spores. Its use is limited to the sterilisation of such albuminous "fluid" media as would coagulate at a higher temperature.

Method.—

1. Fit up a water-bath, heated by a Bunsen flame which is controlled by a thermo-regulator, so that the temperature of the water remains at 56° C.

2. Immerse the tubes or flasks containing the albuminous fluid in the water-bath so that the upper level of such fluid is at least 2 cm. below the level of the water. (The temperature of the bath will now fall somewhat, but after a few minutes will again rise to 56° C).

3. After thirty minutes' exposure to 56° C, extinguish the gas, remove the tubes or flasks from the bath, and subject them to the action of running water so that their contents are rapidly cooled.

4. The vegetative forms of bacteria present in the liquid being killed, stand it for twenty-four hours in a cool, dark place; at the end of that time some at least of such spores as may be present will have germinated and assumed the vegetative form.

5. Destroy these new vegetative forms by a similar exposure to 56° C. on the second day, whilst others, of slower germination, may be caught on the third day, and so on.

6. In order to ensure thorough sterilisation, repeat the process on each of six successive days.

This method of exposing liquids to a temperature of 56° C. in a water-bath for half an hour on each of six successive days is termed *fractional sterilisation*.

Water at 100°C. destroys the vegetative forms of bacteria almost instantaneously, and spores in from five to fifteen minutes. This method of sterilisation is applicable to the metal instruments, such as knives, forceps, etc., used in animal experiments; syringes, rubber corks, rubber and glass tubing, and other small apparatus, and is effected in what is usually spoken of as the "water steriliser" (Fig. 27).

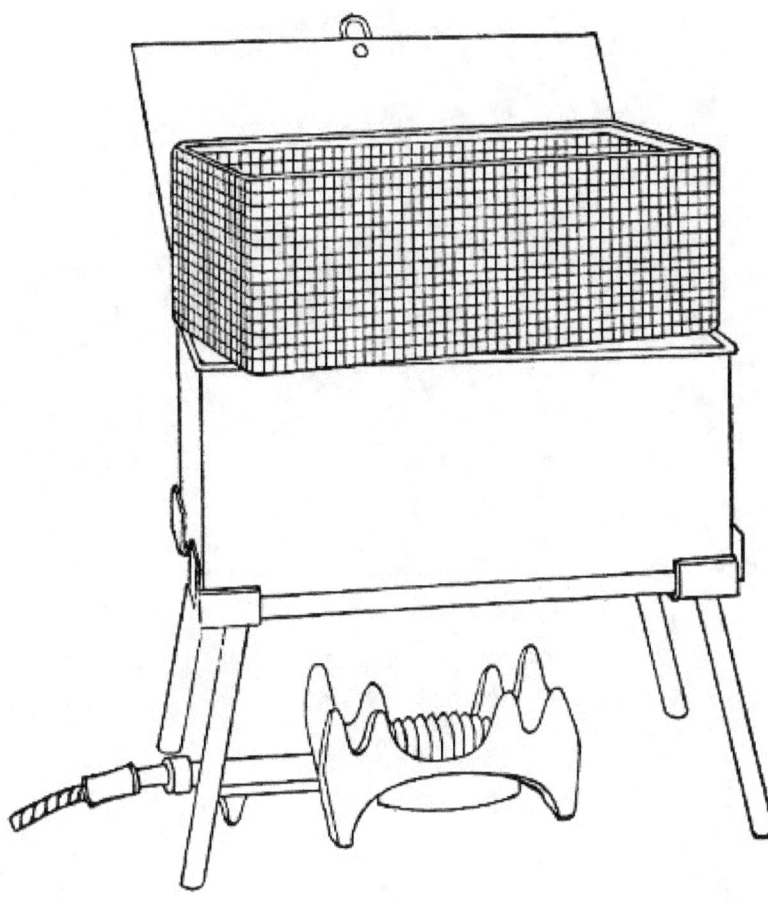

Fig. 27.—Water sterilizer.

This is a rectangular copper box, 26 cm. long, 18 cm. wide, and 12 cm. deep, mounted on legs, heated from below by a Bunsen or radial gas burner, and containing a movable copper wire tray, 2 cm. smaller in every dimension than the steriliser itself, and provided with handles. The top of the steriliser is hinged to form a lid.

Method.—

1. Place the instruments, etc., to be sterilised inside the copper basket, and replace the basket in the steriliser.
2. Pour a sufficient quantity of water into the steriliser, shut down the lid, and light the gas below.

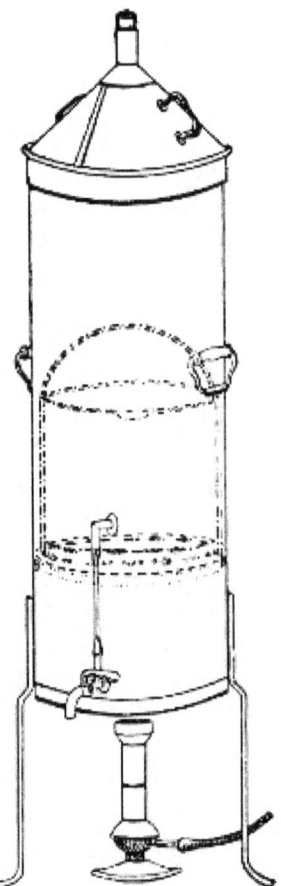

Fig. 28.—Koch's steriliser.

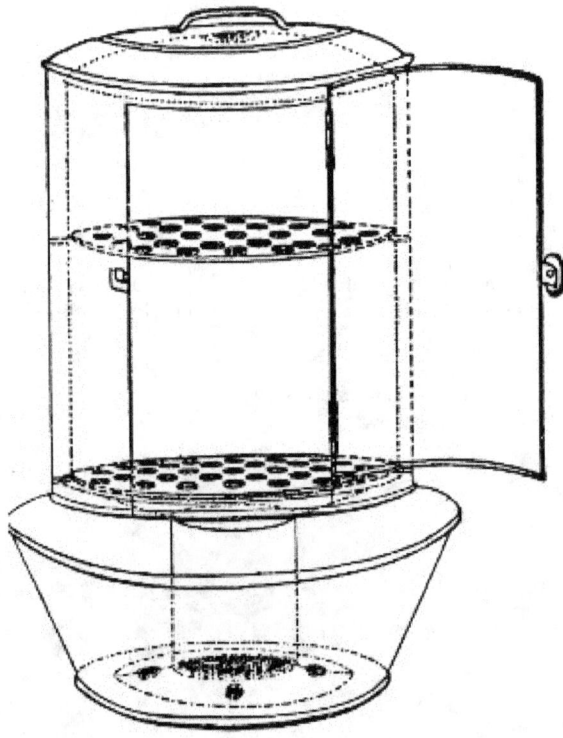

Fig. 29.—Arnold's steriliser.

3. After the water has boiled and steam has been issuing from beneath the lid for at least ten minutes, extinguish the gas, open the lid, and lift out the wire basket by its handles and rest it diagonally on the walls of the steriliser; the contained instruments, etc., are now sterile and ready for use.

4. After use, or when accidentally contaminated, replace the instruments in the basket and return that to the steriliser; completely disinfect by a further boiling for fifteen minutes.

5. After disinfection, and whilst still hot, take out the instruments, dry carefully and at once, and return them to their store cases.

Streaming steam—i. e., steam at 100°C.—destroys the vegetative forms of bacteria in from fifteen to twenty minutes, and the sporing forms in from one to two hours. This method is chiefly used for the sterilisation of the various nutrient media intended for the cultivation of bacteria, and is carried out in a steam kettle of special construction, known as Koch's steam steriliser (Fig. 28) or in one of its many modifications, the most efficient of which is Arnold's (Fig. 29).

The steam steriliser in its simplest form consists of a tall tinned-iron or copper cylindrical vessel, divided into two unequal parts by a movable perforated metal diaphragm, the lower, smaller portion serving for a water reservoir, and the upper part for the reception of wire baskets containing the articles to be sterilised. The vessel is closed by a loose conical lid, provided with handles, and perforated at its apex by a tubulure; it is mounted on a tripod stand and heated from below by a Bunsen burner. The more elaborate steriliser is cased with felt or asbestos board, and provided with a water gauge, also a tap for emptying the water compartment.

To Use the Steam Steriliser.—

1. Fill the water compartment to the level of the perforated diaphragm, place the lid in position, and light the Bunsen burner.

2. After the water has boiled, allow sufficient time to elapse for steam to replace the air in the sterilising compartment, as shown by the steam issuing in a steady, continuous stream from the tubulure in the lid.

3. Remove the lid, quickly lower the wire basket containing media tubes, etc., into the sterilising compartment until it rests on the diaphragm, and replace the lid.

4. After an interval of twenty minutes in the case of fluid media, or thirty minutes in the case of solid media, take off the lid and remove the basket with its contents.

5. Now, but not before, extinguish the gas.

Note.—After removing tubes, flasks, etc., from the steam steriliser, they should be at once separated freely in order to prevent moisture condensing upon the cotton-wool plugs and soaking through into the interior of the tubes.

This treatment will destroy any vegetative forms of bacteria; during the hours of cooling any spores present will germinate, and the young organisms will be destroyed by repeating the process twenty-four hours later; a third sterilisation after a similar interval makes assurance doubly sure.

The method of sterilising by exposure to streaming steam at 100° C. for twenty minutes on each of three consecutive days is termed *discontinuous* or *intermittent sterilisation*.

Exposure to steam at 100° C. for a period of one or two hours, or *continuous sterilisation*, cannot always be depended upon and is therefore not to be recommended.

Superheated steam—i. e., steam under pressure (see Pressure-temperature table, Appendix, page 500) in sealed vessels at a temperature of 115° C.—will destroy both the vegetative and the sporing forms of bacteria within fifteen minutes; if the pressure is increased, and the temperature raised to 120° C., the same end is attained in ten minutes. This method was formerly employed for the sterilisation of media (and indeed is so used in some laboratories still), but most workers now realise that media subjected to this high temperature undergo hydrolytic changes which render them unsuitable for the cultivation of the more delicate micro-organisms. The use of superheated steam should be restricted almost entirely to the disinfection of such contaminated articles, old cultivations, etc., as cannot be dealt with by dry heat or the actual furnace. Sterilisation by means of superheated steam is carried out in a special boiler—Chamberland's autoclave (Fig. 30). The autoclave consists of a stout copper cylinder, provided with a copper or gun-metal lid, which is secured in place by means of bolts and thumbscrews, the joint between the cylinder and its lid being hermetically sealed by the interposition of a rubber washer. The cover is perforated for a branched tube carrying a vent cock, a manometer, and a safety valve. The copper boiler is mounted in the upper half of a cylindrical sheet-iron case—two concentric circular rows of Bunsen burners, each circle having an independent gas-supply, occupying the lower half. In the interior of the boiler is a large movable wire basket, mounted on legs, for the reception of the articles to be sterilised.

To Use the Autoclave.—

1. Pack the articles to be sterilised in the wire basket.

2. Run water into the boiler to the level of the bottom of the basket; also fill the contained flasks and tubes with water.

3. See that the rubber washer is in position, then replace the cover and fasten it tightly on to the autoclave by means of the thumbscrews.

4. Open the vent cock and light both rings of burners.

5. When steam is issuing in a steady, continuous stream from the vent tube, shut off the vent cock and extinguish the outer ring of gas burners.

6. Wait until the index of the manometer records a temperature of 120° C., then regulate the gas and the spring safety valve in such a manner that this temperature is just maintained, and leave it thus for twenty minutes. In the more expensive patterns of autoclave this regulation of the safety valve is carried out automatically, the manometer being fitted with an adjustable pointer which can be set to any required pressure-temperature and so arranged that when the index of the manometer coincides with the adjustable hand the safety valve is opened.

7. Extinguish the gas and allow the manometer index to fall to zero.

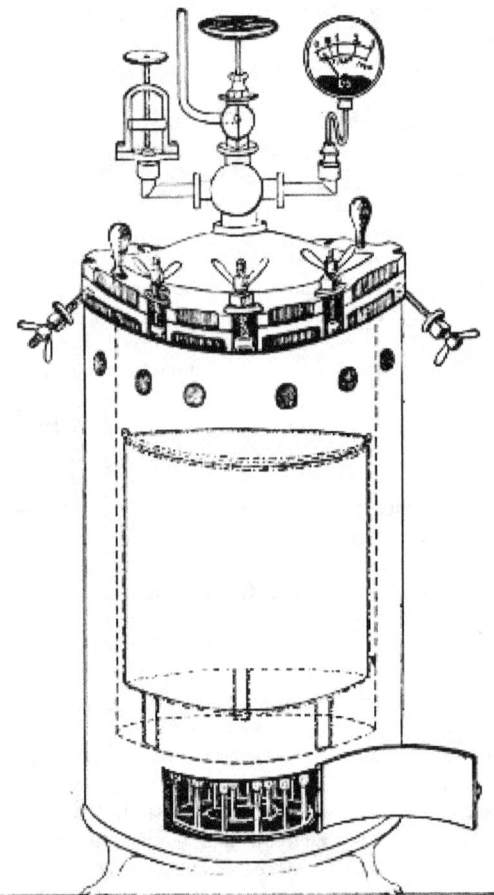

Fig. 30.—Chamberland's Autoclave.

8. Now open the vent cock slowly, and allow the internal pressure to adjust itself to that of the atmosphere.

9. Remove the cover and take out the sterilised contents.

Sterilisation Periods.—An exceedingly useful device for the timing of sterilisation periods (and indeed for many other operations in the laboratory) is the

ELECTRIC SIGNAL TIMING CLOCK.

This is a clock of American type in which the face is surrounded by a metal plate having a series of 60 holes at equal distances apart, corresponding to the minutes on the dial. This plate is connected with one of the poles of a dry battery, the other pole of which is connected to the metal case of the clock for the purpose of actuating an ordinary magnet alarm bell. In the centre of each of the holes in the plate a metal rod is fixed, which then passes through an insulating ring and projects inside the clock face, where it makes contact with the hour hand. The clock is mounted on a heavy base, with a key-board containing 20 numbered plugs. If one of the plugs is inserted in a hole in the plate it makes contact with the rod, and when the hour hand of the clock touches the other end the circuit is completed and the bell starts ringing. The period of this friction contact is approximately 20 seconds. The clock can therefore be used for electrically noting the periods of time from one minute by multiples of one minute up to one hour.

Fig. 31.—Electric signal timing clock.

Filtration.—(*a*) *Cotton-wool Filter.*—Practically the only method in use in the laboratory for the sterilisation of air or of a gas is by filtration through dry cotton-wool or glass-wool, the fibres of which entangle the micro-organisms and prevent their passage.

Perhaps the best example of such a filter is the cotton-wool plug which closes the mouth of a culture tube. Not only does ordinary diffusion take place through it, but if a tube plugged in the usual manner with cotton-wool is removed from the hot incubator, the temperature of the contained air rapidly falls to that of the laboratory, and a partial vacuum is formed; air passes into the tube, through the cotton-wool plug, to restore the equilibrium, and, so long as the plug remains dry, in a germ-free condition. If, however, the plug becomes moist, either by absorption from the atmosphere, or from liquids coming into contact with it, micro-organisms (especially the mould fungi) commence to multiply, and the long thread forms rapidly penetrate the substance of the plug, and gain access to and contaminate the interior of the tube.

Fig. 32.—Cotton-wool air filter.

Method.—

If it is desired to sterilise gases before admission to a vessel containing a pure cultivation of a micro-organism, as, for instance, when forcing a current of oxygen over or through a broth cultivation of the diphtheria bacillus, this can be readily effected as follows:

1. Take a length of glass tubing of, say, 1.5 cm. diameter, in the centre of which a bulb has been blown, fill the bulb with dry cotton-wool (Fig. 32), wrap a layer of cotton-wool around each end of the tube, and secure in position with a turn of thin copper wire or string; then sterilise the piece of apparatus in the hot-air oven.

2. Prepare the cultivation in a Ruffer or Woodhead flask (Fig. 33) the inlet tube of which has its free extremity enveloped in a layer of cotton-wool, secured by thread or wire, whilst the exit tube is plugged in the usual manner.

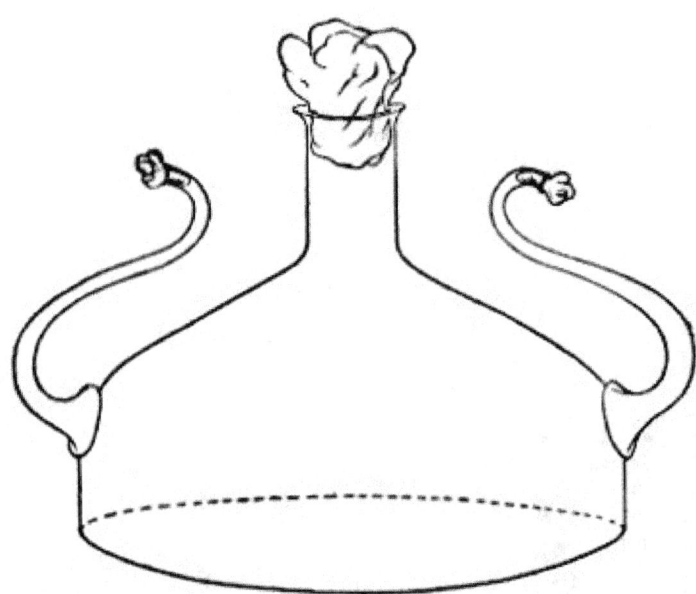

Fig. 33.—Ruffer's flask.

3. Sterilise a short length of rubber tubing by boiling. Transfer it from the boiling water to a beaker of absolute alcohol.

4. When all is ready remove the rubber tube from the alcohol by means of a pair of forceps, drain it thoroughly, and pass through the flame of a Bunsen burner to burn off the last traces of alcohol.

5. Remove the cotton-wool wraps from the entry tube of the flask and from one end of the filter tube and rapidly couple them up by means of the sterile rubber tubing.

6. Connect the other end of the bulb tube with the delivery tube from the gas reservoir.

The gas in its passage through the dry sterile cotton-wool in the bulb of the filter tube will be freed from any contained micro-organisms and will enter the flask in a sterile condition.

(*b*) *Porcelain Filter.*—The sterilisation of liquids by filtration is effected by passing them through a cylindrical vessel, closed at one end like a test-tube, and made either of porous "biscuit" porcelain, hard-burnt and unglazed (Chamberland system), or of Kieselguhr, a fine diatomaceous earth (Berkefeld system), and termed a "bougie" or "candle" (Fig. 34).

Note.—In selecting candles for use in the laboratory avoid those with metal fittings, since during sterilisation cracks develop at the junction of the metal and the siliceous material owing to the unequal expansion.

In this method the bacteria are retained in the pores of the filter while the liquid passes through in a germ-free condition.

It is obvious that to be effective the pores of the filter must be extremely minute, and therefore the rate of filtration will usually be slow. Chamberland filter candles possess finer channels than Berkefeld candles and consequently filter much more slowly. To overcome this disadvantage, either aspiration or pressure, or a combination of these two forces, may be employed to hasten the process.

Doultons white porcelain filters it may be noted are as efficient as the Chamberland candles and filter rather more rapidly.

Apparatus Required.—

1. Separatory funnel containing the unfiltered fluid.

2. Sterile filter candle (Fig. 34), the open end fitted with a rubber stopper (Fig. 34, *a*) perforated to receive the delivery tube of the separatory funnel, and its neck passed through a large rubber washer (Fig. 34, *b*) which fits the mouth of the filter flask.

3. Sterile filter flask of suitable size, for the reception of the filtered fluid, its mouth closed by a cotton-wool plug.

4. Water injector Sprengel (see Fig. 38, *c*) pump, or Geryk's pump (an air pump on the hydraulic principle, sealed by means of low vapor-tension oil, Fig. 35).

If this latter is employed, a Wulff's bottle, fitted as a wash-bottle and containing sulphuric acid, must be interposed between the filter flask and the pump, in order to prevent moist air reaching the oil in the pump.

5. Air filter (*vide* page 40) sterilised.

6. Pressure tubing.

7. Screw clamps (Fig. 36).

Method.—

1. Couple the exhaust pipe of the suction pump with the lateral tube of the filter flask (first removing the cotton-wool plug from this latter), by means of pressure tubing, interposing, if necessary, the wash-bottle of sulphuric acid.

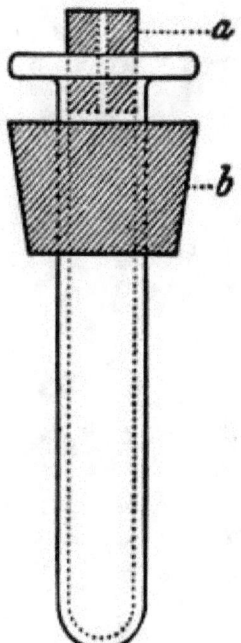

Fig. 34.—Porcelain filter candle.

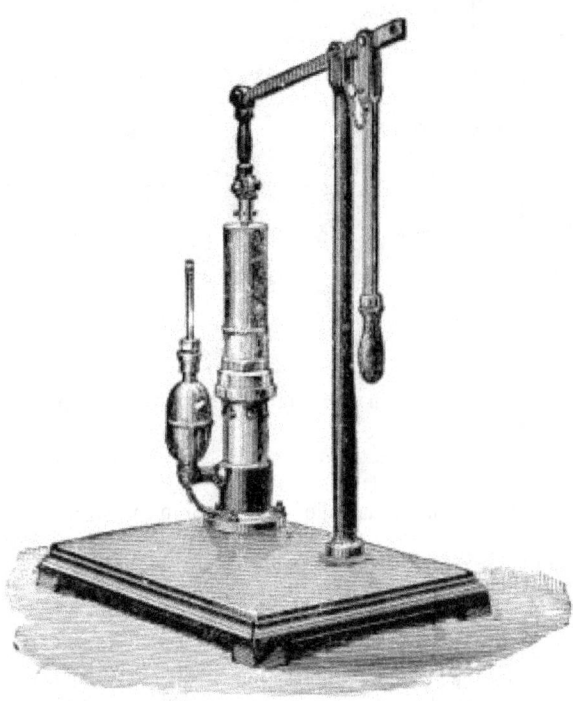

Fig. 35.—Geryk air pump.

2. Remove the cotton-wool plug from the neck of the filter flask and adjust the porcelain candle in its place.

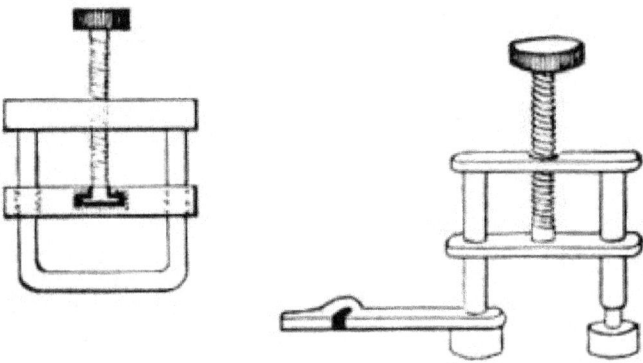

Fig. 36.—Screw clamps.

3. Attach the nozzle of the separatory funnel to the filter candle by means of the perforated rubber stopper (Fig. 37).

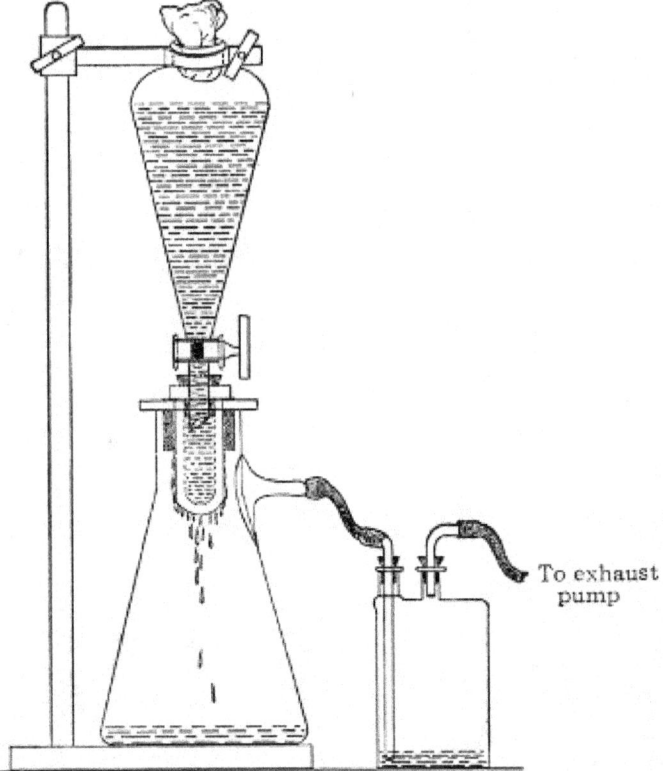

Fig. 37.—Apparatus arranged for filtering—aspiration.

4. Open the tap of the funnel, and exhaust the air from the filter flask and wash-bottle; maintain the vacuum until the filtration is complete.

5. When the filtration is completed close the tap of the funnel; adjust a screw clamp to the pressure tubing attached to the lateral branch of the filter flask; screw it up tightly, and disconnect the acid wash-bottle.

6. Attach the air filter to the open end of the pressure tubing; open the screw clamp gradually, and allow filtered air to enter the flask, to abolish the negative pressure.

7. Detach the rubber tubing from the lateral branch of the flask, flame the end of the branch in the Bunsen, and plug its orifice with sterile cotton-wool.

8. Remove the filter candle from the mouth of the flask, flame the mouth, and plug the neck with sterile cotton-wool.

9. Disinfect the filter candle and separatory funnel by boiling.

If it is found necessary to employ pressure in addition to or in place of suction, insert a perforated rubber stopper into the mouth of the separatory funnel and secure in position with copper wire; next fit a piece of glass tubing through the stopper, and connect the external orifice with an air-pressure pump of some kind (an ordinary foot pump such as is employed for inflating bicycle tyres is one of the most generally useful, for this purpose) or with a cylinder of compressed air or other gas.

In order to filter a large bulk of fluid very rapidly it is necessary to use a higher pressure than glass would stand, and in these cases the metal receptacle designed by Pakes (Fig. 38, *a*), to hold the filter candle itself as well as the fluid to be filtered, should be employed. (A vacuum must also be maintained in the filter flask, by means of an exhaust pump, during the entire process.)

This piece of apparatus consists of a brass cylinder, capacity 2500 c.c., with two shoulders; and an opening in the neck at each end, provided with screw threads.

A nut carrying a pressure gauge fits into the top screw; and into the bottom is fitted a brass cylinder carrying the filter candle and prolonged downwards into a delivery tube. Leakage is prevented by means of rubber washers.

Into the top shoulder a tube is inserted, bent at right angles and provided with a tap. All the brass-work is tinned inside (Fig. 38, *a*). In use the reservoir is generally mounted on a tripod stand.

To Sterilise.—
1. Insert the filter candle into its cylinder and screw this loosely on.

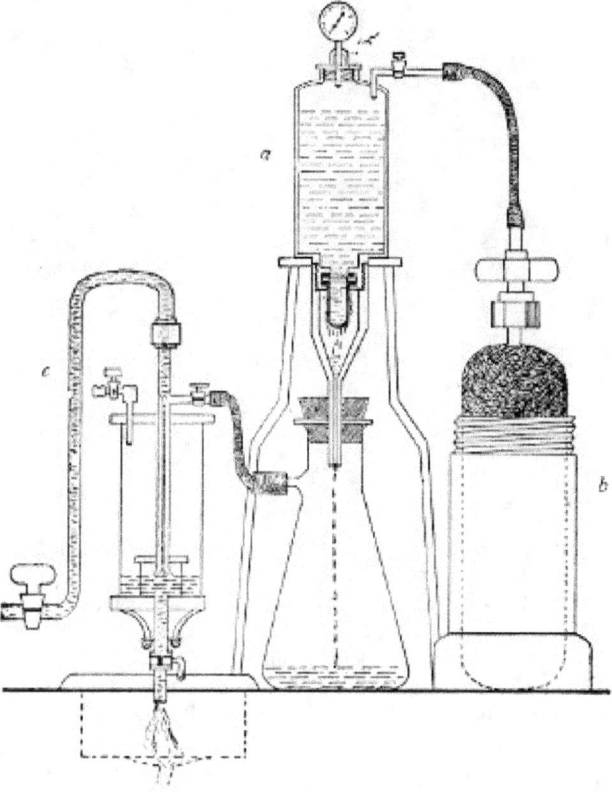

Fig. 38.—Pakes' filtering reservoir—pressure and aspiration.

2. Wrap a layer of cotton-wool around the delivery tube and fasten in position.
3. Remove the nut carrying the pressure gauge and plug the neck with cotton-wool.
4. Heat the whole apparatus in the autoclave at 120° C. for twenty minutes.

Method.—
1. Remove the apparatus from the autoclave, and allow it to cool.
2. Screw home the box carrying the bougie.
3. Set the apparatus up in position, with its delivery tube (from which the cotton-wool wrapping has been removed) passing through a perforated rubber stopper in the neck of a filter flask.

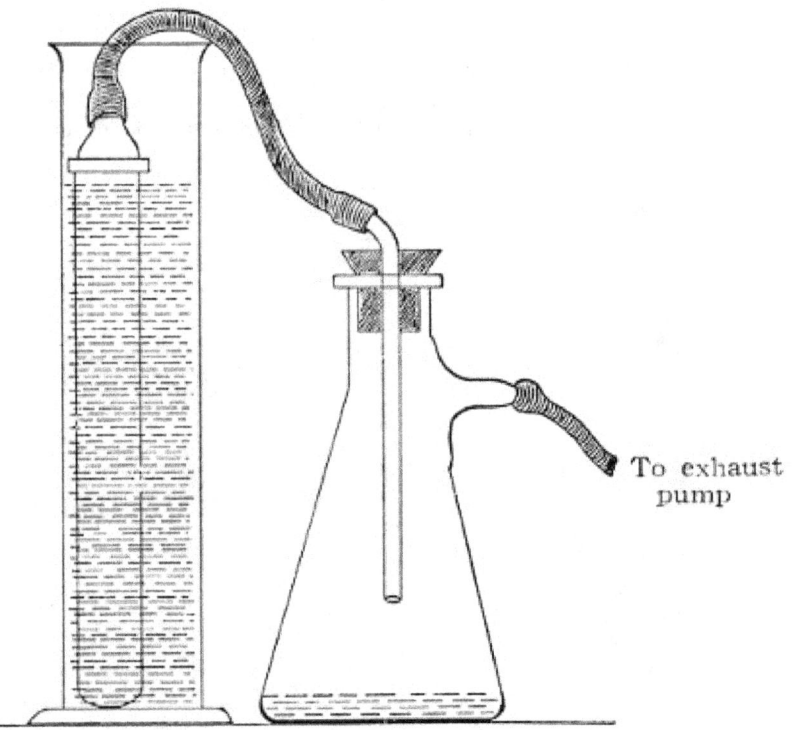

Fig. 39.—Closed candle arranged for filtering.

4. Fill the fluid to be filtered into the cylinder and screw on the nut carrying the pressure gauge. (This nut should be immersed in boiling water for a few minutes previous to screwing on, in order to sterilise it.)

5. Connect the horizontal arm of the entry tube with a cylinder of compressed oxygen (or carbon dioxide, Fig. 38, *b*), by means of pressure tubing.

6. Connect the lateral arm of the filter flask with the exhaust pump (Fig. 38, *c*) and start the latter working.

7. Open the tap of the gas cylinder; then open the tap on the entry tube of the filter cylinder and raise the pressure in its interior until the desired point is recorded on the manometer. Maintain this pressure, usually one or one and a half atmospheres, until filtration is completed, by regulating the tap on the entry tube.

Some forms of filter candle are made with the open end contracted into a delivery nozzle, which is glazed. In this case the apparatus is fitted up in a slightly different manner; the fluid to be filtered is contained in an open cylinder into which the candle is plunged, while its delivery nozzle is connected with the filter flask by means of a piece of flexible pressure tubing (previously sterilised by boiling), as in figure 39.

IV. THE MICROSCOPE.

The essentials of a microscope for bacteriological work may be briefly summed up as follows:

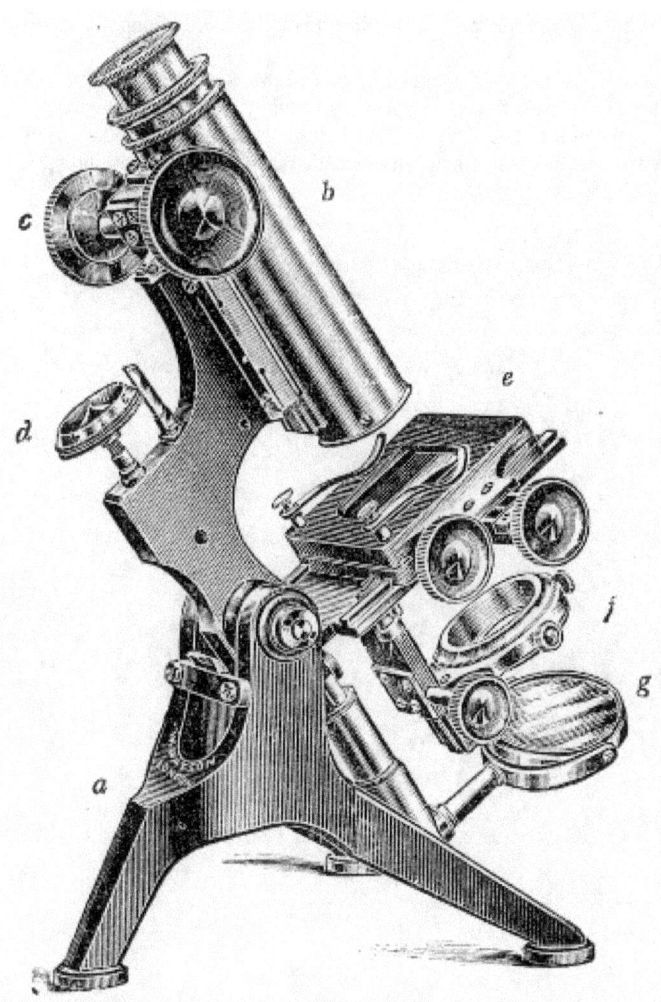

Fig. 40.—Microscope stand.

The instrument, of the monocular type, must be of good workmanship and well finished, rigid, firm, and free from vibration, not only when upright, but also when inclined to an angle or in the horizontal position. The various joints and movements must work smoothly and precisely, equally free from the defects of "loss of time" and "slipping." All screws, etc., should conform to the Royal Microscopical Society's standard. It must also be provided with good lenses and a sufficiently large stage. The details of its component parts, to which attention must be specially directed, are as follows:

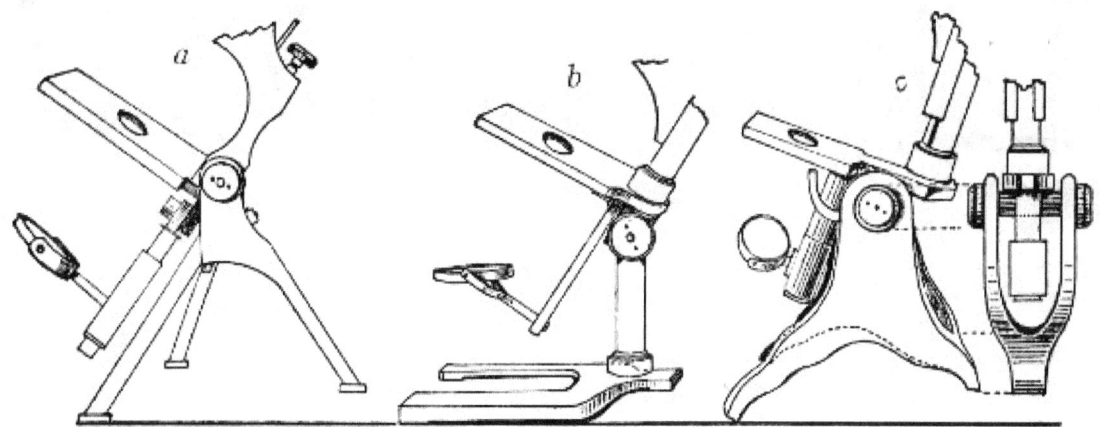

Fig. 41.—Foot, three types.

1. The Base or Foot (Fig. 40, *a*).—Two elementary forms—the tripod (Fig. 41, *a*) and the vertical column set into a plate known as the "horse-shoe" (Fig. 41, *b*)—serve as the patterns for countless modifications in shape and size of this portion of the stand. The chief desiderata—stability and ease of manipulation—are attained in the first by means of the "spread" of the three feet, which are usually shod with cork; in the second, by the dead weight of the foot-plate. The tripod is mechanically the more correct form, and for practical use is much to be preferred. Its chief rival, the Jackson foot (Fig. 41, *c*), is based upon the same principle, and on the score of appearance has much to recommend it.

2. The **body tube** (Fig. 40, *b*) may be either that known as the "long" or "English" (length 250 mm.), or the "short" or "Continental" (length 160 mm.). Neither length appears to possess any material advantage over the other, but it is absolutely necessary to secure objectives which have been manufactured for the particular tube length chosen. In the high-class microscope of the present day the body tube is usually shorter than the Continental, but is provided with a draw tube which, when fully extended, gives a tube length greater than the English, thus permitting the use of either form of objective.

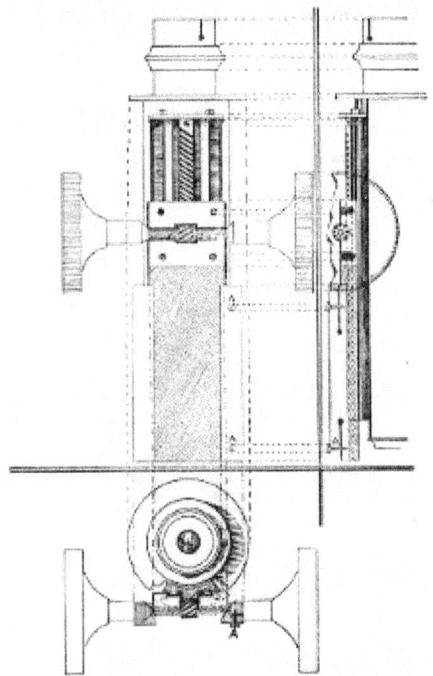

Fig. 42.—Coarse adjustment.

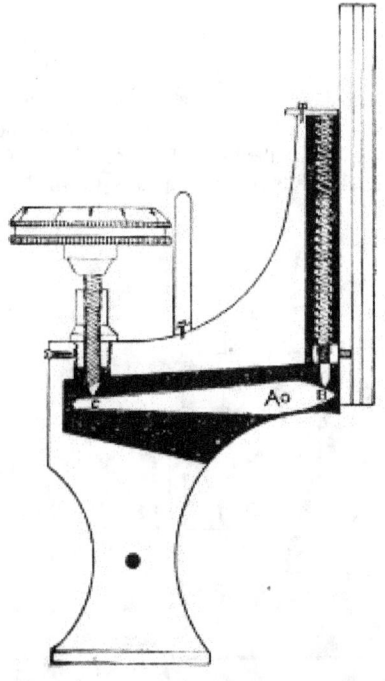

Fig. 43.—Fine adjustment.

For practical purposes the tube length = distance from the end of the nosepiece to the eyeglass of the ocular. This is the measurement referred to in speaking of "long" or "short" tube.

3. The **coarse adjustment** (Fig. 40, *c*) should be a rack-and-pinion movement, steadiness and smoothness of action being secured by means of accurately fitting dovetailed bearings and perfect correspondence between the teeth of the rack and the leaves of the pinion (Fig. 42). Also provision should be made for taking up the "slack" (as by the screws *AA*, Fig. 42).

4. The **fine adjustment** (Fig. 40, *d*) should on no account depend upon the direct action of springs, but should be of the lever pattern, preferably the Nelson (Fig. 43). In this form the unequal length of the arms of the lever secures very delicate movement, and, moreover, only a small portion of the weight of the body tube is transmitted to the thread of the vertical screw actuating the movement.

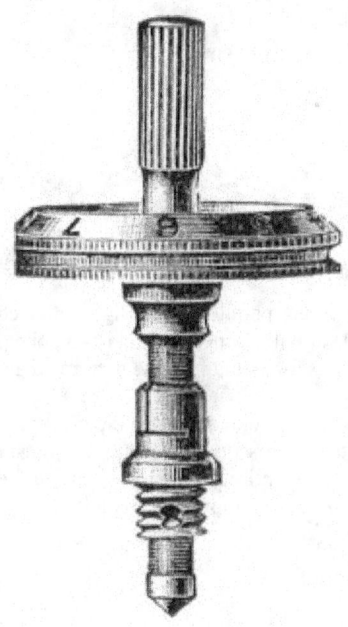

Fig. 44.—Spindle head to fine adjustment.

A spindle milled head (Fig. 44) will be found a very useful device to have fitted in place of the ordinary milled head controlling the fine adjustment. In this contrivance the axis of the milled head is prolonged upward in a short column, the diameter of which is one-sixth of that of the head. The spindle can be rapidly rotated between the fingers for medium power adjustments while the larger milled head can be slowly moved when focussing high powers.

5. The **stage** (Fig. 40, *e*) should be square in shape and large in area—at least 12 cm.—flat and rigid, in order to afford a safe support for the Petri dish used for plate cultivations; and should be supplied with spring clips (removable at will) to secure the 3 by 1 glass slides.

A mechanical stage must be classed as a necessity rather than a luxury so far as the bacteriologist is concerned, as when working with high powers, and especially when examining hanging-drop specimens, it is almost impossible to execute sufficiently delicate movements with the fingers. In selecting a mechanical stage, preference should be given to one which forms an integral part of the instrument (Fig. 45) rather than one which needs to be clamped on to an ordinary plain stage every time it is required, and its traversing movements should be controlled by stationary milled heads (Fig. 45, *AA'*). The shape of the aperture is a not unimportant point; it should be square to allow of free movement over the substage condenser. The mechanical stage should be tapped for three (removable) screw studs to be used in place of the sliding bar, so that if desired the Vernier finder (Fig. 45, *BB'*), such as is usually fitted to this class of stage, or a Maltwood finder, may be employed.

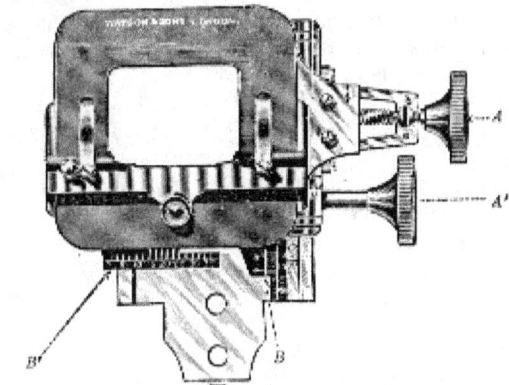

Fig. 45.—Mechanical stage.

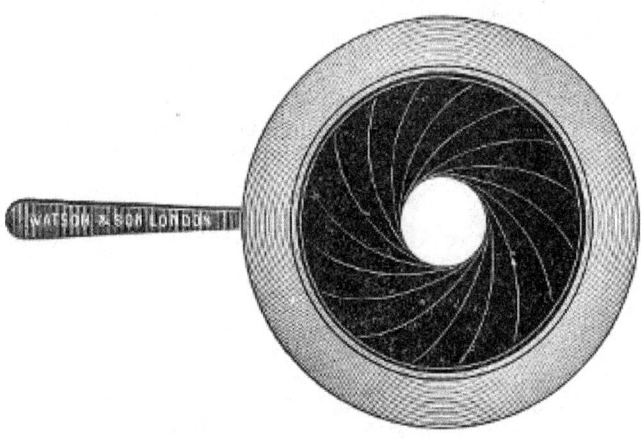

Fig. 46.—Iris diaphragm.

6. Diaphragm.—Separate single diaphragms must be avoided; a revolving plate pierced with different sized apertures and secured below the stage is preferable, but undoubtedly the best form is the "iris" diaphragm (Fig. 46) which enters into the construction of the substage condenser.

7. The **substage condenser** is a necessary part of the optical outfit. Its purpose is to collect the beam of parallel rays of light reflected by the plane mirror, by virtue of a short focus system of lenses, into a cone of large aperture (reducible at will by means of an iris diaphragm mounted as a part of the condenser), which can be accurately focussed on the plane of the object. This focussing must be performed anew for each object, on account of the variation in the thickness of the slides.

The form in most general use is that known as the Abbé (Fig. 47) and consists of a plano-convex lens mounted above a biconvex lens. This combination is carried in a screw-centering holder known as the substage below the stage of the microscope (Fig. 40 *f*), and must be accurately adjusted so that its optical axis coincides with that of the objective. Vertical movement of the entire substage apparatus effected by means of a rack and pinion is a decided advantage, and some means should be provided for temporarily removing the condenser from the optical axis of the microscope.

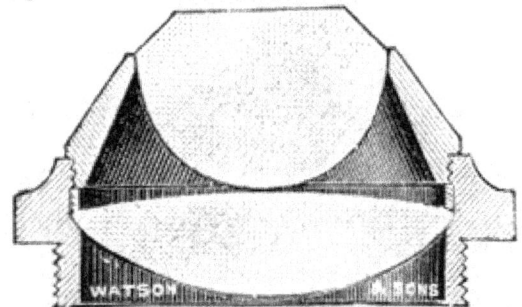

Fig. 47—Optical part of Abbé illuminator.

With the oil immersion objective, however, an **achromatic condenser**, giving an illuminating cone of about 0.9, should be used if the full value of the lens is to be obtained. It is generally assumed that a good objective requires an illuminating cone equivalent to two-thirds of its numerical aperture. The best Abbé condenser transmits a cone of about .45 whilst the aperture of the 1/12 inch immersion lenses of different makers varies from 1.0 to 1.4, hence, the efficiency of these lenses is much curtailed if the condenser is merely the Abbé. These improved condensers must be absolutely centered to the objective and capable of very accurate focussing otherwise much of their value is lost.

8. Mirrors.—Below the substage condenser is attached a gymbal carrying a reversible circular frame with a plane mirror on one side and a concave mirror on the other (Fig. 40, *g*). The plane mirror is that usually employed, but occasionally, as for example when using low powers and with the condenser racked down and thrown out of the optical axis, the concave mirror is used.

9. Oculars, or Eyepieces.—Those known as the Huyghenian oculars (Fig. 48) will be sufficient for all ordinary work without resorting to the more expensive "compensation" oculars. Two or three, magnifying the "real" image (formed by the objective) four, six, or eight times respectively, form a useful equipment.

As an accessory **Ehrlich's Eyepiece** is a very useful piece of apparatus when the enumeration of cells or bacteria has to be carried out. This is an ordinary eyepiece fitted with an adjustable square diaphragm operated by a lever projecting from the side of the mount. Three notches are made in one of the sides of the square and by moving the lever square aperture can be reduced to three-quarters, one-half or one-quarter of the original size.

10. Objectives.—Three objectives are necessary: one for low-power work—*e. g.*, 1 inch, 2/3 inch, or 1/2 inch; one for high-power work—*e. g.*, 1/12 inch oil immersion lens; and an intermediate "medium-power" lens—*e. g.*, 1/6 inch or 1/8 inch (dry). These lenses must be carefully selected, especial attention being paid to the following points:

(*a*) *Correction of Spherical Aberration.*—Spherical aberration gives rise to an ill-defined image, due to the central and peripheral rays focussing at different points.

(*b*) *Correction of Chromatic Aberration.*—Chromatic aberration gives rise to a coloured fringe around the edges of objects due to the fact that the different-coloured rays of the spectrum possess varying refrangibilities and that a simple lens acts toward them as a prism.

(*c*) *Flatness of Field.*—The ideal visual field would be large and, above all, *flat*; in other words, objects at the periphery of the field would be as distinctly "in focus" as those in the centre. Unfortunately, however, this is an optical impossibility and the field is always spherical in shape. Some makers succeed in giving a larger central area that is in focus at one time than others, and although this may theoretically cause an infinitesimal sacrifice of other qualities, it should always be sought for. Successive zones and the entire peripheral ring should come into focus with the alteration of the fine adjustment. This simultaneous sharpness of the entire circle is an indication of the perfect centering of the whole of the lenses in the objective.

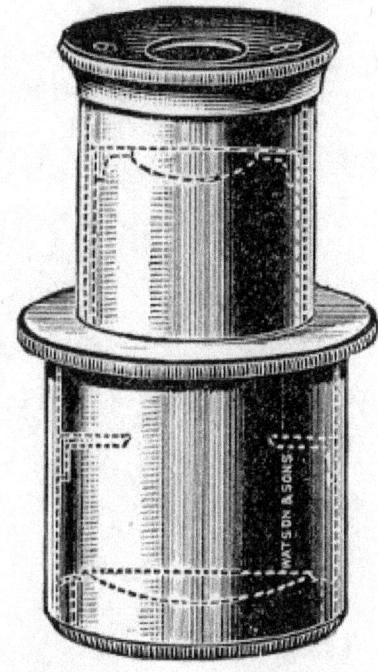

Fig. 48.—Huyghenian eyepiece.

(*d*) *Good Definition.*—Actual magnification is, within limits, of course, of less value than clear definition and high resolving power, for it is upon these properties we depend for our knowledge of the detailed structure of the objects examined.

(*e*) *Numerical Aperture* (*N. A.*).—The numerical aperture may be defined, in general terms, as the ratio of the *effective* diameter of the back lens of the objective to its equivalent focal length. The determination of this point is a process requiring considerable technical skill and mathematical ability, and is completely beyond the powers of the average microscopist.

Although with the increase in power it is correspondingly difficult to combine all these corrections in one objective, they are brought to a high pitch of excellence in the present-day "achromatic" objectives, and so remove the necessity for the use of the higher priced and less durable apochromatic lenses.

In selecting objectives the best "test" objects to employ are:

1. A thin (one cell layer), even   }   { 1", 2/3", 1/2":
"blood film," stained with Jenner's   }   for   { 1/6", 1/8"
or Romanowsky's stain.   }   { 1/12" oil

2. A thin cover-slip preparation   }
of a young cultivation of   }   {1/8" dry
B. diphtheriæ (showing   }   for   {
segmentation) stained with   }   {1/12" oil
methylene-blue.

Accessories.—*Eye Shade* (Fig. 49).—This piece of apparatus consists of a pear-shaped piece of blackened metal or ebonite, hinged to a collar which rotates on the upper part of the body tube of the microscope. It can be used to shut out the image of surrounding objects from the unoccupied eye, and when carrying out prolonged observations will be found of real service.

Nosepiece.—Perhaps the most useful accessory is a nosepiece to carry two of the objectives (Fig. 50), or, better still, all three (Fig. 51). This nosepiece, preferably constructed of aluminium, must be of the covered-in type, consisting of a curved plate attached to the lower end of the body tube—a circular aperture being cut to correspond to the lumen of that tube. To the under surface of this plate is pivoted a similarly curved plate, fitted with three tubulures, each of which carries an objective. By rotating the lower plate each of the objectives can be brought successively in to the optical axis of the microscope.

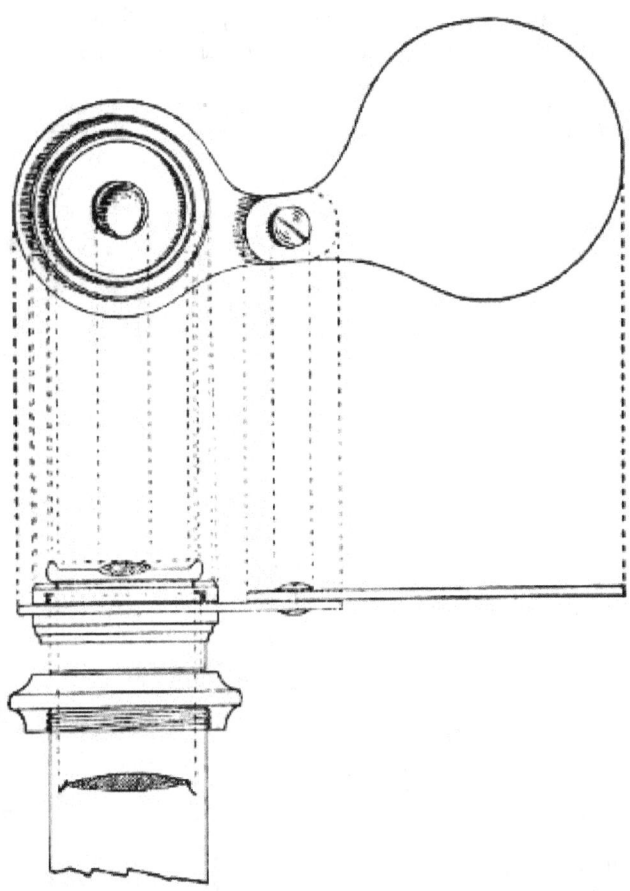

Fig. 49.—Eye shade.

For critical work and particularly for photo-micrography, however, the interchangeable nosepiece is by no means perfect as it is next to impossible to secure accurate centreing of each lens in the optical axis. For special purposes, therefore, it is necessary to employ a special nosepiece such as that made by Zeiss or Leitz into which each objective slides on its own carrier and upon which it is accurately centred.

Fig. 50.—Double nosepiece.

Fig. 51.—Triple nosepiece.

Warm Stage (Fig. 52).—This is a flat metal case containing a system of tubes through the interior of which water of any required temperature can be circulated. It is made to clamp on to the stage of the microscope by the screws $A\ A'$, and is perforated with a large hole coinciding with the optical axis of the microscope; a short tube B, projecting from one end of the warm stage permits water of the desired temperature to be conducted from a reservoir through a length of rubber tubing to the interior of the stage and a similar tube at the other end B' of the stage allows exit to the waste water. By raising the temperature of hanging-drop preparations, etc., placed upon it, above that of the surrounding atmosphere, the warm stage renders possible exact observations on spore germination, hanging-drop cultivations, etc.

Fig. 52.—
Warm stage.

A better form is the electrical hot stage designed by Lorrain Smith; it requires the addition of a lamp resistance and sliding rheostat, also a delicate ammeter reading to .01 of an ampère. It consists of a wooden frame supporting a flat glass bulb with a long neck bent upward at an obtuse angle (Fig. 53). The bulb is filled with liquid paraffin, which rises in the open neck when expanded by heat. The neck also accommodates the thermometer. Two coils of manganin wire run in the paraffin at opposite sides of the bulb (outside the field of vision), coupled to brass terminals on the wooden frame by platinum wire fused into the glass. The resistance of the two coils in series is about 10 ohms. A current of 2-1/2 ampères is needed, and is conducted to the coils in the stage through the rheostat. With the help of the ammeter any desired temperature can be obtained and maintained, up to about 200° C. If immersion oil contact is made between the top lens of the condenser and the lower surface of the bulb, this stage works very well indeed with the 1/12-inch oil immersion lens.

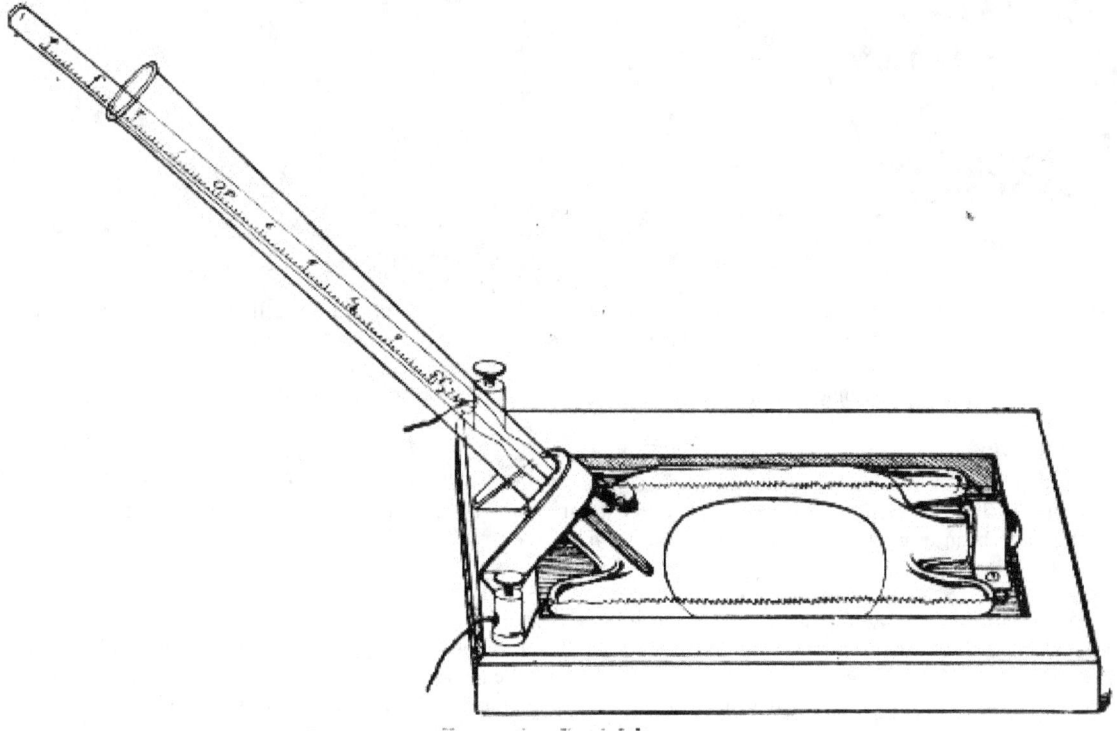

Fig. 53.—
Lorrain Smith's warm stage.

Dark Ground or Paraboloid Condenser.—This is an immersion substage condenser of high aperture by means of which unstained objects such as bacteria can be shown as bright white particles upon a dense black background. The central rays of light are blocked out by means of an opaque stop while the peripheral rays are reflected from the paraboloidal sides of the condenser and refracted by the object viewed. To obtain the best results with this type of condenser a powerful illuminant—such as a small arc lamp or an incandescent gas lamp—is needed, together with picked slides of a certain thickness (specified for the particular make of condenser but generally 1 mm.) and specially thin cover-glasses (not more than 0.17 mm.). The objective must not have a higher NA than 1.0, consequently immersion lenses must be fitted with an internal stop to cut down the aperture.

Micrometer.—Some form of micrometer for the purpose of measuring bacteria and other objects is also essential. Details of those in general use will be found in the following pages.

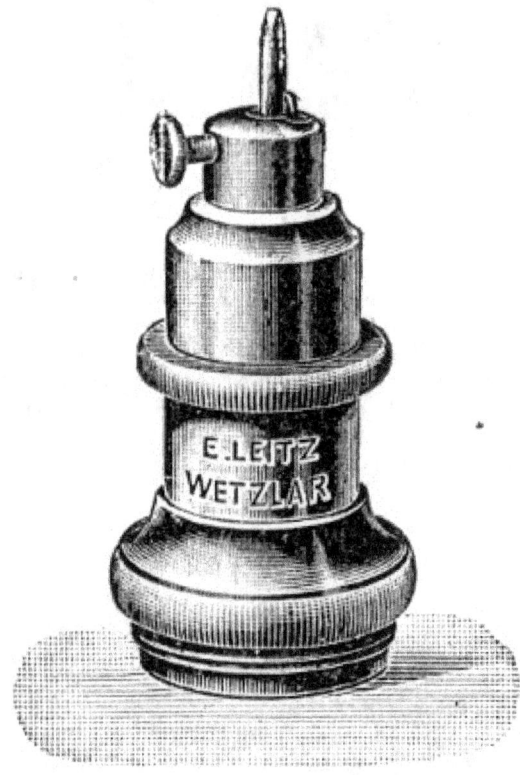

Fig. 54—Diamond Object marker.

Object Marker (Fig. 54).—This is an exceedingly useful piece of apparatus. Made in the form of an objective, the lenses are replaced by a diamond point, set slightly out of the centre, which can be rotated by means of a milled plate. Screwed on to the nosepiece in place of the objective, rotation of the diamond point will rule a small circle on the object slide to permanently record the position of an interesting portion of the specimen. The diamond is mounted on a spring which regulates the pressure, and the size of the circle can be adjusted by means of a lateral screw.

METHODS OF MICROMETRY.

The unit of length as applied to the measurement of microscopical objects is the one-thousandth part of a millimetre (0.001 mm.), denominated a *micron* (sometimes, and erroneously, referred to as a micro-millimetre), and indicated in writing by the Greek letter μ. Of the many methods in use for the measurement of bacteria, three only will be here described, viz.:

(*a*) By means of the Camera Lucida.
(*b*) By means of the ocular or Eyepiece Micrometer.
(*c*) By means of the Filar Micrometer (Ramsden's micrometer eyepiece).

For each of these methods a **stage micrometer** is necessary. This is a 3 by 1 inch glass slip having engraved on it a scale divided to hundredths of a millimetre (0.01 mm.), every tenth line being made longer than the intervening ones, to facilitate counting; and from these engraved lines the measurement in every case is evaluated. A cover-glass is cemented over the scale to protect it from injury.

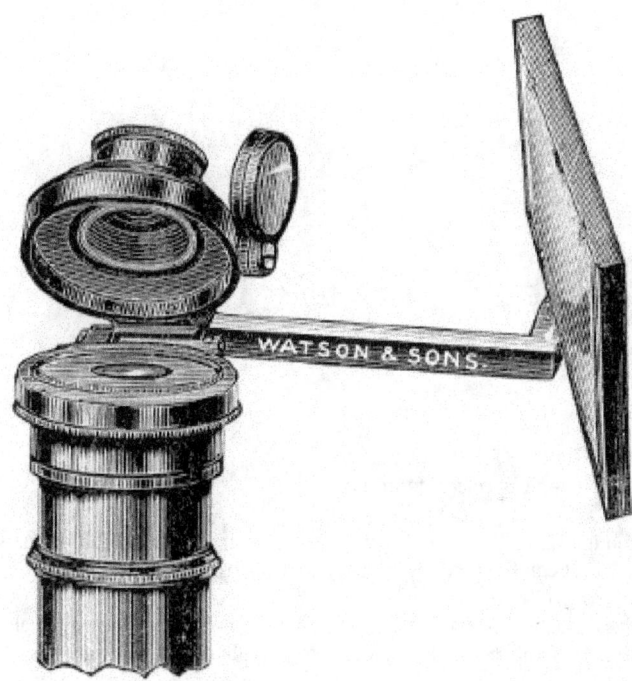

Fig. 55.—Camera lucida, Abbé pattern.

(*a*) By means of the Camera Lucida.

1. Attach a camera lucida (of the Wollaston, Beale, or Abbé pattern) (Fig. 55) to the eyepiece of the microscope.
2. Adjust the micrometer on the stage of the microscope and accurately focus the divisions.
3. Project the scale of the stage micrometer on to a piece of paper and with pen or pencil sketch in the magnified image, each division of which corresponds to 10μ. Mark on the paper the optical combination (ocular objective and tube length) employed to produce this particular magnification.
4. Repeat this procedure for each of the possible combinations of oculars and objectives fitted to the microscope supplied, and carefully preserve the scales thus obtained.

To measure an object by this method simply project the image on to the scale corresponding to the particular optical combination in use at the moment. Read off the number of divisions it occupies and express them as *micra*.

In place of preserving a scale for each optical combination, the object to be measured and the micrometer scale may be projected and sketched, in turn, on the same piece of paper, taking particular care that the centre of the eyepiece is 25 cm. from the paper on which the divisions are drawn.

Fig. 56.—Eyepiece micrometer, ordinary.

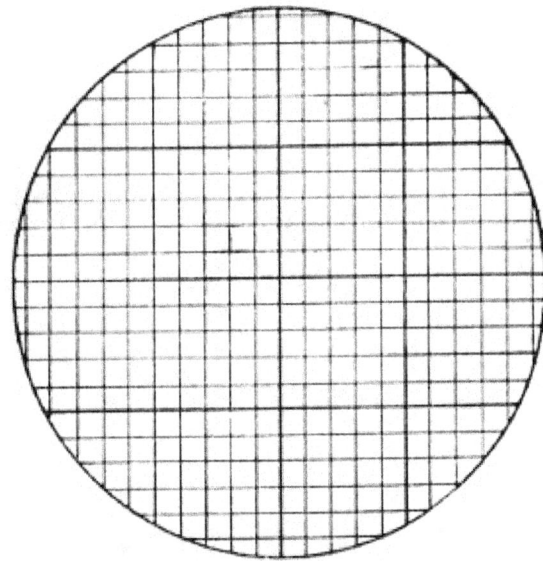

Fig. 57.—Eyepiece micrometer, net.

(*b*) By means of the Eyepiece Micrometer.

The **eyepiece micrometer** is a circular glass disc having engraved on it a scale divided to tenths of a millimetre (0.1 mm.) (Fig. 56), or the entire surface ruled in 0.1 mm. squares (the net micrometer) (Fig. 57). It can be fitted inside the mount of any ocular just above the aperture of the diaphragm and must be adjusted exactly in the focus of the eye lens.

Some makers mount the glass disc together with a circular cover-glass in such a way that when placed in position in any Huyghenian eyepiece of their own manufacture, the scale is exactly in focus for normal vision. Special eyepieces are also obtainable having a sledging adjustment to the eye lens for focussing the micrometer.

The value of one division of the micrometer scale must first be ascertained for each optical combination by the aid of the stage micrometer, thus:

1. Insert the eyepiece micrometer inside the ocular and adjust the stage micrometer on the stage of the microscope.
2. Focus the scale of the stage micrometer accurately; the lines will appear to be immediately below those of the eyepiece micrometer. Make the lines on the two micrometers parallel by rotating the ocular.
3. Make two of the lines on the ocular micrometer coincide with those bounding one division of the stage micrometer; this is effected by increasing or diminishing the tube length; and note the number of included divisions.
4. Calculate the value of each division of the eyepiece micrometer in terms of μ, by means of the following formula:

$x = 10\, y.$

Where x = the number of included divisions of the eyepiece micrometer.

y = the number of included divisions of the stage micrometer.

5. Note the optical combination employed in this experiment and record it with the calculated micrometer value.

Repeat this process for each of the other combinations. Carefully record the results.

To measure an object by this method read off the number of divisions of the eyepiece micrometer it occupies and express the result in *micra* by a reference to the standard value for the particular optical combination employed.

Zeiss prepares a compensating eyepiece micrometer for use with his apochromatic objectives, the divisions of which are so computed that (with a tube length of 160 mm.) the value of each is equivalent to as many *micra* as there are millimetres in the focal length of the objective employed.

Wright's Eikonometer is really a modification of the eyepiece micrometer for rapidly measuring microscopical objects by direct inspection, having previously determined the magnifying power of the particular optical combination employed. It is a small piece of apparatus resembling an eyepiece, with a sliding eye lens, which can be accurately focussed on a micrometer scale fixed within the instrument. When placed over the microscope ocular the divisions of this scale measure the actual size of the virtual image in millimetres.

In order to use this instrument for direct measurement, it is first necessary to determine the magnifying power of each combination of ocular, tube length and objective.

Place a stage micrometer divided into hundredths of a millimetre on the microscope stage and focus accurately.

Rest the eikonometer on the eyepiece. Observation through the eikonometer shows its micrometer scale superposed on the image of the stage micrometer.

Rotate the eikonometer until the lines on the two scales are parallel, and make the various adjustments to ensure that two lines on the eikonometer scale coincide with two lines on the stage micrometer.

For the sake of illustration it may be assumed that five of the divisions on the stage micrometer accurately fill one of the divisions of the eikonometer scale; this indicates a magnifying power of 500 as the constant for that particular optical combination, and a record should be made of the fact.

The magnification constants of the various other optical combinations should be similarly made and recorded.

To measure any object subsequently it should be first focussed carefully in the ordinary way.

The eikonometer should then be applied to the eyepiece and the size of the object read off on the eikonometer scale as millimetres, and the actual size calculated by dividing the observed size by the magnification constant for the particular optical combination employed in the observation.

(*c*) By means of the filar micrometer.

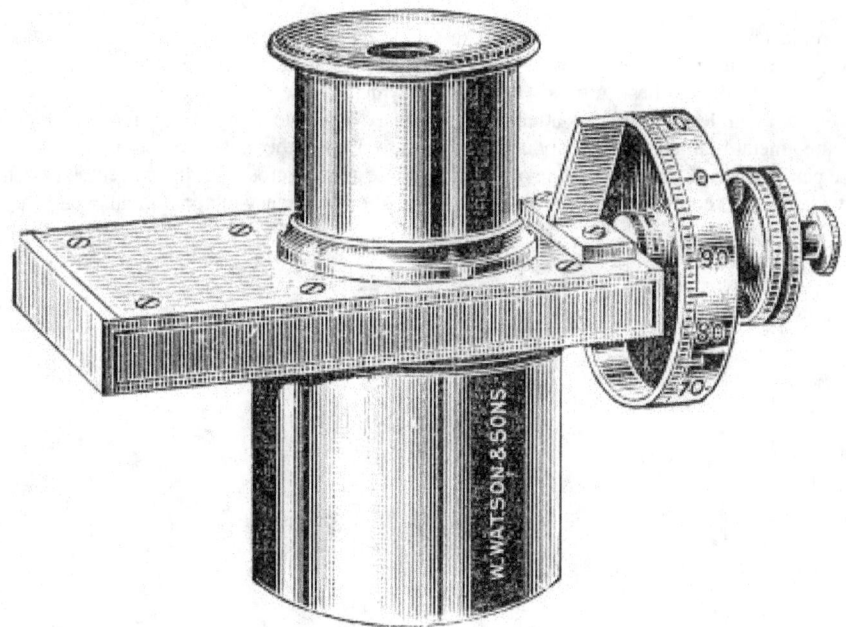

Fig. 58.—Ramsden's Filar micrometer.

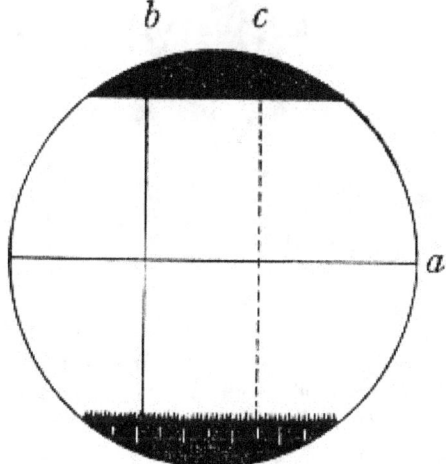

Fig. 59.—Ramsden's micrometer field, a, fixed wire; b, reference wire (fixed); c, travelling wire.

The **Filar** or cobweb Micrometer (Ramsden's micrometer) eyepiece (Fig. 58) consists of an ocular having a fine "fixed" wire stretching horizontally across the field (Fig. 59), a vertical reference wire—fixed—adjusted at right angles to the first; and a fine wire, parallel to the reference wire, which can be moved across the field by the action of a micrometer screw; the drum head is divided into one hundred parts, which successively pass a fixed index as the head is turned. In the lower part of the field is a comb with the intervals between its teeth corresponding to one complete revolution of this screw-head.

As in the previous method, the value of each division of the micrometer scale (*i. e.*, the comb) must first be determined for each optical combination. This is effected as follows:

1. Place the filar micrometer and the stage micrometer in their respective positions.
2. Rotate the screw of the filar micrometer until the movable wire coincides with the fixed one, and the index marks zero on the drum head. (If when the drum head is at zero the two wires do not exactly coincide they must be adjusted by loosening the drum screw and resetting the drum.)
3. Focus the scale of each micrometer accurately, and make the lines on them parallel.
4. Rotate the head of the micrometer screw until the movable line has transversed one division of the stage micrometer. Note the number of complete revolutions (by means of the recording comb) and the fractions of a revolution (by means of scale on the head of the micrometer screw), which are required to measure the 0.01 mm.

5. Make several such estimations and average the results.
6. Note the optical combination employed in this experiment and record it carefully, together with the micrometer value in terms of μ.
7. Repeat this process for each of the different optical combinations and record the results.

To measure an object by this method, simply note the number of revolutions and fractions of a revolution of the screw-head required to traverse such object from edge to edge, and express the result as *micra* by reference to the recorded values for that particular optical combination.

Microscope Illuminant.—In tropical and subtropical regions diffuse daylight is the best illuminant. In temperate climes however daylight of the desirable quantity is not always available, and recourse must be had to oil lamps, gas lamps—preferably those with incandescent mantles—and electricity; and of these the last is undoubtedly the best. A handy lamp holder which can be manufactured in the laboratory is shown in Fig. 60. It consists of a base board weighted with lead to which is attached the ordinary domestic lamp holder, and behind this is fastened a curved sheet-iron reflector. An obscured metal filament lamp of about 16 candle power gives the most suitable light, and if monochromatic light is needed, the blue grease pencil is streaked over the side of the lamp nearest the microscope; the current is switched on and when the glass bulb is warm, rubbing with a wad of cotton-wool will readily distribute the blue greasy material in an even film over the ground glass.

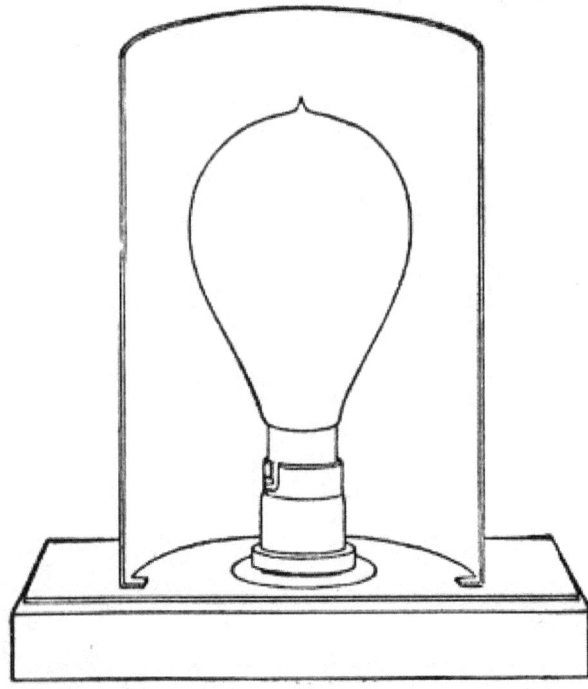

Fig. 60.—Electric microscope lamp.

FOOTNOTES:

Its importance will be realised, however, when it is stated in the words of the late Professor Abbé: "The numerical aperture of a lens determines all its essential qualities; the brightness of the image increases with a given magnification and other things being equal, as the square of the aperture; the resolving and defining powers are directly related to it, the focal depth of differentiation of depths varies inversely as the aperture, and so forth."

Made by Mr. Otto Baumbach, 10, Lime Grove, Manchester.

V. MICROSCOPICAL EXAMINATION OF BACTERIA AND OTHER MICRO-FUNGI.

APPARATUS AND REAGENTS USED IN ORDINARY MICROSCOPICAL EXAMINATION.

The following comprises the essential apparatus and reagents for routine work with which each student should be provided.
1. India-rubber "change-mat" upon which cover-glasses may be rested during the process of staining.
2. Squares of blotting paper about 10 cm., for drying cover-slips and slides.
(The filter paper known as "German lined"—a highly absorbent, closely woven paper, having an even surface and no loose "fluff" to adhere to the specimens—is the most useful for this purpose.)

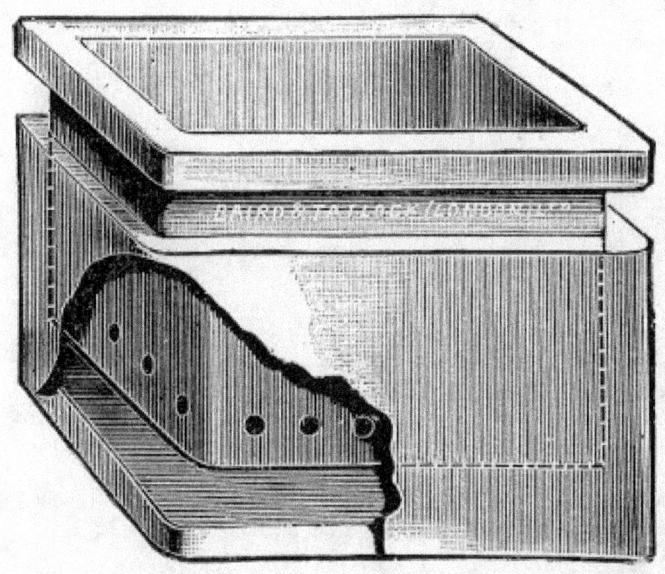

Fig. 61.—Disinfectant Jar.

3. Glass jar filled with 2 per cent. lysol solution for the reception of infected cover-glasses and infected pipettes, etc.

4. A square glazed earthenware box with a loose lining containing 2 per cent. lysol solution for the reception of infected material and used slides. The bottom of the lining is perforated so that when full the lining and its contents can be lifted bodily out of the box, when the disinfectant solution drains away and the slides, etc., can easily be emptied out. The empty lining is then returned to the box with its disinfectant solution (Fig. 61).

5. Bunsen burner provided with "peep-flame" by-pass.

6. Porcelain trough holding five or six hanging-drop slides (Fig. 62).

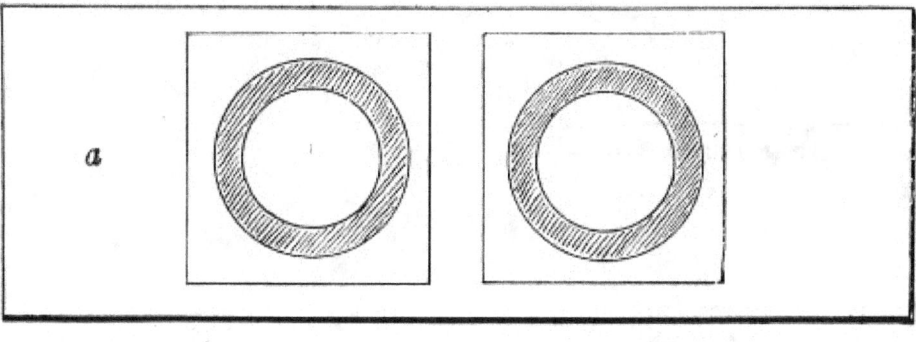

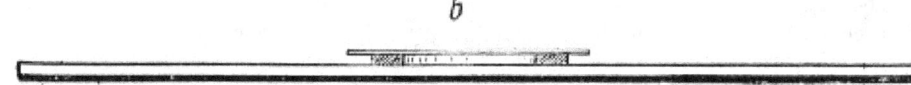

Fig. 62.—Hanging-drop slides: a, Double cell seen from above; b, single cell seen from the side.

The best form of hanging-drop slide is a modification of Boettcher's glass ring slide, and is prepared by cementing a circular cell of tin, 13 to 15 mm. diameter, and 1 to 2 mm. in height, to the centre of a 3 by 1 slip by means of Canada balsam. It is often extremely convenient to have two of these cells cemented close together on one slide (Fig. 62, a).

Another form of hanging-drop slide is made in which a circular or oval concavity or "cell" is ground out of the centre of a 3 by 1 slip. These are more expensive, less convenient to work with, and are more easily contaminated by drops of material under examination, and should be carefully avoided.

7. Three aluminium rods (Fig. 63), each about 25 cm. long and carrying a piece of 0.015 gauge platino-iridium wire 7.5 cm. in length. The end of one of the wires is bent round to form an oval loop, of about 1 mm. in its short diameter, and is termed a loop or an oese; the terminal 3 or 4 mm. of another wire is flattened out by hammering it on a smooth iron surface to form a "spatula"; the third is left untouched or is pointed by the aid of a file. These instruments are used for inoculating culture tubes and preparing specimens for microscopical examination.

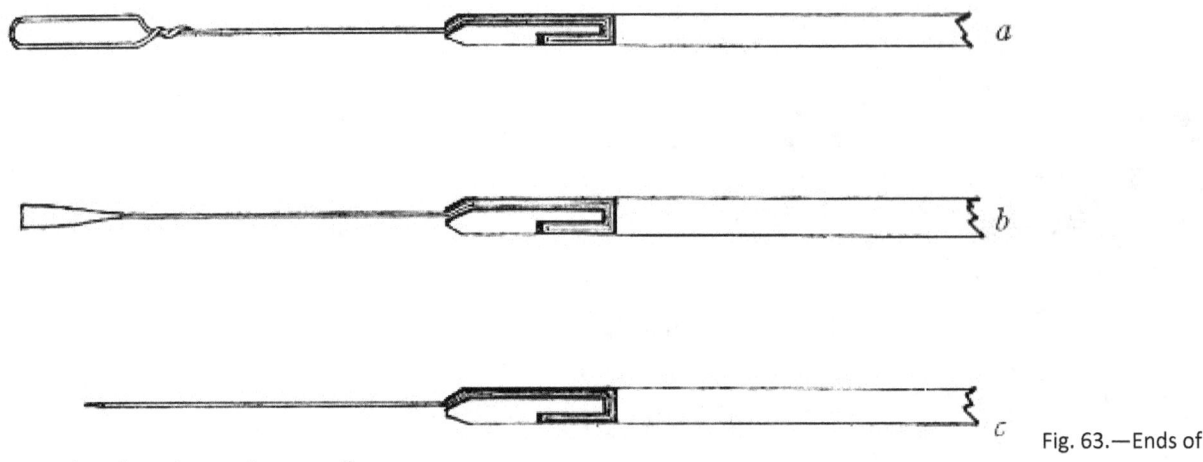

Fig. 63.—Ends of platinum rods. a, loop; b, spatula; c, needle.

The method of mounting these wires may be described as follows:

Take a piece of aluminium wire 25 cm. long and about 0.25 cm. in diameter, and drill a fine hole completely through the wire about a centimetre from one end. Sink a straight narrow channel along one side of the wire, in its long axis, from the hole to the nearest end, shallow at first, but gradually becoming deeper.

On the opposite side of the wire make a short cut, 2 mm. in length, leading from the hole in the same direction.

Now pass one end of the platinum wire through the hole, turn up about 2 mm. at right angles and press the short piece into the short cut. Turn the long end of the wire sharply, also at right angles, and sink it into the long channel so that it emerges from about the centre of the cut end of the aluminium wire (Fig. 63). A few sharp taps with a watch maker's hammer will now close in the sides of the two channels over the wire and hold it securely.

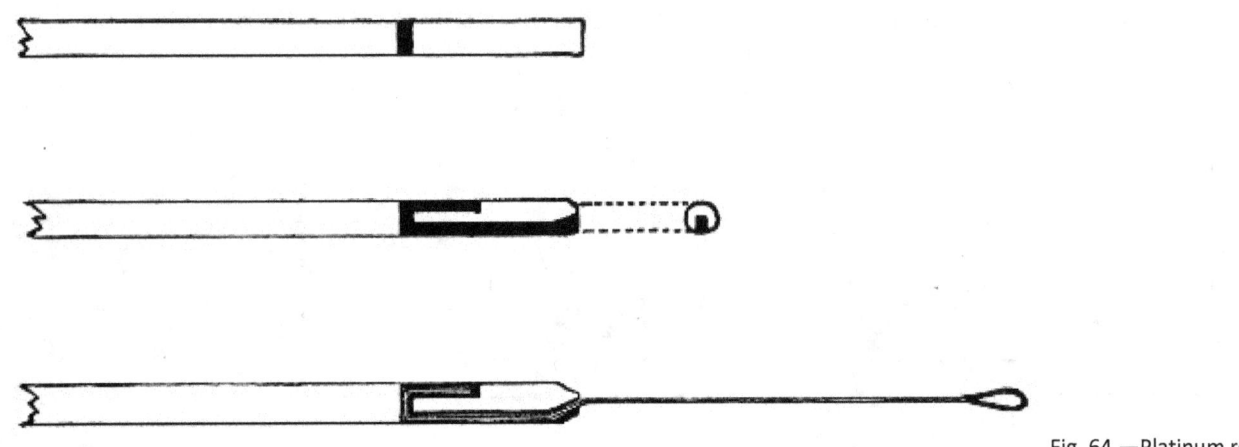

Fig. 64.—Platinum rod in aluminium handle—method of mounting. The platinum wire may be fused into the end of a piece of glass rod, but such a handle is vastly inferior to aluminium and is not to be recommended.

8. Two pairs of sharp-pointed spring forceps (10 cm. long), one of which must be kept perfectly clean and reserved for handling clean cover-slips, the other being for use during staining operations.

9. A box of clean 3 by 1 glass slips.

10. A glass capsule with tightly fitting (ground on) glass lid, containing clean cover-slips in absolute alcohol.

11. One of Faber's "grease pencils" (yellow, red, or blue) for writing on glass.

12. A wooden rack (Fig. 65) with twelve drop-bottles (Fig. 66) each 60 c.c. capacity, containing

Aniline water.
Gentian violet, saturated alcoholic solution.
Lugol's (Gram's) iodine.
Absolute alcohol.
Methylene-blue, } Fuchsin, basic, } saturated alcoholic solution.
Neutral red, 1 per cent. aqueous solution.
Leishman's modified Romanowsky stain.
Carbolic acid, 5 per cent. aqueous solution.
Acetic acid, 1 per cent. solution.
Sulphuric acid, 25 per cent. solution.
Xylol.

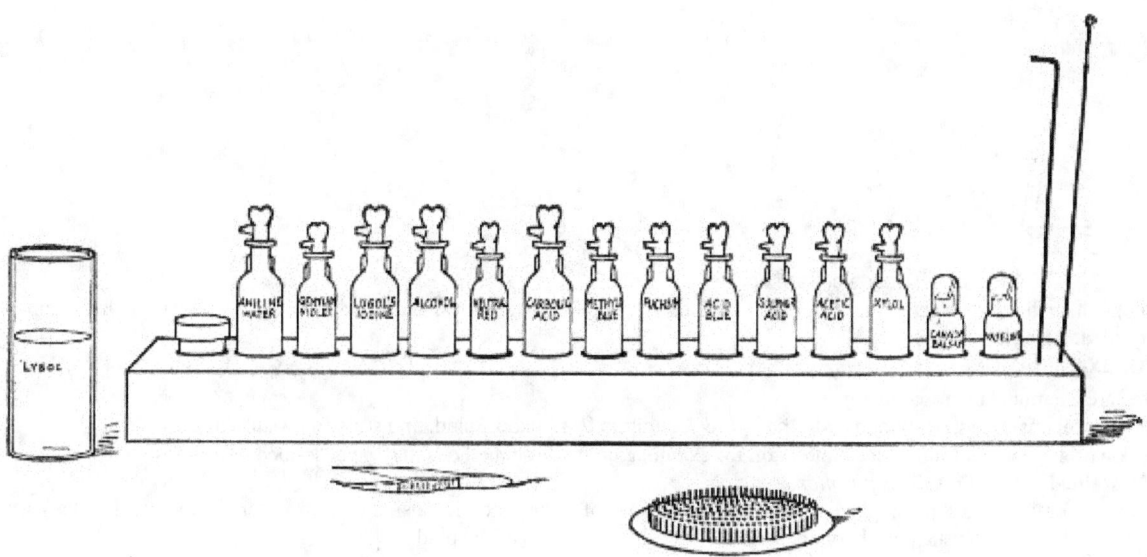

Fig. 65.—Staining rack, rubber change mat and lysol pot.

Fig. 66.—Drop bottle.

The Elements of Bacteriological Technique

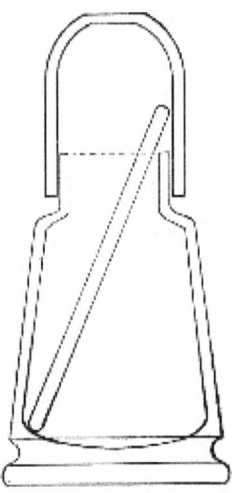

Fig. 67.—Canada balsam pot.

And two pots with air-tight glass caps (Fig. 67), each provided with a piece of glass rod and filled respectively with Canada balsam dissolved in xylol, and sterile vaseline.

METHODS OF EXAMINATION.

Bacteria, etc., are examined microscopically.

1. In the living state, unstained, or stained. 2. In the "fixed" condition (*i. e.*, fixed, killed, and stained by suitable methods).

The preparation of a specimen from a tube cultivation for examination by these methods may be described as follows:

1. Living, Unstained.—(*a*) *"Fresh" Preparation.*—

1. Clean and dry a 3 by 1 glass slip and place it on one of the squares of filter paper. Deposit a drop of water (preferably distilled) or a drop of 1 per cent. solution of caustic potash, on the centre of the slip, by means of the platinum loop.

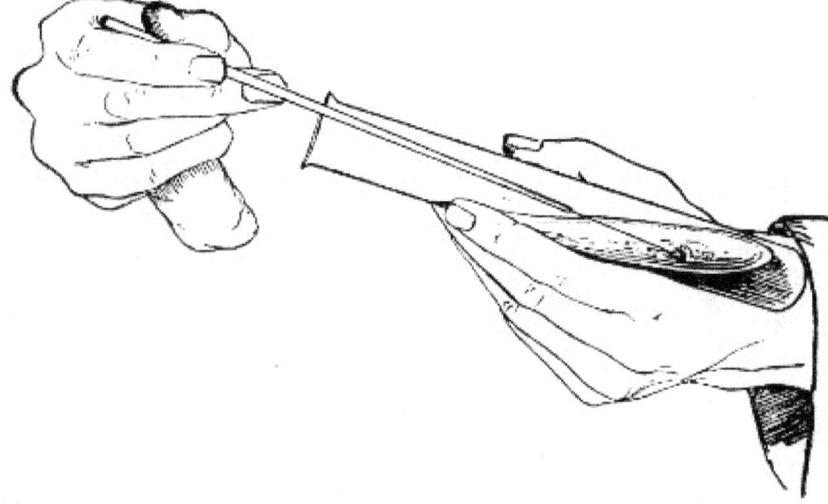

Fig. 68.—Holding tubes for removing bacterial growth, as seen from the front.

Technique of Opening and Closing a Culture Tube.

2. Remove the tube cultivation from its rack or jar with the left hand and ignite the cotton-wool plug by holding it to the flame of the Bunsen burner. Extinguish the flame by blowing on the plug, whilst rotating the tube on its long axis, its mouth directed vertically upward, between the thumb and fingers. (This operation is termed "flaming the plug," and is intended to destroy any micro-organisms that may have become entangled in the loose fibres of the cotton-wool, and which, if not thus destroyed, might fall into the tube when the plug is removed and so accidentally contaminate the cultivation.)

3. Hold the tube at or near its centre between the ends of the thumb and first two fingers of the left hand, and allow the sealed end to rest upon the back of the hand between the thumb and forefinger, the plug pointing to the right. Keep the tube as nearly in the horizontal position as is consistent with safety, to diminish the risk of the accidental entry of organisms (Fig. 68).

4. Take the handle of the loop between the thumb and forefinger of the right hand, holding the instrument in a position similar to that occupied by a pen or a paint-brush, and sterilise the platinum portion by holding it in the flame of a Bunsen burner until it is red hot. Sterilise the adjacent portion of the aluminium handle by passing it rapidly twice or thrice through the flame. After sterilising it, the loop must not be allowed to leave the hand or to touch against anything but the material it is intended to examine, until it is finished with and has been again sterilised.

5. Grasp the cotton-wool plug of the test-tube between the little finger and the palm of the right hand (whilst still holding the loop as directed in step 4), and remove it from the mouth of the tube by a "screwing" motion of the right hand.

6. Introduce the platinum loop into the tube and hold it in this position until satisfied that it is quite cool. (The cooling may be hastened by touching the loop on one of the drops of moisture which are usually to be found condensed on the interior of the glass tube, or by dipping it into the condensation water at the bottom; at the same time care must be taken in the case of cultures on solid media to avoid touching either the medium or the growth.)

7. Remove a small portion of the growth by taking up a drop of liquid, in the case of a fluid culture, in the loop; or by touching the loop on the surface of the growth when the culture is on solid medium; and withdraw the loop from the tube without again touching the medium or the glass sides of the tube.

8. Replace the cotton-wool plug in the mouth of the tube.

9. Replace the tube cultivation in its rack or jar.

10. Mix the contents of the loop thoroughly with the drop of water on the 3 by 1 slide.

11. Again sterilise the loop as directed in step 4, and replace it in its stand.

12. Remove a cover-slip from the glass capsule by means of the cover-slip forceps, rest it for a moment on its edge, on a piece of filter paper to remove the excess of alcohol, then pass it through the flame of the Bunsen burner. This burns off the remainder of the alcohol, and the cover-slip so "flamed" is now clean, dry, and sterile.

13. Lower the cover-slip, still held in the forceps, on to the surface of the drop of fluid on the 3 by 1 slip, carefully and gently, to avoid the inclusion of air bubbles.

14. Examine microscopically (*vide infra*).

During the microscopical examination, stains and other reagents may be run in under a cover-slip by the simple method of placing a drop of the reagent in contact with one edge of the cover-glass and applying the torn edge of a piece of blotting paper to the opposite side. The reagent may then be observed to flow across the field and come into contact with such of the micro-organisms as lie in its path.

The non-toxic basic dyes most generally employed for the intra-vitam staining of bacteria are

Neutral red, }

Quinoleine blue }
                    in 0.5 per cent. aqueous solutions.

Methylene green }

Vesuvin, }

Negative Stain (Burri).—By this method of demonstration the appearances presented by dark ground illumination (by means of a paraboloid condenser) are closely simulated, since minute particles, bacteria, blood or pus cells etc. stand out as brilliantly white or colourless bodies on a dark grey-brown background.

Reagent required:

Any one of the liquid waterproof black drawing inks (Chin-chin, Pelican, etc.). This is prepared for use as follows:
Measure out and mix:

Liquid black ink,   25 c.c.

Tincture of iodine   1 c.c.

Allow the mixture to stand 24 hours, centrifugalise thoroughly, pipette off the supernatant liquid to a clean bottle and then add a crystal of thymol or one drop of formalin as a preservative.

Method.—

1. With the sterilised loop deposit one drop of the liquid ink close to one end of a 3 by 1 slide.

2. With the sterilised loop deposit a drop of the fluid culture (or of an emulsion from a solid culture) by the side of the drop of ink (Fig. 69, *a*); mix the two drops thoroughly by the aid of the loop.

3. Sterilise the loop.

4. Hold the slide firmly on the bench with the thumb and forefinger of the left hand applied to the end nearest the drop of fluid.

5. Take another clean 3 by 1 slide in the right hand and lower its short end obliquely (at an angle of about 60°) transversely on to the mixed ink and culture on the first slide, and allow the fluid to spread across the slide and fill the angle of incidence.

6. Maintaining the original angle, draw the second slide firmly and evenly along the first toward the end farthest from the left hand (Fig. 69, *b*).

7. Throw the second slide into a pot of disinfectant; allow the first slide to dry in the air.

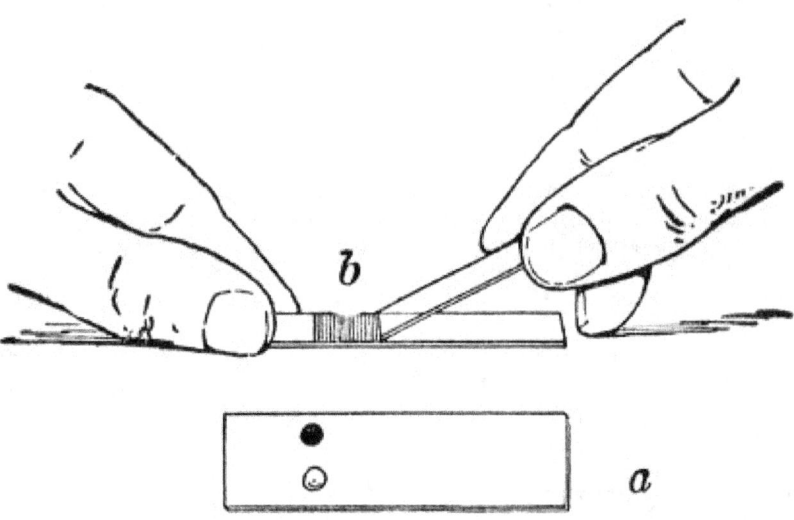

Fig. 69.—Spreading negative film.

8. Place a drop of immersion oil on the centre of the film, lower the 1/12-inch objective into the oil and examine microscopically without the intervention of a cover-slip.

(The film of ink may be covered with a long cover-glass and xylol balsam as a permanent preparation.)

(b) *Hanging-drop Preparation.*—

1. Smear a layer of sterile vaseline on the upper surface of the ring cell of a hanging-drop slide by means of the glass rod provided with the vaseline bottle, and place the slide on a piece of filter paper.

2. "Flame" a cover-slip and place it on the filter paper by the side of the hanging-drop slide.

3. Place a drop of water on the centre of the cover-slip by means of the platinum loop.

4. Obtain a small quantity of the material it is desired to examine, in the manner detailed above (pages 74-76, steps 2 to 11 must be followed in their entirety and with the strictest exactitude whenever tube contents are being handled), and mix it with the drop of water on the cover-slip.

5. Raise the cover-slip in the points of the forceps and rapidly invert it on to the ring cell of the hanging-drop slide, so that the drop of fluid occupies the centre of the ring. (Carefully avoid contact between the drop of fluid and either the ring cell or the layer of vaseline. Should this happen, the now *infected* hanging-drop slide and its cover-slip must be dropped into the pot of lysol and a new preparation made.)

6. Press the cover-slip firmly down into the vaseline on to the top of the ring cell. (This spreads out the vaseline into a thin layer, and besides ensuring the adhesion of the cover-slip, seals the cells and so retards evaporation.)

7. Examine microscopically.

The examination of a "fresh" specimen or a "hanging-drop" preparation is directed to the determination of the following data:

1. The nature of the bacteria present—*e. g.*, cocci, bacilli, etc.

2. The purity of the cultivation; this can only be determined when gross morphological differences exist between the organisms present.

3. The presence or absence of spores; when present, spores show their typical refrangibility exceedingly well by this method.

4. The presence or absence of mobility. In a hanging-drop specimen some form of movement can practically always be observed, and its character must be carefully determined by noting the relative positions of adjacent micro-organisms.

(*a*) Brownian or molecular movement. Minute particles of solid matter (including bacteria), when suspended in a fluid, will always show a vibratory movement affecting the entire field, but never altering the relative positions of the bacteria. (Cocci exhibit this movement, but with the exception of the Micrococcus agilis, the cocci are non-motile.)

(*b*) Streaming movement. This is due to currents set up in the hanging drop as a result of jarring of the specimen or of evaporation, or to the fact that the cover-slip is not perfectly level, and although the relative positions of the bacteria may vary, still the flowing movement of large numbers of organisms in some one direction will usually be sufficient to demonstrate the nature of this motion.

(*c*) Locomotive movement, or **true motility**, is determined by observing some one particular bacillus changing its position in the field independently of, and in a direction contrary to, other organisms present.

When the examination is completed and the specimen finished with, the "fresh specimen"—*i. e.*, the slide with the cover-slip attached—must be dropped into the lysol pot. In the hanging-drop specimen, however, the cover-slip only is infected, and this may be raised from the ring cell by means of forceps and dropped into the disinfectant.

Permanent Staining of the Hanging-drop Specimen.—Occasionally it is necessary to fix and stain a hanging-drop preparation. This may be done as follows:

1. Remove the cover-slip from the cell by the aid of the forceps.

2. If the drop is small, fix it by dropping it face downward, whilst still wet, on to the surface of some Gulland's solution or corrosive sublimate solution (*vide* page 82) in a watch-glass. If the drop is large, place it face upward on the rubber mat, cover it with an inverted watch-glass, and allow it to dry. Then fix it in the alcohol and ether solution (*vide*, page 82).

3. Dip the cover-glass into a beaker containing hot water in order to remove some of the vaseline adhering to it.

4. Wash successively in alcohol, xylol, ether, and alcohol, to remove the last traces of grease.

5. Wash in water.

6. Stain, wash, dry, and mount as for an ordinary cover-slip film preparation (*vide* pages 83-85).

2. Killed, Stained.—In this method three distinct processes are necessary:
"Preparing" and "fixing" the film. Staining. Mounting.

Preparing the Film.—
1. Flame a cover-slip and place it on a piece of filter paper.
2. Place a drop of water on the centre of the cover-slip by means of platinum loop.
3. Obtain a small quantity of the material to be examined upon a sterilised platinum loop (see pages 74-76, steps 2 to 11) and mix it with the drops of water on the cover-slip.
4. Spread the drop of emulsion evenly over the cover-slip in the form of a square film to within 1 mm. of each edge of the cover-slip.
5. Allow it to dry completely in the air.

Fixing.—Fix by passing the cover-slip, held in the fingers, three or four times through the flame of a Bunsen burner.

In some instances (*e. g.*, when the films after staining are intended for micrometric observations) it is almost essential to fix by exposure to a uniform temperature of 115° C., for twenty minutes. This is best done in a carefully regulated hot-air oven.

Fixation may also be effected by immersing in some fixative fluid, such as one of the following:
1. Absolute alcohol, for five to fifteen minutes.
2. Absolute alcohol, Ether, equal parts, for five to thirty minutes (*e. g.*, for blood or milk).
3. Osmic acid, 1 per cent. aqueous solution, for thirty seconds.
4. Corrosive sublimate, saturated aqueous solution, for five minutes.
5. Corrosive sublimate (Lang), for five minutes. This solution is prepared by dissolving:

Sodium chloride	0.75 gramme
Hydrarg. perchloride	12.00 grammes
Acetic acid	5.00 grammes
In distilled water	100.00 c.c.

Filter.

6. Gulland's solution, for five minutes. This solution is prepared by mixing:

Absolute alcohol	25.0 c.c.
Ether	25.0 c.c.
Corrosive sublimate, 20 per cent. alcoholic solution	0.4 c.c.

7. Formalin 10 per cent. aqueous solution (= 4 per cent. aqueous solution of formaldehyde since formalin is a 40 per cent. solution of the gas in water).

Either of these methods of fixation coagulates the albuminous material and ensures perfect adhesion of the film to the cover-slip.

Clearing.—Wash the cover-slip thoroughly in running water and proceed with the staining.

If the film has been prepared from broth, liquefied gelatine, or pus or other morbid exudations, saturate the film after fixation with acetic acid 2 per cent. and allow it to act for two minutes.

Wash with alcohol, then let the alcohol remain on the cover-slip for two minutes. (This will "clear" the groundwork and give a much sharper and cleaner film than would otherwise be obtained.)

If the film has been prepared from blood or bloodstained fluid, treat with acetic acid 2 per cent. for two minutes after fixation. Wash with water, dry, and proceed with the staining. (This will remove the hæmoglobin and facilitate examination.)

Staining.—
1. Rest the cover-slip, film side uppermost, on the rubber mat.
2. By means of a drop-bottle, cover the film side of the cover-slip with the selected stain, allow it to act for a few minutes, then wash off the excess in running water.

The penetrating power of stains is increased by (*a*) physical means—*e. g.*, heating the stain; (*b*) chemical means—*e. g.*, by the addition of carbolic acid, 5 per cent. aqueous solution; caustic alkalies, 2 per cent. aqueous solutions; water saturated with aniline oil; borax, 0.5 per cent. aqueous solution.

The most commonly used dyes for cover-slip film preparations are the aniline dyes.

(A) Basic: (a) Methylene-blue. (b) Gentian violet. (c) Fuchsin.

These dyes are kept in saturated alcoholic (90 per cent.) solutions so that decomposition may be retarded.

Two or three drops of alcoholic solution of these dyes to, say, 4 c.c. water, usually makes a sufficiently strong staining fluid for cover-slip film preparations.

Carbolic methylene-blue (C.M.B.) and carbol fuchsin (C.F.) are prepared by covering the cover-slip with 5 per cent. solution of carbolic acid and adding a few drops of the saturated alcoholic solution of methylene-blue or fuchsin respectively to it. For aniline gentian violet (A.G.V.) the stain is added to a saturated solution of aniline oil in water.

(d) Thionine blue. (e) Bismarck brown. (f) Neutral red. (B) Acid: (a) Eosin, aqueous yellowish. (b) Safranine.

These dyes are kept in 1 per cent. aqueous solution to which is added 5 per cent. of alcohol, as a preservative. They are generally used in this form.

A few nuclear stains (carmine, hæmatoxylin) are occasionally used more especially in "section" work.

Decolourisation.—After overstaining, films may be decolourised by washing for a longer or shorter time in one of the following reagents arranged in ascending order of power

1. Water.
2. Chloroform.
3. Acetic acid, 1 per cent.
4. Alcohol.
5. Alcohol absolute, equal parts. Acetic acid, 1 per cent., Hydrochloric, 1 per cent. aqueous solution. Hydrochloric, 1 per cent. Alcoholic (90 per cent.) solution.
6. Mineral acids: Sulphuric, 25 per cent. aqueous solution. Nitric, 33 per cent. aqueous solution.

Counterstaining.—Use colours which will contrast with the first stain; *e. g.*,

Vesuvin, }

Neutral red, } for films stained by methylene-blue or Gram's method.

Eosin, }

Fuchsin, }

Methylene-blue, } for films stained by fuchsin.

Gentian violet, }

8. *Mounting.*—
1. Wash the film carefully in running water.
2. Blot off the superfluous water with the filter paper, or dry more completely between two folds of blotting paper.
3. Complete the drying in the air, or by holding the cover-slip in the fingers at a safe distance above the flame of the Bunsen burner.
4. Place a drop of xylol balsam on the centre of a clean 3 by 1 glass slide and invert the cover-slip over the balsam, and lower it carefully to avoid the inclusion of air bubbles.

Note.—Xylol is used in preference to chloroform to dissolve Canada balsam, as it does not decolourise the specimen.

Impression films (*Klatschpraeparat*) are prepared from isolated colonies of bacteria in order that their characteristic formation may be examined by higher powers than can be brought to bear on the living cultivation. They are prepared from plate cultivations (*vide* page 230) in the following manner.

1. Remove a clean cover-slip from the alcohol pot with sterile forceps and burn off the spirit.
2. Open the plate and rest one edge of the cover-slip on the surface of the medium a little to one side of the selected colony. Lower it cautiously over the colony until horizontal. Avoid any lateral movement or the inclusion of bubbles of air.
3. Make gentle vertical pressure on the centre of the cover-slip with the points of the forceps to ensure perfect contact with the colony.
4. Steady one edge of the cover-slip with the forceps and pass the point of a mounted needle just under the opposite edge and raise the cover-slip carefully; the colony will be adherent to it. When nearly vertical, grasp the cover-slip with the forceps and remove it from the plate. Re-cover the plate.
5. Place the cover-slip, film uppermost, on the rubber mat, and cover it with an inverted watch-glass until dry.
6. Fix by immersing in one of the fixing fluids previously mentioned (*vide* page 82).
7. Clear with acetic acid and alcohol.
8. Stain and mount as an ordinary cover-slip film preparation, being careful to perform all washing operations with extreme gentleness.

Microscopical Examination of the Unstained Specimens.—
1. Place the body tube of the microscope in the vertical position.
2. Arrange the hanging-drop slide on the microscope stage so that the drop of fluid is in the optical axis of the instrument, and secure it in that position by means of the spring clips.
3. Use the 1/6-inch objective, rack down the body tube until the front lens of the objective is almost in contact with the cover-slip—that is, well within its focal distance. This is best done whilst bending down the head to one side of the microscope, so that the eyes are on a level with the stage.
4. Apply the eye to the ocular and adjust the plane mirror to the position which secures the best illumination.
5. Rack the condenser down slightly and cut down the aperture of the iris diaphragm so that the light, although even, is dim.
6. Rack up the body tube by means of the coarse adjustment until the bacteria come into view; then focus exactly by means of the fine adjustment.

Some difficulty is often experienced at first in finding the hanging drop, and if the first attempt is unsuccessful, the student must not on any account, whilst still applying his eye to the ocular, rack the body tube down (for by so doing there is every likelihood of the front lens of the objective being forced through the cover-glass, and not only spoiling the specimen, but also contaminating the objective); but, on the contrary, withdraw his eye, rack the tube up, and commence again from step 2.

Dark Ground Illumination.—
1. Set up the microscope stand in the vertical position and insert the highest eyepiece available.
2. Remove the nosepiece from the microscope tube and fit the 2/3 inch objective in place.
3. Remove the substage condenser and replace it by the dark ground condenser.
4. Fit up the source of illumination some 30-50 cm. distant from the microscope. (This should be the Liliput Arc Lamp (Leitz), Nernst Lamp or incandescent gas lamp; if either of the two latter are employed, a bull's eye condenser to produce parallel rays must be interposed between light and microscope); and adjust illuminant and microscope so that the substage plane mirror is completely filled with light.

5. Focus the two concentric rings engraved upon the upper surface of the condenser and centre them accurately by means of the centring screws.

6. Prepare a "fresh" specimen (see pages 74-76) of the material it is desired to observe, using selected, new, 3 by 1 glass slips of less than 1 mm. thickness, and No. 1 cover-glasses (0.17 mm. thick), which should be cleaned with a piece of soft washleather and not with the emery paper, as scratches on the glass produce haziness in the preparation.

7. Deposit a large drop of immersion oil (or pure water) on the upper surface of the condenser and rack it down a few millimetres.

8. Adjust the fresh preparation on the microscope stage and fasten it in position with the stage clips.

9. Rack up the condenser until the immersion fluid makes contact with the under surface of the slide; avoid the formation of air bubbles.

10. Adjust the substage mirror so that the light is reflected upward. A bright spot will be seen on the fresh preparation near the centre of the field.

11. Replace the 2/3-inch objective by the 1/12-inch oil immersion lens which has been fitted with the special stop to reduce its N. A.; place a drop of immersion oil upon the centre of the cover-glasses of the fresh preparation and lower the microscope tube until the front lens of the objective has entered the oil drop.

12. Focus the bright spot referred to in step 10. If it no longer occupies the centre of the field, alter the angle of the substage mirror until it does.

13. Now focus the lens accurately on the film, cautiously vary the height of the dark ground condenser until the best position is found. The intensely illuminated bacteria will stand out in vivid contrast to the dark background.

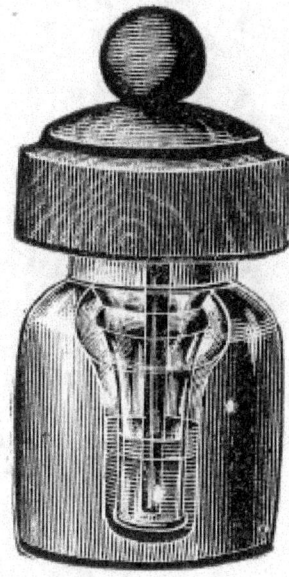

Fig. 70.—Immersion oil bottle.

Microscopical Examination of the Stained Specimen.—(The body tube of the microscope may be vertical or inclined to an angle.)

1. Secure the slide on the stage of the microscope by means of the spring clips.
2. Place a drop of cedarwood oil on the centre of the cover-slip.

The immersion oil is pure cedarwood oil, and is kept in a small bottle of stout glass (Fig. 70), the cavity of which is shaped like an inverted cone, and is provided with a safety funnel (so that the oil does not escape if the bottle is accidentally overturned) and a dust cap of boxwood fitted with a wooden rod with which the drop of oil is applied to the cover-glass or lens.

3. Use the 1/12-inch oil immersion lens of the microscope. Rack down the body tube till the front lens of the objective is in contact with the oil and nearly touching the cover-slip.
4. Rack up the condenser until it is in contact with the under surface of the slide.
5. Apply the eye to the ocular and arrange the plane mirror so as to obtain the greatest possible amount of light.
6. Rack up the body tube until the stained film comes into view.
7. Focus the condenser accurately on the film.
8. Focus the film accurately by means of the fine adjustment.

VI. STAINING METHODS.

In the following pages are collected the various "stock" stains in everyday use in the bacteriological laboratory, together with a selection of the most convenient and generally useful staining methods for demonstrating particular structures or differentiating groups of bacteria. The stains employed should either be those prepared by Gruebler, of Leipzig, or Merck, of Darmstadt. The methods printed in ordinary type are those which a long experience has shown to be the most reliable, and to give the best results—those relegated to small type comprise such as are not so generally useful, but give excellent results in the hands of the experienced worker.

BACTERIA STAINS.

Methylene-blue.—
1. *Saturated Aqueous Solution.*
Weigh out

Methylene-blue   1.5 grammes

Place in a stoppered bottle having a capacity of from 150 to 200 c.c. and add

Distilled water   100.0 c.c.

Allow the water to remain in contact with the dye for two weeks, shaking the contents of the bottle vigourously for a few moments every day. Filter.

2. *Saturated Alcoholic Solution.*
Weigh out

Methylene-blue   1.5 grammes

Place in a stoppered bottle of 150 c.c. capacity and add

Alcohol, 90 per cent   100.0 c.c.

Allow the alcohol to remain in contact with the dye for two hours, shaking vigourously every few minutes. Filter.

3. *Carbolic Methylene-blue* (Kuehne).
Weigh out

Methylene-blue   1.5 grammes

Carbolic acid   5.0 grammes

and dissolve in

Distilled water   100.0 c.c.

and add

Absolute alcohol   10.0 c.c.

Filter.

4. *Alkaline Methylene-blue* (Loeffler).
Measure out and mix

Methylene-blue, saturated alcoholic solution   30.0 c.c.

Caustic potash, 0.1 per cent. aqueous solution   100.0 c.c.

Filter.

Gentian Violet.—
5. *Saturated Aqueous Solution.*
Weigh out

Gentian violet   2.25 grammes

and proceed as in preparing the corresponding solution of methylene-blue.

6. *Saturated Alcoholic Solution.*
Weigh out

Gentian violet   5.0 grammes

and proceed as in preparing the corresponding solution of methylene-blue.

7. *Carbolic Gentian Violet* (Nicollé).
Measure out and mix

Gentian violet, saturated alcoholic solution   10.0 c.c.

Carbolic acid, 1 per cent. aqueous solution   100.0 c.c.

Filter.

8. *Anilin Water Solution* (Koch-Ehrlich).
Measure out

Distilled water   100 c.c.

Add anilin oil drop by drop (shaking well after the addition of each drop) until the solution is opaque.
Filter until clear.
and add

Absolute alcohol   10 c.c.

Saturated alcoholic solution gentian violet   11 c.c.

Filter.

Note.—This solution will not keep longer than 14 days.
Thionine Blue (or Lauth's Violet).—
9. *Carbolic Thionine Blue* (Nicollé).
Weigh out

Thionine blue   1.0 gramme

Carbolic acid   2.5 grammes

and dissolve in

Distilled water   100.0 c.c.

Filter.
Before use dilute with equal quantity of distilled water and again filter.
Fuchsin (Basic).—
10. *Saturated Aqueous Solution.*
Weigh out

Basic fuchsin   1.5 grammes

and proceed as in preparing the corresponding solution of methylene-blue (*q. v.*).
11. *Saturated Alcoholic Solution.*
Weigh out

Basic fuchsin   3.5 grammes

and proceed as in preparing the corresponding solution of methylene-blue.
12. *Carbolic Fuchsin* (Ziehl).
Weigh out

Basic fuchsin   1.0 gramme

Carbolic acid   5.0 grammes

dissolve in

Distilled water   100.0 c.c.

and add

Absolute alcohol   10.0 c.c.

Filter.
CONTRAST STAINS.

Eosin.—There are several commercial varieties of eosin, which, from the bacteriological point of view, possess very different values. Gruebler lists four varieties, of which two only are useful for bacteriological work:
Eosin, aqueous yellowish. Eosin, aqueous bluish.
13. *Eosin Aqueous Solution* (Yellowish or Bluish Shade), 1 per cent.
Weigh out

Eosin, aqueous   1.0 gramme

dissolve in

Distilled water   100.0 c.c.

and add

Absolute alcohol   5.0 c.c.

Filter.
14. *Eosin Alcoholic Solution*, 0.5 per cent.
Weigh out

Eosin, alcoholic   0.5 gramme

and dissolve in

Alcohol (70 per cent.)   100.0 c.c.

Filter.
Safranine.—
15. *Aqueous Solution.*
Weigh out.

Safranine   0.5 gramme

and dissolve in

Distilled water   100.0 c.c.

Filter.

Neutral Red.—
16. *Aqueous Solution.*
Weigh out

Neutral red   1.0 gramme

and dissolve in

Distilled water   100.0 c.c.

Filter.

Vesuvin (or Bismarck Brown).—
17. *Saturated Aqueous Solution.*
Weigh out

Vesuvin   0.5 gramme

and dissolve in

Distilled water   100.0 c.c.

Filter.

TISSUE STAINS.

Aniline Gentian Violet (For Weigert's Fibrin Stain).—
Weigh out

Gentian violet   1.0 gramme

and dissolve in

Absolute alcohol   15.0 c.c.

Distilled water   80.0 c.c.

then add

Aniline oil   3.0 c.c.

Shake well and filter before use.

Hæmatoxylin (Ehrlich).—
1. Weigh out

Hæmatoxylin   2.0 grammes

and dissolve in

Absolute alcohol   100.0 c.c.

2. Weigh out

Ammonium alum   2.0 grammes

and dissolve in

Distilled water   100.0 c.c.

3. Mix 1 and 2, allow the mixture to stand forty-eight hours, then filter.
4. Add

Glycerine   85.0 c.c.

Acetic acid, glacial   10.0 c.c.

5. Allow the stain to stand for one month exposed to light; then filter again ready for use.

Hæmatin (Mayer's).—
A. Weigh out

Hæmatin   1.0 gramme

and dissolve in

Alcohol 90 per cent. (warmed to 37°C.)   50 c.c.

B. Weigh out

Potash alum   50 grammes

and dissolve in

Distilled water   100 c.c.

Prepare these two solutions in separate flasks. Take a clean flask of 250 c.c. capacity and insert a large funnel in its neck. Pour the solutions A and B simultaneously and slowly into the funnel to mix thoroughly. Store for future use.

Note.—If acid hæmatin is required, introduce glacial acetic acid (3 c.c.) into the mixing flask before adding the solutions A and B.

Alum Carmine (Mayer).—

Weigh out

Alum   2.5 grammes

Carmine   1.0 gramme

and place in a glass beaker.
Measure out in a measuring cylinder,

Distilled water   100.0 c.c.

Place the beaker on a sand-bath, add the water in successive small quantities, and keep the mixture boiling for twenty minutes. Measure the solution and make up to 100 c.c. by the addition of distilled water. Filter.

Lithium Carmine (Orth).—
Weigh out

Carmine   2.5 grammes

and dissolve in

Lithium carbonate, cold saturated solution   100.0 c.c.

Filter.

Picrocarmine.—
Weigh out

Picrocarmine   2.0 grammes

and dissolve in

Distilled water   100.0 c.c.

BLOOD STAINS

When watery solutions of medicinal methylene blue and water soluble eosins are mixed a precipitate is formed which is soluble only in alcohol, and solutions of this precipitate impart a peculiar reddish-purple colour to chromatin. This compound was first used by Romanowsky to demonstrate malarial parasites, but various modifications are now employed for staining blood films generally, and also for bacteria and protozoa. The best modifications of the original Romanowsky are those of Jenner and Leishman—Jenner being most suitable for the histological study of the blood, and Leishman for the demonstration of protozoa.

Jenner's Stain.—
A. Weigh out:

Eosin aqueous yellow   6.0 grammes

Dissolve in

Distilled water (non-alkaline)   250 c.c.

This will make a thick solution.
B. Weigh out:

Methylene blue (medicinally pure) Hoechst   5.0 grammes

Dissolve in

Distilled water (non-alkaline)   250 c.c.

1. Add B to A very slowly, stirring all the time. A viscous precipitate forms which frequently loses its viscosity when heat is applied. (This explains the necessity of mixing slowly).
2. Evaporate slowly in a porcelain basin, stirring occasionally, on a water bath at 55° C. When a paste begins to form scrape and break up occasionally. (On no account must the paste be allowed to fuse.)
3. Grind the resulting mass into an amorphous powder.
4. Weigh out:

Amorphous powder   0.5 grammes

Dissolve in

Methylic alcohol (Merck's puriss, for analysis)   100 c.c.

Allow time for true solution. (About three days is sufficient.)
Method.—
1. Prepare film, dry, but *do not fix*.
2. Flood the unfixed film with the stain, allow it to act for 3 minutes (the methylic alcohol of the stain fixes the film).
3. Pour off the stain and wash in distilled water until the film presents a pink colour.
4. Dry and mount.

Leishman's Stain.—
A. Weigh out:

Methylene blue (medicinal)   1 gramme

Dissolve in

Sodium carbonate, 0.5 per cent. aqueous solution   100 c.c.

Keep at 65° C. for 12 hours in either a hot incubator or a water-bath; then stand in dark place at room temperature (20°C.) for ten days.

B. Weigh out:

Eosin, extra B. A.   0.1 gramme

Dissolve in

Distilled water   100 c.c.

1. Mix the two solutions A and B in equal volumes, and allow the mixture to stand for 12 hours with occasional stirring.
2. Filter, and collect precipitate on filter paper.
3. Wash precipitate thoroughly with distilled water, and dry.
4. Weigh out 0.15 gramme of the dried precipitate; rub up in a mortar with 5 c.c. of methylic alcohol (Merck's puriss, for analysis). Allow undissolved powder to settle, then decant the supernatant fluid to a clean 100 c.c. measuring cylinder.
5. Add further 5 c.c. alcohol to sediment in mortar and repeat the process, and so on until all the sediment has been dissolved.
6. Now make up the fluid in the measuring cylinder to 100 c.c. by the addition of more methylic alcohol.

Method.—
1. Prepare film, dry, but *do not fix*.
2. Flood the unfixed film with stain, allow it to act 30 seconds.
3. Add double the volume of distilled water to the stain on the film, and mix with glass rod or platinum loop.
4. Allow this diluted stain to act five minutes.
5. Wash off with distilled water.
6. Leave some water on film for thirty seconds to intensify the colour contrasts.
7. Dry and mount.

METHODS OF DEMONSTRATING STRUCTURE OF BACTERIA, ETC.

To Demonstrate Capsules.

1. MacConkey.—

Stain.—

Weigh out

Dahlia   0.5 gramme

Methyl green (00 crystals)   1.5 grammes

rub up in a mortar with

Distilled water   100.0 c.c.

Add

Fuchsin, saturated alcoholic solution   10.0 c.c.

and make up to 200 c.c. by the addition of

Distilled water   90.0 c.c.

Filter.

Allow the stain to stand for two weeks before use; keep in a dark place or in an amber glass bottle. Owing to the unstable character of the methyl green, this stain deteriorates after about six months.

Method.—
1. Prepare and fix film in the usual manner.
2. Flood the cover-slip with the stain and allow it to act for five to ten minutes.
3. Wash very thoroughly in water; if necessary, direct a powerful stream of water on the film from a wash-bottle.
4. Dry and mount.

2. Muir's Method.—
1. Prepare, dry and fix film in the ordinary manner.
2. Flood the film with carbolic fuchsin, warm until steam begins to rise. Allow the stain to act for thirty seconds.
3. Wash quickly with methylated spirit.
4. Wash thoroughly with water.
5. Subject the film to the action of the following mordant for five seconds:

Corrosive sublimate, saturated aqueous solution   2 c.c.

Tannic acid, 20 per cent. aqueous solution   2 c.c.

Potash alum saturated aqueous solution   5 c.c.

6. Wash thoroughly in water.
7. Treat with methylated spirit for about sixty seconds. (The preparation should now be pale red.)
8. Wash thoroughly in water.
9. Counterstain in methylene blue, aqueous solution thirty seconds.
10. Wash in water.

11. Dehydrate in alcohol.
12. Clear in xylol and mount in xylol balsam.

3. Welch's Method.—
1. Prepare and fix film in the usual manner.
2. Flood the slide with acetic acid 2 per cent.; allow the acid to remain in contact with the film for two minutes. This swells up and fixes the capsule and enables it to take the stain.
3. Blow off the acetic acid by the aid of a pipette.
4. Immerse in aniline gentian violet, five to thirty seconds.
5. Wash in water.
6. Dry and mount.

4. Ribbert's Method.—
Stain.—
Measure out and mix:

Acetic acid, glacial	12.5 c.c.
Alcohol, absolute	50.0 c.c.
Distilled water	100.0 c.c.

Warm to 36° C. (*e. g.*, in the "hot" incubator) and saturate with dahlia. Filter.
Method.—
1. Prepare and fix films in the usual manner.
2. Cover the film with the stain and allow it to act for one or two seconds only.
3. Wash thoroughly in water.
4. Dry and mount.

To Demonstrate Flagella.
1. Muir's Modified Pitfield.—This is the best method and gives the most reliable results, for not only is the percentage of successful preparations higher than with any other, but the bacilli and flagella retain their relative proportions.
(a) **Mordant.—**

Tannic acid, 10 per cent. aqueous solution	10 c.c.
Corrosive sublimate, saturated aqueous solution	5 c.c.
Alum, saturated aqueous solution	5 c.c.
Carbolic fuchsin (Ziehl)	5 c.c.

Mix thoroughly.
A precipitate forms which must be allowed to settle for a few hours.
Decant off the clear fluid into tubes and centrifugalise thoroughly.
This solution is at its best some four or five days after manufacture; it keeps for about a couple of weeks, but must be re-centrifugalised each time, before use.

(b) *Stain.—*

Alum, saturated aqueous solution	25 c.c.
Gentian violet, saturated alcoholic solution	5 c.c.

Filter.
This stain must be freshly prepared.
Method.—The cultivations employed should be smear agar cultures, twelve to eighteen hours old if incubated at 37°C, twenty-four to thirty hours if incubated at 22°C.
1. Remove a very small quantity of the growth by means of the platinum spatula.
2. Emulsify it with a few cubic centimetres of distilled water in a watch-glass, by gently moving the spatula to and fro in the water. Do not rub up the growth on the side of the watch-glass. Some workers prefer to use tap water, others employ normal saline solution, but distilled water gives the best emulsion.
3. Spread a thin film of the emulsion on a newly flamed cover-slip, using no force, but rather *leading* the drop over the cover-slip with the platinum loop.
4. Allow the film to dry in the air, properly protected from falling dust.
5. Fix by passing thrice through the Bunsen flame, holding the cover-slip whilst doing so by one corner between the finger and thumb.
6. Pour on the film as much of the mordant as the cover-glass will hold. Grasp the cover-slip with the forceps and hold it, high above the flame, until steam rises. Allow the steaming mordant to remain in contact with the film two minutes.
7. Wash well in water and dry carefully.
8. Pour on the film as much of the stain as the cover-glass will hold. Steam over the flame as before for two minutes.
9. Wash well in water.
10. Dry and mount.

2. "Pitfield" Original Method.—
(a) *Mordant.—*

Tannic acid	1 gramme
Water	10 c.c.

(b) *Stain.—*

Saturated aqueous solution of alum	10 c.c.
Saturated alcoholic solution of gentian violet	1 c.c.
Distilled water	5 c.c.

Mix equal parts of *a* and *b* before using.
1. Prepare and fix the film in the manner described above.
2. Boil the mixture and immerse the cover-slip in it, whilst still hot, for one minute.
3. Wash in water.
4. Examine in water; if satisfactory, dry and mount in Canada balsam.

3. MacCrorrie's Method.—
Mordant-Stain.—
Measure out and mix.

Night blue, saturated alcoholic solution	10 c.c.
Potash alum, saturated aqueous solution	10 c.c.
Tannin, 10 per cent. aqueous solution	10 c.c.

Note.—The addition of gallic acid, 0.1 to 0.2 gramme, may improve the solution, but is not necessary.
Method.—
1. Prepare and fix the films as above.
2. Pour some of the mordant-stain on the film and warm gently, high above the flame, for two minutes (or place in the "hot" incubator for a like period).
3. Wash thoroughly in water.
4. Dry and mount.

4. Loeffler's Method.—
(a) *Mordant.—*

Tannic acid, 20 per cent. aqueous solution	10 c.c.
Ferrous sulphate, saturated aqueous solution	5 c.c.
Hæmatoxylin solution	3 c.c.
Carbolic acid, 1 per cent. aqueous solution	4 c.c.

This solution must be freshly prepared.
Hæmatoxylin solution is prepared by boiling 1 gramme logwood

with 8 c.c. distilled water, filtering and replacing the loss from evaporation.
Alternative Mordant (Bunge's Mordant).—

Tannic acid, 20 per cent. aqueous solution	10 c.c.
Ferrous sulphate, saturated aqueous solution	5 c.c.
Fuchsin, saturated alcoholic solution	1 c.c.

(b) *Stain.—*
Weigh out

Methylene-blue }
Or methylene-violet } 4 grammes
Or fuchsin }

and dissolve in

Aniline water, freshly saturated and filtered	100 c.c.

Method.—
1. Prepare and fix films as above.
2. Pour the mordant on to the film and warm cautiously over the flame till steam rises; keep the mordant gently steaming for one minute.
3. Wash well in distilled water till no more colour is discharged; if necessary, wash carefully with absolute alcohol.
4. Filter a few drops of the stain on to the film, warm as before, and allow the steaming stain to act for one minute.
5. Wash well in distilled water.
6. Dry and mount.

Note.—The flagella of some organisms can be demonstrated better by means of an alkaline stain or an acid stain—a point to be determined for each. Speaking generally, those bacilli which give rise to an acid reaction in the culture medium require an alkali; those which form alkali in cultivation require an acid. According to requirements, therefore, Loeffler recommends the addition of sodium hydrate, 1 per cent. aqueous solution, 1 c.c.; or an equal quantity of an exactly comparable solution of sulphuric acid.

5. Van Ermengem's Method.—This method, being merely a precipitation of a silver salt on the micro-organisms and not a true stain, creates a false impression as to the relative proportions of bacteria and flagella.

(*a*) *Fixing Fluid.*—

Osmic acid, 2 per cent. aqueous solution	10 c.c.
Tannic acid, 20 per cent. aqueous solution	20 c.c.
Acetic acid, glacial	1 c.c.

The fixing fluid should be prepared some days before use and filtered as required. In colour it should be distinctly violet.

(*b*) *Sensitising Solution.*—

Silver nitrate, 0.5 per cent. aqueous solution.

This solution must be kept in a dark blue glass bottle or in a dark cupboard.

Filter immediately before use.

(*c*) *Reducing Solution.*—

Weigh out

Gallic acid	5 grammes
Tannic acid	3 grammes
Potassium acetate, fused	10 grammes

and dissolve in

Distilled water	350 c.c.

Filter.

This solution will keep active for several days, but fresh solution must be used for each preparation.

Method.—

1. Prepare emulsion, make and fix films as above in the preceding method, steps 1 to 4.

2. Pour on the film as much of the fixing solution as the cover-glass will hold, heat carefully over the flame till steam rises, and allow the steaming fixing fluid to act for five minutes.

3. Wash well in water.

4. Wash in absolute alcohol.

5. Wash in distilled water.

6. Pour some of the sensitising solution on the film and allow it to act for from thirty seconds to one minute; blot off the excess of fluid with filter paper.

7. Without washing, transfer the film to a watch-glass containing the reducing solution and allow it to remain therein for from thirty seconds to one minute; blot off the excess of fluid with filter paper.

8. Without washing, again treat the film with the sensitising solution, this time until the film commences to turn black.

9. Wash in distilled water.

10. Dry and mount.

To Stain Nuclei of Yeast Cells.

1. Prepare and fix film in the usual manner.

2. Soak in ferric ammonia sulphate 3 per cent. aqueous solution for two hours.

3. Wash thoroughly in water.

4. Stain in hæmatoxylin solution (see page 95) for thirty minutes.

5. Wash in water.

6. Differentiate in ferric ammonia sulphate solution for 1-1/2-2 minutes, examining wet under microscope during the process.

To Stain Spores.

1. Single Stain.—

1. Prepare cover-slip film in the usual way.

2. In fixing, pass the cover-slip film fifteen or thirty times through the flame instead of only three. This destroys the resisting power of the spore membrane and allows the stain to reach the interior.

3. Stain in the usual way with methylene-blue or fuchsin.

4. Wash in water.

5. Dry and mount.

2. Double Stain.—

1. Prepare and fix film in the usual way—*i. e.*, pass three times through flame to fix.

2. Cover the film with hot carbol-fuchsin and hold in the forceps above a small flame until the fluid begins to steam. Set the cover-slip down and allow it to cool. Repeat the process when the stain ceases to steam and continue to repeat until the stain has been in contact with the film for twenty minutes. (This stains both spores and bacteria.)

3. Wash in water.
4. Decolourise in alcohol, 2 parts; acetic acid, 1 per cent., 1 part. (This removes the stain from everything but the spores.)
5. Wash in water.
6. Mount the cover-slip in water and examine microscopically with the 1/6-inch objective. (Spores should be red, and the rest of the film colourless or a very light pink.) If satisfactory, pass on to section 7; if unsatisfactory, repeat steps 2 to 5.
7. Counterstain in weak methylene-blue. (Now spores red, bacilli blue.)
8. Wash in water.
9. Dry and mount.

The spores of different bacilli differ greatly in their resistance to decolourising reagents; even the spores of the same species of organisms vary according to their age. Young spores are more easily decolourised than those more mature.

Sulphuric acid, 1 per cent. aqueous solution, and hydrochloric acid, 0.5 per cent. alcoholic (90 per cent.) solution, are useful decolourising reagents.

3. Moeller's Method.—
1. Prepare and fix films in the usual manner.
2. Immerse in absolute alcohol for two minutes, then in chloroform for two minutes; wash in water. This dissolves out any fat or crystals that might otherwise retain the "spore" stain.
3. Immerse in chromic acid, 5 per cent. aqueous solution, for one minute; wash in water.
4. Pour Ziehl's carbolic fuchsin on the film, warm as in previous methods, and allow it to act for ten minutes.
5. Wash in water.
6. Decolourise in sulphuric acid, 5 per cent. aqueous solution, for five seconds.
7. Wash in water.
8. Counterstain with Kuehne's carbolic methylene-blue for one or two minutes.
9. Wash in water.
10. Dry and mount.
(Spores red, bacilli blue.)

4. Abbott's Method.—
1. Prepare and fix films in the usual manner.
2. Pour Loeffler's alkaline methylene-blue on the film; warm cautiously over the flame till steam rises and allow the hot steam to act for one to five minutes.
3. Wash thoroughly in water.
4. Decolourise in nitric acid, 2 per cent. alcoholic (alcohol 80 per cent.) solution.
5. Wash thoroughly in water.
6. Counterstain in eosin, 1 per cent. aqueous solution.
7. Wash.
8. Dry and mount.
(Spores blue, bacilli red.)

DIFFERENTIAL METHODS OF STAINING.

Gram's Method.—This method depends upon the fact that the protoplasm of some bacteria permits aniline gentian violet and Lugol's iodine solution, when applied consecutively, to enter into a chemical combination which results in the formation of a new blue-black pigment, only very sparingly soluble in absolute alcohol. Such organisms are said to "stain by Gram," or to be "Gram positive."
1. Prepare a cover-slip film and fix in the usual way.
2. Stain in aniline gentian violet three to five minutes. Filter as much aniline water on to the cover-slip as it will hold; then add the smallest quantity of alcoholic solution of gentian violet which suffices to saturate the aniline water and form a "bronze scum" upon its surface—if too much of the alcoholic gentian violet is added the alcohol present redissolves this scum.

To prepare aniline water, pour 4 or 5 c.c. aniline oil into a stoppered bottle and add distilled water, 100 c.c. Shake vigourously and filter immediately before use. The excess of oil sinks to the bottom of the bottle and may be used again.
3. Wash in water.
4. Treat with Lugol's iodine solution until the film is black or dark brown.

To do this treat with iodine solution for a few seconds, wash in water, and examine the film over a piece of white filter paper. Note the colour. Repeat this process until the film ceases to darken with the fresh application of iodine solution.

Lugol's solution is prepared by dissolving

Iodine	1 gramme
Iodide of potassium	3 grammes
In distilled water	300 c.c.

5. Wash in water.
6. Wash with alcohol until no more colour is discharged and the alcohol runs away clear and colourless.

The following mixture may be substituted for absolute alcohol as a decolouriser

Acetone	10 c.c.

Absolute alcohol   100 c.c.

7. Wash in water.

8. Counterstain very lightly with aqueous solution of Neutral Red. Other counterstains may be used such as dilute eosin, dilute fuchsin, or vesuvin.

Note.—This section may be omitted when dealing with films prepared from pure cultivations.

9. Wash in water.

10. Dry and mount.

Gram-Claudius Method.—

1. Prepare a cover-slip film and fix in the usual way.

2. Stain in methyl violet, 1 per cent. aqueous solution for three to five minutes.

3. Treat with two lots picric acid, saturated aqueous solution.

4. Wash in water and dry.

5. Decolourise with clove oil.

6. Wash off clove oil with xylol.

7. Mount in xylol balsam.

Gram-Weigert Method.—

1-5. Proceed as for the corresponding sections of Gram's method (*quod vide*).

6. Dry in the air.

7. Wash in aniline oil, 1 part, xylol, 2 parts, until no more colour is discharged.

8. Wash in xylol.

9. Mount in xylol balsam.

Modified Gram-Weigert Method.—(To demonstrate trichophyta in hair.)

1. Soak the hairs in ether for ten minutes to remove the fat.

2. Stain thirty minutes in a tar-like solution of aniline gentian violet (prepared by adding 15 drops of the alcoholic solution of gentian violet to 3 drops of aniline water).

3. Dry the hairs between pieces of blotting paper.

4. Treat with perfectly fresh iodine solution.

5. Again dry between blotting paper.

6. Treat with aniline oil to remove excess of stain. (If necessary, add a drop or two of nitric acid to the oil.)

7. Again treat with aniline oil.

8. Treat with aniline oil and xylol, equal parts.

9. Clear with xylol.

10. Mount in xylol balsam.

To obtain the best differentiation the preparation should be repeatedly examined microscopically (with a 1/6-inch objective) between steps 5 and 9, as the actual time involved varies with different specimens.

Ziehl-Neelsen's Method.—(To demonstrate tubercle and other acid-fast bacilli.)

1. Smear a thin, even film of the specimen on the cover-slip by means of the platinum loop. (In the case of sputum, if it is a very watery specimen, allow the film to dry, then spread a second and even a third layer over the first.)

2. Fix by passing three times through the flame.

3. Stain in hot carbol-fuchsin (as in staining for spores) for five to ten minutes. (This stains everything on the film.) Avoid over-heating.

4. Decolourise by dipping in sulphuric acid, 25 per cent. (This removes stain from everything but acid-fast bacilli; *e. g.*, tubercle, leprosy, and smegma bacilli and the film turns yellow.)

5. Wash in water. (A pale red colour returns to the film).

6. Wash in alcohol till no more colour is discharged. (This often, but not invariably, removes the stain from acid-fast bacilli other than tubercle; *e. g.*, smegma bacillus.)

7. Wash in water.

8. Counterstain in weak methylene-blue. (Stains non-acid-fast bacilli, leucocytes, epithelial cells, etc.)

9. Wash in water, dry, and mount.

Pappenheim's Method.—

This method is supposed to differentiate between B. tuberculosis and other acid-fast micro-organisms.

1. Prepare and fix film in the usual way.

2. Stain in carbol-fuchsin *without heat* for three minutes.

3. Without previously washing in water treat the film with three or four successive applications of corallin (Rosolic acid) solution.

Corallin	1 gramme
Methylene-blue (saturated alcoholic solution)	100 c.c.
Glycerine	20 c.c.

4. Wash in water.

5. Dry and mount.

Neisser's Method—Modified.—(To demonstrate diphtheroid bacilli.)

Stain I.—

Measure out and mix

Methylene-blue, saturated alcoholic solution	4.0 c.c.
Acetic acid, 5 per cent. aqueous solution	96.0 c.c.

Filter.
Stain II.—
Weigh out

Neutral red	2.5 grammes

and dissolve in

Distilled water	1000 c.c.

Filter.
Method.—
1. Prepare and fix films in the usual way.
2. Pour stain I on the film and allow it to act for two minutes.
3. Wash thoroughly in water.
4. Treat with Lugol's iodine for ten seconds.
5. Wash thoroughly in water.
6. Pour stain II on to the film and allow it to act for thirty seconds.
7. Wash thoroughly in water.
8. Dry and mount.

Note.—The cultivation from which the films are prepared must be upon blood-serum which has been incubated at 37°C. for from nine to eighteen hours.

The bacilli are stained a light red by the neutral red, which contrasts well with the two or three black spots, situated at the poles and occasionally one in the centre representing protoplasmic aggregations (? metachromatic granules) stained by the acid methylene-blue.

Wheal and Chown (Oxford) Method.—(To demonstrate actinomyces.)
1. Stain briefly with Ehrlich's hæmatoxylin (until nuclei are faint blue after washing with tap water).
2. Wash in tap water.
3. Stain in hot carbol-fuchsin (as for tubercle bacilli) for five to ten minutes.
4. Wash in tap water.
5. Decolourise with Spengler's picric acid alcohol. This is prepared by mixing:

Alcohol, absolute	20 c.c.
Picric acid, saturated aqueous solution	10 c.c.
Distilled water	10 c.c.

During the progress of steps 1-5 the preparation must be repeatedly examined microscopically with the 1/6-inch objective.

When properly differentiated the clubs appear brilliant red on greenish ground.
6. Dehydrate in alcohol.
7. Clear in xylol.
8. Mount in xylol balsam.

This method serves equally well for films and for sections.

VII. METHODS OF DEMONSTRATING BACTERIA IN TISSUES.

For bacteriological purposes, sections of tissue are most conveniently prepared by either the **freezing method** or the **paraffin method**.
The latter is decidedly preferable, but as it is of greater importance to demonstrate the bacteria, if such are present, than to preserve the tissue elements unaltered, the "frozen" sections are often of value.

Whichever method is selected, it is necessary to take small pieces of the tissue for sectioning,—2 to 5 mm. cubes when possible, but in any case not exceeding half a centimetre in thickness. Post-mortem material should be secured as soon after the death of the animal as possible.

The tissue is prepared for cutting by—
(*a*) Fixation; that is, by causing the death of the cellular elements in such a manner that they retain their characteristic shape and form.

The fixing fluids in general use are: Absolute alcohol; corrosive sublimate, saturated aqueous solution; corrosive sublimate, Lang's solution (*vide* page 82); formaldehyde, 4 per cent. aqueous solution. (Of these, Lang's corrosive sublimate solution is decidedly the best all-round "fixative.")

(*b*) **Hardening**; that is, by rendering the tissue of sufficient consistency to admit of thin slices or "sections" being cut from it. This is effected by passing the tissue successively through alcohols of gradually increasing strength: 30 per cent. alcohol, 50 per cent. alcohol, 75 per cent. alcohol, 90 per cent. alcohol, absolute alcohol.

In both these processes a large excess of fluid should always be used.

FREEZING METHOD.

1. **Fixation.** Place the pieces of tissue in a wide-mouthed glass bottle and fill with absolute alcohol. Allow the tissues to remain therein for twenty-four hours.

2. **Hardening.** Remove the alcohol (no longer absolute, as it has taken up water from the tissues) from the bottle and replace it with fresh absolute alcohol. Allow the tissues to remain therein for twenty-four hours.

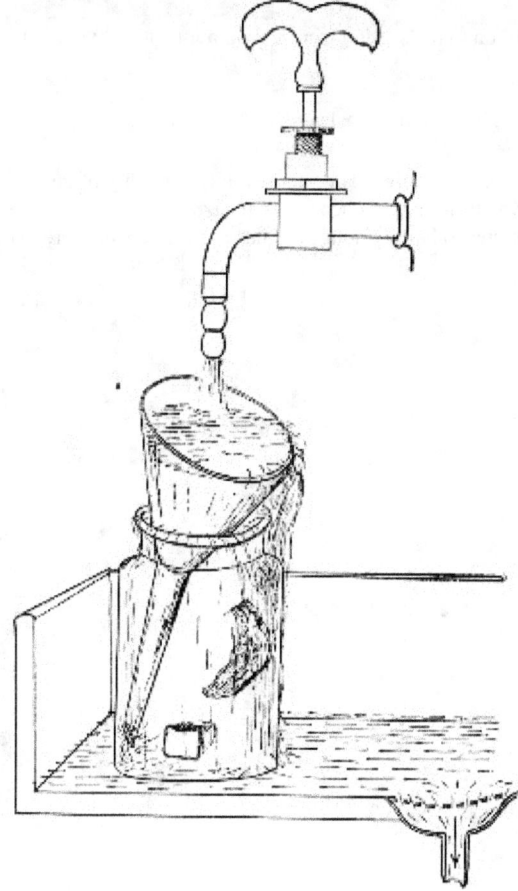

Fig. 71.—Washing tissues.

Note.—If not needed for cutting immediately, the hardened tissues can be stored in 75 per cent. alcohol.

3. Remove the alcohol from the tissues by soaking in water from one to two hours. Remove the stopper from the bottle; rest a glass funnel in the open mouth and place under a tap of running water. The water of course, overflows, but the tissues remain in the bottle (Fig. 71).

4. Impregnate the tissues with mucilage for twelve to twenty-four hours, according to size. Transfer the pieces of tissue to a bottle containing sterilised gum mixture.

Formula.—

Gum arabic	5 grammes
Saccharose	1 gramme
Boric acid	1 gramme
Water	100 c.c.

5. Place the tissue on the plate of a freezing microtome (Cathcart's is perhaps the best form), cover and surround with fresh gum mixture; freeze with ether, or for preference, carbon dioxide, and cut sections.

6. Float the sections off the knife into a glass dish containing tepid water and allow them to remain therein for about an hour to dissolve out the gum.

(If not required at once, store in 90 per cent. alcohol.)

7. Transfer to a glass capsule containing the selected staining fluid, by means of a section lifter.

8. Transfer the sections in turn to a capsule containing absolute alcohol (to dehydrate) and to one containing xylol or oil of cloves (to clear).

9. Mount in xylol balsam.

Alternative Rapid Method.—

1. Cut very small blocks of the tissue.
2. Fix in formalin 10 per cent. aqueous solution (fixation fluid No. 7, page 82) for 24 hours.
3. Transfer block to plate of freezing microtome and freeze with carbon dioxide vapour.
4. Float the sections off the knife into a glass dish of tepid water.
5. Stain the sections in glass capsules containing selected stains.
6. Place the stained section in a dish of clean water and introduce a glass slide obliquely beneath the section; with a mounted needle draw the section on to the slide and hold it there; gently remove the slide from the water, taking care that any folds in the section are floated out before the slide is finally removed from the water.
7. Drain away as much water as possible from the section. Drop absolute alcohol on to the section from a drop bottle, to dehydrate it.
8. Double a piece of blotting paper and gently press it on the section to dry it.
9. Drop on xylol to clear the section.
10. Place a large drop of xylol balsam on the section and carefully lower a cover-glass on to the balsam.

PARAFFIN METHOD.

1. **Fixation.** Place the pieces of tissue, resting on cotton-wool, in a wide-mouthed glass bottle. Pour on a sufficient quantity of the corrosive sublimate fixing fluid; allow the tissue to remain therein for twelve to twenty-four hours according to size.
2. Pour off the fixing fluid and wash thoroughly in running water for twenty minutes to half an hour to remove the excess of corrosive sublimate.

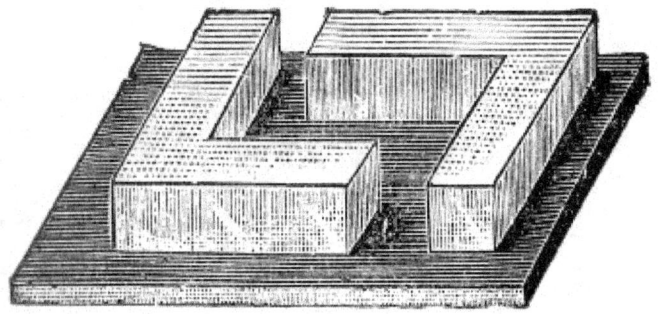

Fig. 72.—L-shaped brass moulds.

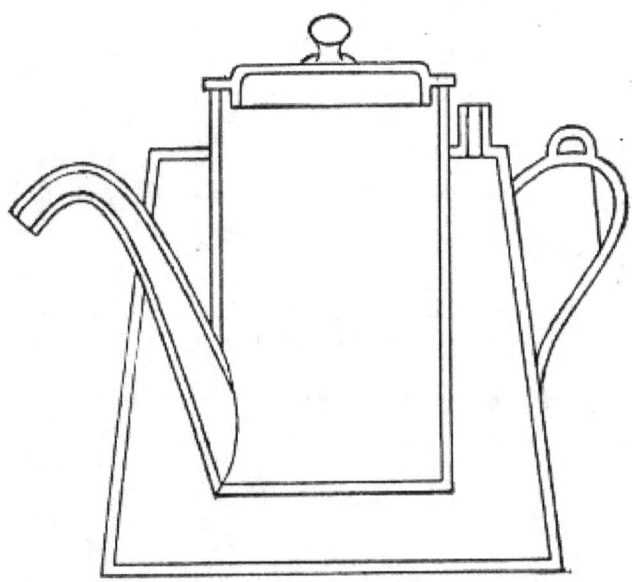

Fig. 73.—Paraffin kettle.

3. **Hardening.** Place the tissues in each of the following strengths of alcohol in turn for from twelve to twenty-four hours: 50 per cent., 75 per cent., 90 per cent., absolute.

4. **Dehydration** is effected by transferring the tissues to fresh absolute alcohol.

5. **Clearing.** Half fill a wide-mouthed bottle with chloroform. On the surface of the chloroform float a layer of absolute alcohol about five to ten millimetres in depth. Place the pieces of tissue in the layer of alcohol and when they have sunk through this layer, transfer them to pure chloroform for from six to twenty-four hours according to the size of the pieces. When "cleared," the tissue becomes more or less transparent.

6. **Infiltration.** Place the cleared tissues in fresh chloroform with several pieces of paraffin wax and stand in a warm place, such as on the top of the warm incubator. The warmth gradually melts the paraffin and the tissues should remain in the mixture about twenty-four hours.

7. Transfer the tissues to a vessel containing pure melted paraffin. Place this vessel in a paraffin water-bath regulated for 2° C. above the melting-point of the paraffin used, and allow the tissues to soak for some four to six hours to ensure complete impregnation. The paraffin used should have a melting-point of not more than 58° C. For all ordinary purposes 54°C. will be found quite high enough.

8. Imbed in fresh paraffin in a metal (or paper) mould.

(*a*) Arrange a pair of **L**-shaped pieces of metal on a plate of glass to form a rectangular trough (Fig. 72).

(*b*) Pour fresh melted paraffin into the mould from a special vessel (Fig. 73).

(*c*) Lift the piece of tissue from the paraffin bath and arrange it in the mould.

(*d*) Blow gently on the surface of the paraffin in the mould, and as soon as a film of solid paraffin has formed, carefully lift the glass plate on which the mould is set and lower plate and mould together into a basin of cold water.

(*e*) When the block is cold, break off the metal **L**'s; trim off the excess of paraffin from around the tissue with a knife, taking care to retain the rectangular shape, and store the block in a pill-box.

When several pieces of tissue have to be imbedded at one time, shapes of stout copper, 10 cm., 5 cm., and 2.5 cm. square respectively, and 0.75 cm. deep (Fig. 74) will be found extremely useful. These placed upon plates of glass replace the pair of L's in the above process. When the paraffin has set firmly the screw *a* should be loosened to allow the two halves of the flange *b* to separate slightly—this facilitates removal of the paraffin block.

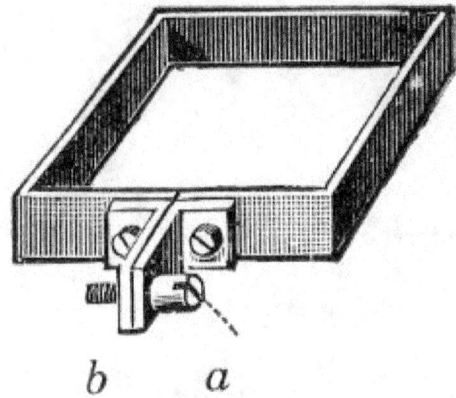

Fig. 74.—Paraffin mould.

8. Cement the block on the carrier of a "paraffin" microtome (the Minot, the Jung, or the Cambridge Rocker) with a little melted paraffin. Greater security is obtained if the paraffin around the base of the block is melted by means of a hot metal or glass rod.

9. Cut sections—thin, and if possible in ribbands.

Mounting Paraffin Sections.—

1. Place a large drop of 30 per cent. alcohol on the centre of a slide (or cover-slip) and float the section on to the surface of the drop, from a section lifter.

2. Hold the slide in the fingers of one hand and warm cautiously over the flame of a Bunsen burner, touching the under surface of the glass from time to time on the back of the other hand. As soon as the slide feels distinctly warm to the skin, the paraffin section will flatten out and all wrinkles disappear.

(The slide with the section floating on it may be rested on the top of the paraffin bath for two or three minutes, instead of warming over the flame as here described.)

3. Cautiously tilt up the slide and blot off the excess of spirit with blotting paper, leaving the section attached to the centre of the slide.

4. Place the slide in a wire rack (Fig. 75), section downward, in the "hot" incubator for twelve to twenty-four hours. At the end of this time the section is firmly adherent to the glass, and is treated during the subsequent steps as a "fixed" cover-glass film preparation.

Note.—If large, thick sections have to be manipulated, or if time is of importance or acids are used during the staining process, it is often advisable to add a trace of Mayer's albumin to the alcohol before floating out the section. If this substance is employed, a sojourn of twenty minutes to half an hour in the "hot" incubator will be found ample to ensure firm adhesion of the section to the slide. The albuminous fluid is prepared as follows:

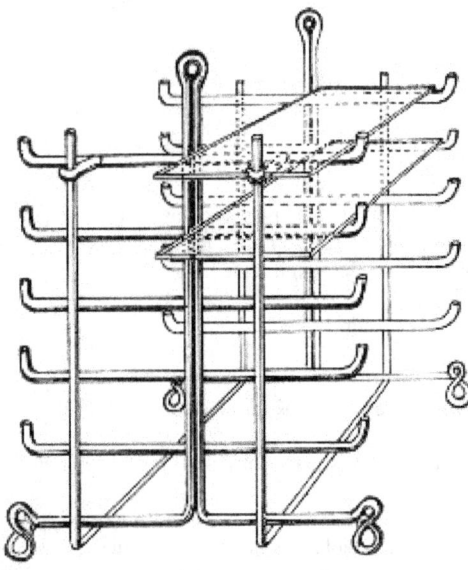

Fig. 75.—Section rack.

Mayer's Albumin.—
Weigh out

Salicylate of soda   1 gramme

and dissolve in

Glycerine   50 c.c.

Add

White of egg   50 c.c.

Mix thoroughly by means of an egg whisk.
Filter into a clean bottle.
As an alternative method paint a thin layer of Schallibaum's solution on the slide with a camel's hair pencil; lay the section carefully on this film and heat gently to fix the section.

Schallibaum's solution:

Clove oil   30 c.c.

Collodion   10 c.c.

Keep in a dark blue bottle in a cool place.

Staining Paraffin Sections.—
1. Warm paraffin section over the Bunsen flame to soften (*but not to melt*) the paraffin, then dissolve out the wax with xylol poured on from a drop bottle.
2. Remove xylol by flushing the section with alcohol.
3. If the tissue was originally "fixed" in a corrosive sublimate solution, the section must now be treated with Lugol's iodine solution for two minutes and subsequently immersed in 90 per cent. alcohol to remove all traces of yellow staining.
4. Wash in water.
5. Stain deeply, if using a single stain, as the subsequent processes decolourise.
6. Wash in water, decolourise if necessary.
7. Flood with several changes of absolute alcohol to dehydrate the section.
8. Clear in xylol. (Oil of cloves is not usually employed, as it decolourises the section.)
9. Mount in xylol balsam.

SPECIAL STAINING METHODS FOR SECTIONS.

Double-staining Carmine and Gram-Weigert.—
1. Prepare the section for staining as above, sections 1 to 3.
2. Stain in lithium carmine (Orth's) or picrocarmine for ten to thirty minutes, in a porcelain staining pot (Fig. 76).
3. Wash in picric acid solution until yellow. At this stage cell nuclei are red, protoplasm is yellow, and bacteria are colourless.
Picric acid solution is prepared by mixing

Picric acid, saturated aqueous solution   40 c.c.

Hydrochloric acid   1 c.c.

Alcohol (90 per cent.)   160 c.c.

4. Wash in water.
5. Wash in alcohol.
6. Stain in aniline gentian violet.
7. Wash in iodine solution till dark brown or black.
8. Wash in water.
9. Dip in absolute alcohol for a second.
10. Decolourise with aniline oil till no more colour is discharged.

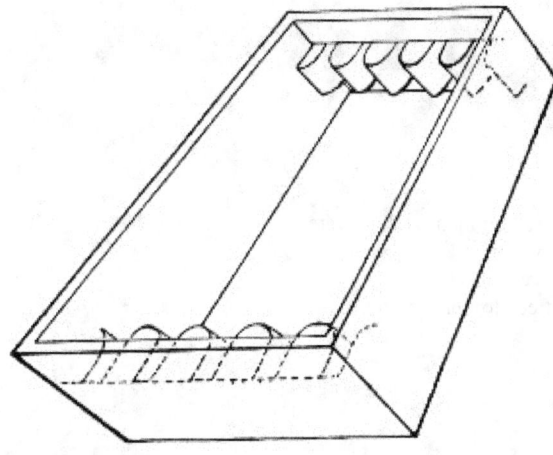

Fig. 76.—Staining pot.

11. Wash with aniline oil, 2 parts, xylol, 1 part.
12. Clear with xylol.
13. Mount in xylol balsam.

Alternative Gram-Weigert Method for Sections.—
1. Fix paraffin section on slide and prepare for staining in the usual manner.
2. Stain in alum carmine for about fifteen minutes.
3. Wash thoroughly in water.
4. Filter aniline gentian violet solution on to the section on the slide and allow to stain about twenty-five minutes.
5. Wash thoroughly in water.
6. Treat with Lugol's iodine until section ceases to become any blacker.
7. Wash thoroughly in water.
8. Treat with a mixture of equal parts of aniline oil and xylol until no more colour comes away.
9. Wash thoroughly with xylol.
10. Decolourise and dehydrate rapidly with absolute alcohol until there remains only a very faint bluish tint.
11. Clear with xylol.
12. Mount in xylol balsam.
(Then fibrin and hyaline tissue are stained deep blue, whilst bacteria which "stain Gram" appear of a deep blue-violet colour.)

Unna-Pappenheim Method.—
Stain.—
Weigh out and mix

Methylene green   0.15 gramme

Pyronin   0.25 gramme

and dissolve in
Carbolic acid 0.5 per cent. aqueous solution 78 c.c.
Measure out

Alcohol   2.5 c.c.   }

Glycerine   20.0 c.c.   } and add to the stain.

Method.—
1. Place tissue in the above stain for ten minutes.
2. Differentiate and dehydrate with absolute alcohol.
3. Clear in xylol.
4. Mount in xylol balsam.

To Demonstrate Capsules.—
1. *MacConkey's Method.*—Stain precisely as for cover-slip films (*vide* page 100).
2. *Friedländer's Method.—*
Stain.—

Gentian violet, saturated alcoholic solution	50 c.c.
Acetic acid, glacial	10 c.c.
Distilled water	100 c.c.

Method.—
1. Prepare the sections for staining, *secundum artem*.
2. Stain sections in the warm (*e. g.*, in the hot incubator) for twenty-four hours.
3. Wash with water.
4. Decolourise lightly with acetic acid, 1 per cent.
5. Dehydrate rapidly with absolute alcohol.
6. Clear with xylol.
7. Mount in xylol balsam.

To Demonstrate Acid-fast Bacilli.—
1. Prepare the sections for staining in the usual way.
2. Stain with hæmatin solution ten to twenty seconds, to obtain a pure nuclear stain; then wash in water.
3. Stain with carbolic fuchsin twenty to thirty minutes at 47°C.; then wash in water.
4. Treat with aniline hydrochlorate, 2 per cent. aqueous solution, for two to five seconds.
5. Decolourise in 75 per cent. alcohol till section appears free from stain—fifteen to thirty minutes.
6. Dehydrate with absolute alcohol.
7. Clear very rapidly with xylol.
8. Mount in xylol balsam.

To Demonstrate Spirochætes in Tissues.
Piridin Method (Levaditi).—
1. Cut slices of tissue 1 mm. thick.
2. Fix in 10 per cent. formalin solution for twenty-four hours.
3. Wash in water for one hour.
4. Place in 96 per cent. alcohol for twenty-four hours.
5. Measure into a dark green or amber bottle 100 c.c. silver nitrate solution 1 per cent., and 10 grammes pyridin puriss. Transfer slices of tissue to this. Stopper and keep at room temperature three hours, then in thermostat at 50° C. for four to six hours.
6. Wash quickly in 10 per cent. pyridin solution.
7. Reduce silver by transferring slices of tissue to following solution for forty-eight hours.

Pyrogallic acid	4 grammes
Acetone	10 c.c.
Pyridin puriss	15 grammes
Distilled water	100 c.c.

8. Wash well in water.
Take through alcohols of increasing strength up to absolute, keeping in each strength for twenty-four hours.
9. Clear, embed, cut very thin sections, mount, remove paraffin, again clear and mount in xylol balsam.
The spirochætes if present are black and show up against the pale yellow color of the background.
Weak carbol fuchsin, neutral red or toluidin blue can also be used to stain the background if desired, after the removal of the paraffin in step 9.

To Demonstrate Protozoa in Sections (Leishman).—
Reagents required:

Leishman's Polychrome stain.

Acetic acid 1 in 1500 aqueous solution.

Caustic soda 1 in 7000 aqueous solution.

Distilled water.

1. Mount section, remove paraffin and take into distilled water as usual (*vide* page 121).
2. Drain off the excess of water.
3. Cover the section with diluted Leishman (1 part stain, 2 parts distilled water) and allow to act for five to ten minutes (until tissue appears a deep blue).
4. Decolourise with acetic acid solution until only the nuclei appear blue (examine the section wet, with low power objective).
5. If the eosin colour is too well marked treat with the caustic soda solution until the desired tint is obtained (as seen with the 1/6-inch objective).
6. Wash with distilled water.
7. Rapidly dehydrate with alcohol.
8. Clear with xylol.

9. Mount in xylol balsam.

VIII. CLASSIFICATION OF FUNGI.

For practical purposes Fungi may be divided into:
1. Hymenomycetes (including the mushrooms, etc.). **2. Hyphomycetes** (moulds). **3. Blastomycetes** (yeasts and torulæ). **4. Schizomycetes** (bacteria).

Note.—Formerly myxomycetes were included in the fungi; they are now recognized as belonging to the animal kingdom, and are termed "mycetozoa."

MORPHOLOGY OF THE HYPHOMYCETES.

At the commencement of his studies, the attention of the student is directed to the various non-pathogenic moulds and yeasts, not only that he may gain the necessary technique whilst handling cultivations of harmless organisms, but also because these very species are amongst the commonest of those that may accidentally contaminate his future preparations.

The hyphomycetes are composed of a mycelium of short jointed rods or "hyphæ" springing from an axis or germinal tube which develops from the spore. Hyphæ are—

(*a*) Nutritive or submerged.
(*b*) Reproductive or aerial.

The protoplasm of these cells contains granules, pigment, oil globules, and sometimes crystals of calcium oxalate.

Reproduction.—Apical spore formation—asexual; zoospores—sexual.

Mucorinæ.—*Mucor* (Fig. 77).—Note the branching filaments—"mycelium" (*a*), "hyphæ" (*b*).

Note the asexual reproduction.

1. A filament grows upward. At its apex a septum forms, then a globular swelling appears—"sporagium" (*d*). This possesses a definite membrane.
2. From the septum grows a club-shaped mass of protoplasm—"columella" (*c*).

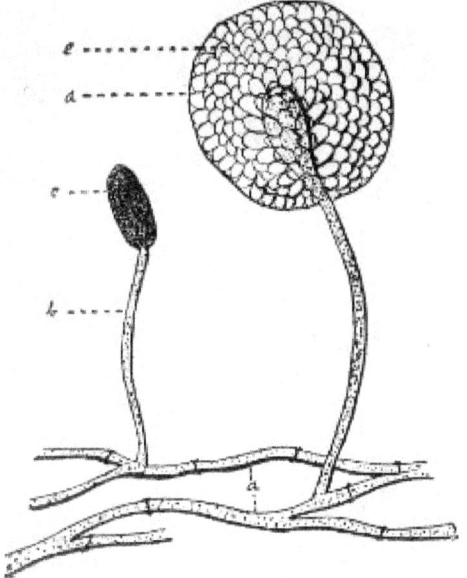

Fig. 77.—Mucor mucedo.

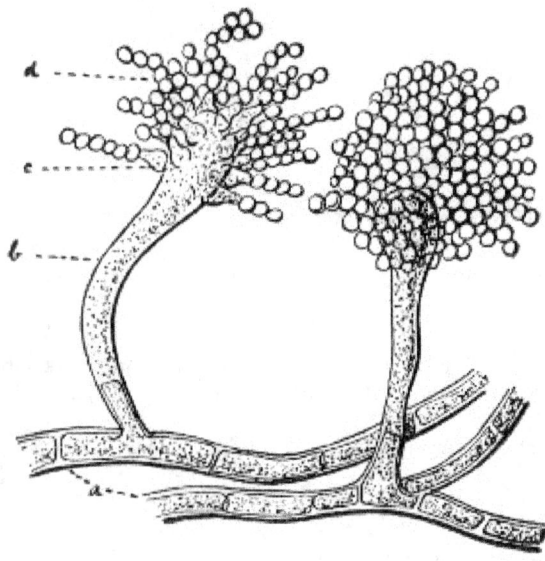

Fig. 78.—Aspergillus

3. The rest of the contained protoplasm breaks up into "swarm spores" (*e*).
Finally the membrane ruptures and spores escape.
Perisporaceæ.—*Aspergillus* (Fig. 78).—Note the branching filaments—"mycelium" (*a*).

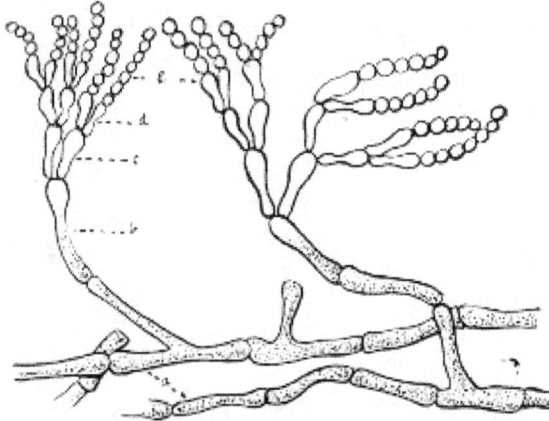

Fig. 79.—Penicillium.

Note the asexual reproduction.
1. A filament (*b*) grows upward, its termination becomes clubbed; on the clubbed extremity flask-shaped cells appear—"sterigmata" (*c*).
2. At free end of each sterigma is formed an oval body—a spore or "gonidium" (*d*), which, when ripe, is thrown off from the sterigma. Two or more gonidia may be supported upon each sterigma.
Penicillium (Fig. 79).—Note the branching filaments—"mycelium" (*a*) (frequently containing globules).
Note the asexual reproduction.
1. A filament grows upward—"goniodophore" (*b*)—and its apex divides up into several branches—"basidia" (*c*).
2. At the apex of each basidium a flask-shaped cell, "sterigma" (*d*), appears.
3. At the apex of each sterigma appears a row of oval cells—"spores" or "conidia" (*e*). These, when ripe, are cast off from the sterigmata.

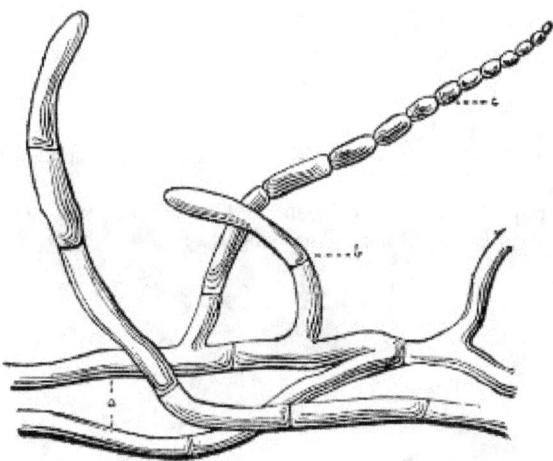

Fig. 80.—Oïdium.

Ascomycetæ.—*Oïdium* (Fig. 80).—(This family is perhaps as nearly related to the blastomycetes as it is to the hyphomycetes.)

Note the branching filaments—"pseudomycelium" (*a*). Here and there filaments are broken up at their ends into oval or rod-shaped segments, "oïdia," and behave as spores.

Note the asexual reproduction. From the pseudomycelium arise true hyphæ (*b*), each of which in turn ends in a chain of spores (*c*).

MORPHOLOGY OF THE BLASTOMYCETES.

The blastomycetes are composed of spherical or oval cells (8 to 9.5μ in diameter), which, when rapidly multiplying by budding, may form a spurious mycelium. A thin cell-wall encloses the granular protoplasm, in which vacuoles and sometimes a nucleus may be noted. This latter is best seen when stained with hæmatoxylin (see page 105).

During their growth and multiplication the blastomycetes split up solutions containing sugar into alcohol and CO_2.

Saccharomyces (Fig. 81).—Note the round or oval cells of granular protoplasm (*a*) containing solid particles and vacuoles (*c*), and surrounded by a definite envelope.

Reproduction.—Budding; ascospores—asexual.

Note the asexual *reproduction*.

1. "Gemmation"—that is, the budding out of daughter cells (*b*) from various parts of the gradually enlarging mother cell. These are eventually cast off and in turn become mother cells and form fresh groups of buds.

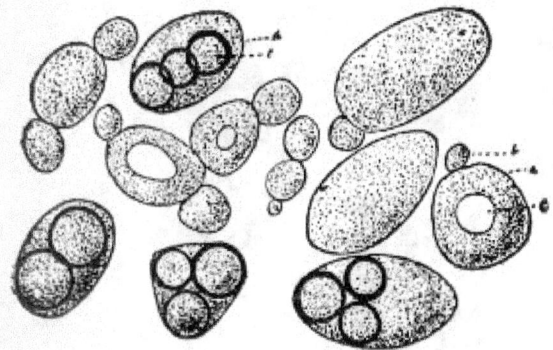

Fig. 81.—Saccharomyces with ascospores.

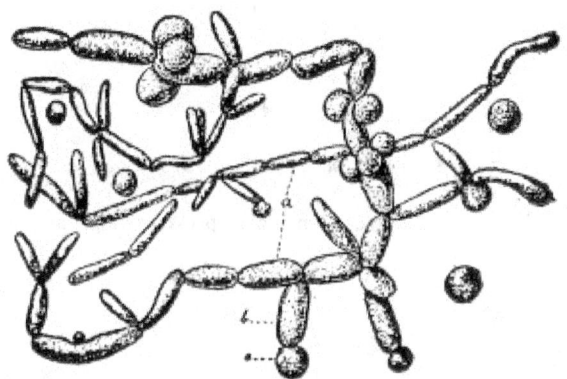

Fig. 82.—Torula.

2. Spore formation—"ascospores" (*e*). These are formed at definite temperatures and within well-defined periods; *e. g.*, Saccharomyces cerevisiæ, thirty hours at 25° to 37°C., or ten days at 12°C.

Torulæ (Fig. 82).—Torulæ, whilst resembling yeasts in almost every other respect, never form endo-spores. Note the elongated, sausage-shaped cells (*a*) the larger oval cells (*b*) and the globular cells (*c*) the former two often interlacing and growing as a film.

Note the absence of ascospore formation.

IX. SCHIZOMYCETES.

Classification and Morphology.—Bacteria are often classified, in general terms, according to their life functions, into—
Saprogenic, or putrefactive bacteria; *Zymogenic*, or fermentative bacteria; *Pathogenic*, or disease-producing bacteria;
or according to their food requirements into—
Prototrophic, requiring no organic food (*e. g.*, nitrifying bacteria); *Metatrophic*, requiring organic food (*e. g.*, saprophytes and facultative parasites); *Paratrophic*, requiring living food (obligate parasites);
or according to their metabolic products into—
Chromogenic, or pigment-producing bacteria; *Photogenic*, or light-producing bacteria; *Aerogenic*, or gas-producing bacteria;
and so on.

Such broad groupings as these have, however, but little practical value when applied to the systematic study of the fission fungi. On the other hand, no really scientific classification of the schizomycetes has yet been drawn up, and the varying morphological appearances of the members of the family are still utilised as a basis for classification, as under—

1. Cocci. (Fig. 83).—Rounded or oval cells, subdivided according to the arrangement of the individuals after fission, into—
Diplococci and *Streptococci*, where division takes place in one plane only, and the individuals remain attached (*a*) in pairs or (*b*) in chains.

Tetrads, *Merismopedia*, or *Pediococci*, where division takes place alternately in two planes at right angles to each other, and the individuals remain attached in flat tablets of four, or its multiples.

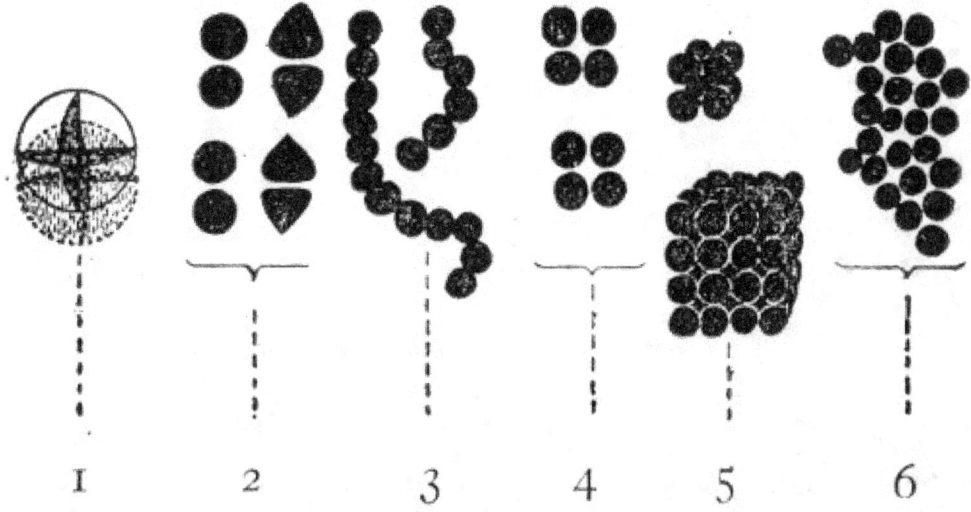

Fig. 83.—Types of bacteria—cocci: 1, Diagram of sphere indicating planes of fission; 2, diplococci; 3, streptococci; 4, tetrads; 5, sarcinæ; 6, staphylococci.

Sarcinæ, where division takes place in three planes successively, and the individuals remain attached in cubical packets of eight and its multiples.

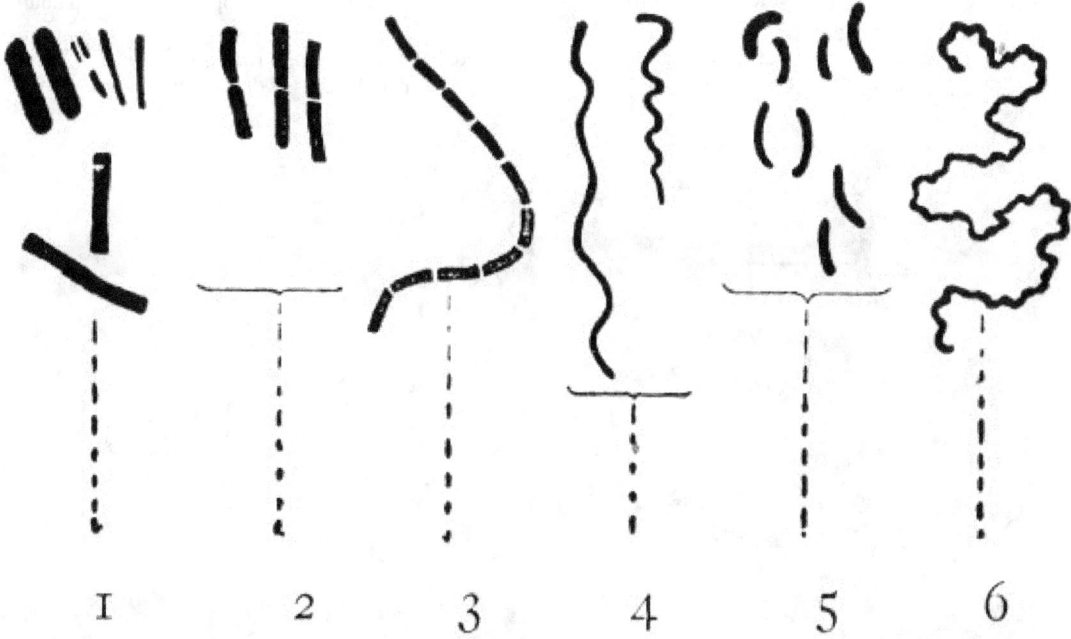

Fig. 84.—Types of bacteria—bacilli, etc.: 1, Bacilli; 2, diplobacilli; 3 streptobacilli; 4, spirilla; 5, vibrios; 6, spirochætæ.

Micrococci or *Staphylococci*, where division takes place in three planes, but with no definite sequence; consequently the individuals remain attached in pairs, short chains, plates of four, cubical packets of eight, and irregular masses containing numerous cocci.

2. Bacili (Fig. 84, 1 to 3).—Rod-shaped cells. A bacillus, however short, can usually be distinguished from a coccus in that two sides are parallel. Some bacilli after fission retain a characteristic arrangement and may be spoken of as *Diplobacilli* or *Streptobacilli*.

Leptothrix is a term that in the past has been loosely used to signify a long thread, but is now restricted to such forms as belong to the leptothriciæ (*vide infra*).

3. *Spirilla* (Fig. 84, 4 to 6).—Curved and twisted filaments. Classified, according to shape, into—

Spirillum. Vibrio (comma). Spirochæta.

Many Spirochætes appear to belong to the animal kingdom and are grouped under protozoa; other organisms to which this name has been given are undoubtedly bacteria.

Higher forms of bacteria are also met with, which possess the following characteristics: They are attached, unbranched, filamentous forms, showing—

(*a*) Differentiation between base and apex;
(*b*) Growth apparently apical;
(*c*) Exaggerated pleomorphism;
(*d*) "Pseudo-branching" from apposition of cells; and are classified into—

1. Beggiotoa. } Free swimming forms, which 2. Thiothrix. } contain sulphur granules.
3. Crenothrix. } 4. Cladothrix. } These forms do not contain 5. Leptothrix. } sulphur granules.
6. Streptothrix. A group which exhibits true but not dichotomous branching, and contains some pathogenic species.

The morphology of the same bacterium may vary greatly under different conditions.

For example, under one set of conditions the examination of a pure cultivation of a bacillus may show a short oval rod as the predominant form, whilst another culture of the same bacillus, but grown under different conditions, may consist almost entirely of long filaments or threads. This variation in morphology is known as "pleomorphism."

Some of the factors influencing pleomorphism are:

1. The composition, reaction, etc., of the *nutrient medium* in which the organism is growing.
2. *The atmosphere* in which it is cultivated.
3. *The temperature* at which it is incubated.
4. Exposure to or protection from *light*.

The various points in the anatomy morphology and physiology of bacteria upon which stress is laid in the following pages should be studied as closely as is possible in preparations of the micro-organisms named in connection with each.

ANATOMY.

1. *Capsule* (Fig. 85, *b*).—A gelatinous envelope (probably akin to mucin in composition) surrounding each individual organism, and preventing absolute contact between any two. In some species the capsule (*e. g.*, B. pneumoniæ) is well marked, but it cannot be demonstrated in all. In very well marked cases of gelatinisation of the cell wall, the individual cells are cemented together in a coherent mass, to which the term "zoogloea" is applied (*e. g.*, Streptococcus mesenteroides). In some species colouring matter or ferric oxide is stored in the capsule.

2. *Cell Wall* (Fig. 85, *c*).—A protective differentiation of the outer layer of the cell protoplasm; difficult to demonstrate, but treatment with iodine or salt solution sometimes causes shrinkage of the cell contents—"plasmolysis"—and so renders the cell wall apparent (*e. g.*, B.

megatherium) in the manner shown in figure 85. Stained bacilli, when examined with the polarising microscope, often show a doubly refractile cell wall (*e. g.*, B. tuberculosis and B. anthracis).

In some of the higher bacteria the cell wall exhibits this differentiation to a marked degree and forms a hard sheath within which the cell protoplasm is freely movable; and during the process of reproduction the cell protoplasm may be extruded, leaving the empty tube unaltered in shape.

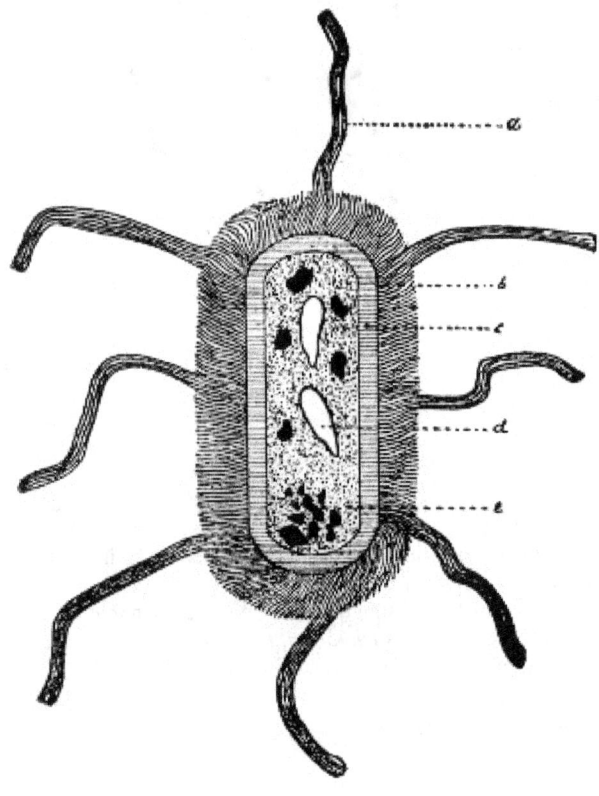

Fig. 85.—Dragrammatic sketch of composite bacterium to illustrate details of anatomical structure.

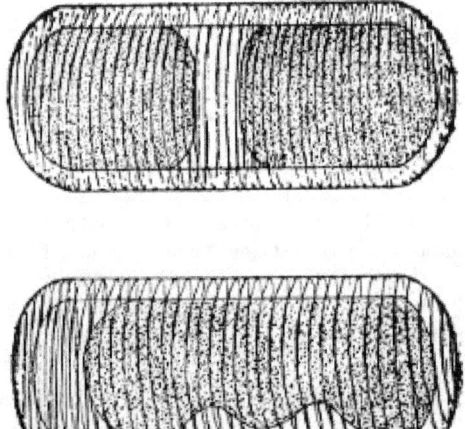

Fig. 86.—Plasmolysis.

3. *Cell Contents.*—Protoplasm (mycoprotein) contains a high percentage of nitrogen, but is said to differ from proteid in that it is not precipitated by C_2H_6O. It is usually homogeneous in appearance—sometimes granular—and may contain oil globules or sap vacuoles (Fig. 85, *d*), chromatin granules, and even sulphur granules. Sap vacuoles must be distinguished from spores, on the one hand, and the vacuolated appearance due to plasmolysis, on the other.

The cell contents may sometimes be differentiated into a parietal layer, and a central body (*e. g.*, beggiotoa) when stained by hæmatoxylin.

4. *Nucleus.*—This structure has not been conclusively proved to exist, but in some bacteria chromatin particles have been observed near the centre of the bacterial cell and denser masses of protoplasm situated at the poles which exhibit a more marked affinity than the rest of the cell protoplasm for aniline dyes. These latter are termed polar granules or *Polkoerner* (Fig. 85, *e*). Occasionally these aggregations of protoplasm alter the colour of the dye they take up. They are then known as metachromatic bodies or *Ernstschen Koerner* (*e. g.*, B. diphtheriæ).

5. *Flagella* (Organs of Locomotion, Fig. 85, *a*).—These are gelatinous elongations of the cell protoplasm (or more probably of the capsule), occurring either at one pole, at both poles, or scattered around the entire periphery. Flagella are not pseudopodia. The possession of flagella was at one time suggested as a basis for a system of classification, when the following types of ciliation were differentiated (Fig. 87):

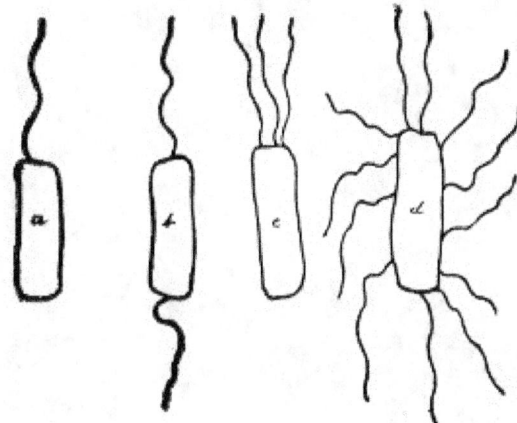

Fig. 87.—Types of ciliation.

1. Polar: (*a*) *Monotrichous* (a single flagellum situated at one pole; *e. g.*, B. pyocyaneus).
(*b*) *Amphitrichous* (a single flagellum at each pole; *e. g.*, Spirillum volutans).
(*c*) *Lophotrichous* (a tuft or bunch of flagella situated at each pole; *e. g.*, B. cyanogenus).
2. Diffuse: *Peritrichous* (flagella scattered around the entire periphery *e. g.*, B. typhosus).

PHYSIOLOGY.

Reproduction.—*Active Stage.*—Vegetative, *i. e.*, by the division of cells, or "fission."
1. The cell becomes elongated and the protoplasm aggregated at opposite poles.
2. A circular constriction of the organism takes place midway between these aggregations, and a septum is formed in the interior of the cell at right angles to its length.
3. The division deepens, the septum divides into two lamellæ, and finally two cells are formed.

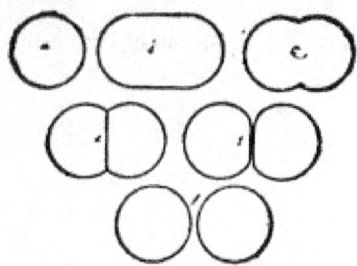

Fig. 88.—Fission of cocci.

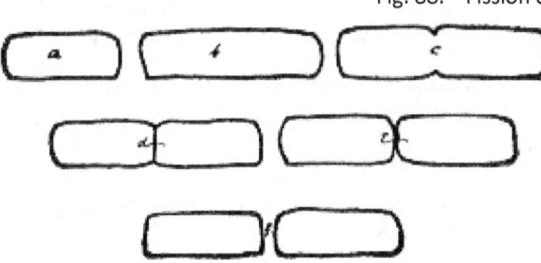

Fig. 89.—Fission of bacteria.

4. The daughter cells may remain united by the gelatinous envelope for a variable time. Eventually they separate and themselves subdivide.

Cultures on artificial media, after growing in the same medium for some time—*i. e.*, when the pabulum is exhausted—show "involution forms" (Fig. 90), well exemplified in cultures of B. pestis on agar two days old, B. diphtheriæ on potato four to six days old.

Fig. 90.—Involution forms.

They are of two classes, viz.:

(*a*) Involution forms characterised by alterations of shape (Fig. 90). (Not necessarily dead.)

(*b*) Involution forms characterised by loss of staining power. (Always dead.)

Resting Stage.—Spore Formation.—Conditions influencing spore formation: In an old culture nothing may be left but spores. It used to be supposed that spores were *always* formed, so that the species might not become extinct, when

(*a*) The supply of nutrient was exhausted.

(*b*) The medium became toxic from the accumulation of metabolic products.

(*c*) The environment became unfavourable; *e. g.*, change of temperature.

This is not altogether correct; *e. g.*, the temperature at which spores are best formed is constant for each bacterium, but varies with different species; again, aerobes require oxygen for sporulation, but anaerobes will not spore in its presence.

(A) Arthrogenous: Noted only in the micrococci. One complete element resulting from ordinary fission becomes differentiated for the purpose, enlarges, and develops a dense cell wall. One or more of the cells in a series may undergo this alteration.

This process is probably not real spore formation, but merely relative increase of resistance. These so-called arthrospores have never been observed to "germinate," nor is their resistance very marked, as they fail to initiate new cultures, after having been exposed to a temperature of 80° C. for ten minutes.

(B) Endogenous: The cell protoplasm becomes differentiated and condensed into a spherical or oval mass (very rarely cylindrical). After further contraction the outer layers of the mass become still more highly differentiated and form a distinct spore membrane, and the spore itself is now highly refractile. It has been suggested, and apparently on good grounds, that the spore membrane consists of two layers, the exosporium and the endosporium. Each cell forms one spore only, usually in the middle, occasionally at one end (some exceptions, however, are recorded; *e. g.*, B. inflatus). The shape of the parent cell may be unaltered, as in the anthrax bacillus, or altered, as in the tetanus bacillus, and these points serve as the basis for a classification of spore-bearing bacilli, as follows:

(A) Cell body of the parent bacillus unaltered in shape (Fig. 91, *a*).

(B) Cell of the parent bacillus altered in shape.

1. *Clostridium* (Fig. 91, *b*): Rod swollen at the centre and attenuated at the poles; spindle shape; *e. g.*, B. butyricus.

2. *Cuneate* (Fig. 91, *c*): Rods swollen slightly at one pole and more or less pointed at the other; wedge-shaped.

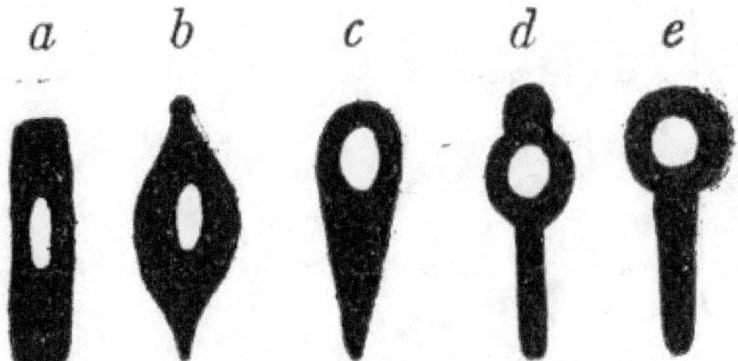

Fig. 91—Types of spore-bearing bacilli.

3. *Clavate* (Fig. 91, *d*): Rods swollen at one pole and cylindrical (unaltered) at the other; keyhole-shaped; *e. g.*, B. chauvei.

4. *Capitate* (Fig. 91, *e*): Rods with a spherical enlargement at one pole; drumstick-shaped; *e. g.*, B. tetani.

The endo-spores remain within the parent cell for a variable time (in one case it is stated that germination of the spore occurs within the interior of the parent cell—"endo-germination"), but are eventually set free, as a result of the swelling up and solution of the cell membrane of the parent bacillus in the surrounding liquid, or of the rupture of that membrane. They then present the following characteristics:

1. Well-formed, dense cell membranes, which renders them extremely difficult to stain, but when once stained equally difficult to decolourise.

2. High refractility, which distinguished them from vacuoles.

3. Higher resistance than the parent organism to such lethal agents as heat, desiccation, starvation, time, etc., this resistance being due to

(*a*) Low water contents of plasma of the spore.

(*b*) Low heat-conducting power   } of the spore membrane.

(*c*) Low permeability   }

This resistance varies somewhat with the particular species—*e. g.*, some spores may resist boiling for a few minutes—but practically all are killed if the boiling is continued for ten minutes.

Germination.—When transplanted to suitable media and placed under favourable conditions, the spores germinate, usually within twenty-four to thirty-six hours, and successively undergo the following changes which may be followed in hanging-drop cultures on a warm stage:

1. Swell up slowly and enlarge, through the absorption of water.

2. Lose their refrangibility.

3. At this stage one of three processes (but the particular process is always constant for the same species) may be observed:

(*a*) The spore grows out into the new bacillus without discarding the spore membrane (which in this case now becomes the cell membrane); *e. g.*, B. leptosporus.

(*b*) It loses its spore membrane by solution; *e. g.*, B. anthracis.

(*c*) It loses its spore membrane by rupture.

In this process the rupture may be either polar (at one pole only *e. g.*, B. butyricus), or bipolar (*e. g.*, B. sessile), or equatorial; (*e. g.*, B. subtilis).

In those cases where the spore membrane is discarded the cell membrane of the new bacillus may either be formed from—

(*a*) The inner layer of the spore membrane, which has undergone a preliminary splitting into parietal and visceral layers; *e. g.*, B. butyricus.

(*b*) The outer layers of the cell protoplasm, which become differentiated for that purpose; *e. g.*, B. megatherium.

The new bacillus now increases in size, elongates, and takes on a vegetative growth—*i. e.*, undergoes fission—the bacilli resulting from which may in their turn give rise to spores.

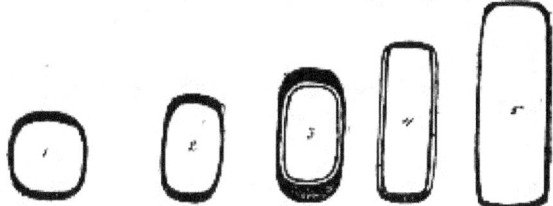

Fig. 92. Simple.

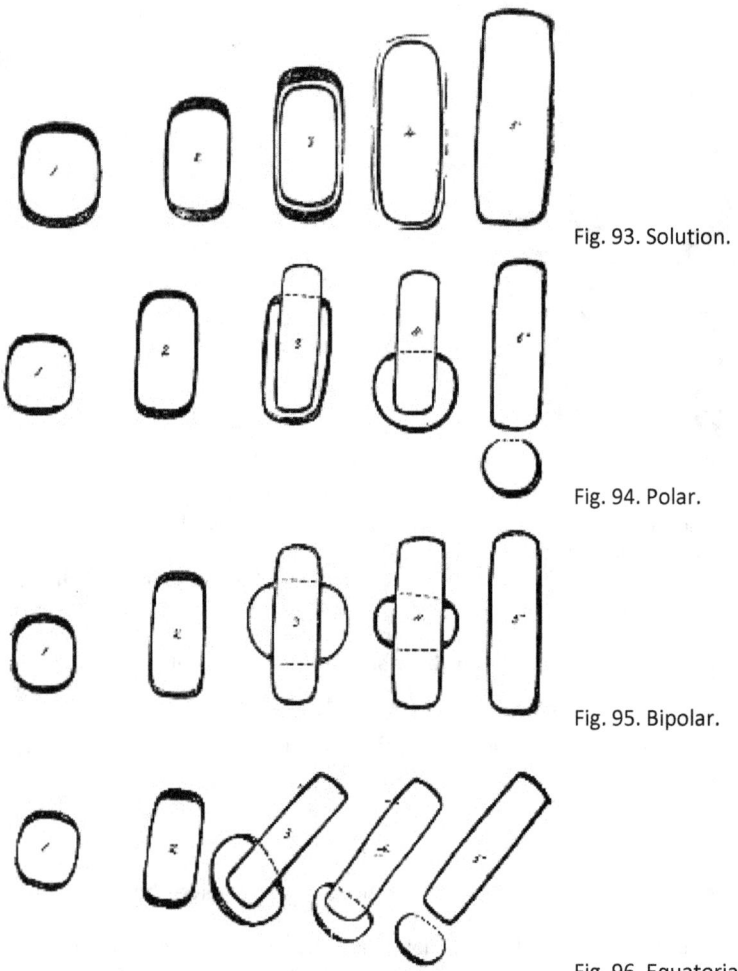

Fig. 93. Solution.

Fig. 94. Polar.

Fig. 95. Bipolar.

Fig. 96. Equatorial.

Food Stuffs.—1. *Organic Foods.—*
(*a*) The pure parasites (*e. g.*, B. lepræ) will not live outside the living body.
(*b*) Both saprophytic and facultative parasitic bacteria agree in requiring non-concentrated food.
(*c*) The facultative parasites need highly organised foods; *e. g.*, proteids or other sources of nitrogen and carbon, and salts.
(*d*) The saprophytic bacteria are more easily cultivated; *e. g.*,
1. Some bacteria will grow in almost pure distilled water.
2. Some bacteria will grow in pure solutions of the carbohydrates.
3. *Water* is absolutely essential to the *growth* of bacteria.

Food of a definite reaction is needed for the growth of bacteria. As a general rule growth is most active in media which react slightly acid to phenolphthalein—that is, neutral or faintly alkaline to litmus. Mould growth, on the other hand, is most vigourous in media that are strongly acid to phenolphthalein.

Environment.—The influence of physical agents upon bacterial life and growth is strongly marked.

1. *Atmosphere.*—The presence of *oxygen* is necessary for the growth of some bacteria, and death follows when the supply is cut off. Such organisms are termed *obligate aerobes*.

Some bacteria appear to thrive equally well whether supplied with or deprived of oxygen. These are termed *facultative anaerobes*.

A third class will only live and multiply when the access of free oxygen is completely excluded. These are termed *obligate anaerobes*.

2. *Temperature.*—Practically no bacterial growth occurs below 5°C, and very little above 40° C. 30°C. to 37° C is the most favorable for the large majority of micro-organisms.

The maximum and minimum temperatures at which growth takes place, as well as the optimum, are fairly constant for each bacterium.
Bacteria have been classified, according to their optimum temperature, into—

	Min.	Opt.	Max.
1. Psychrophilic bacteria (chiefly water organisms)	0° C.	15° C.	30°C.
2. Mesophilic bacteria (includes pathogenic bacteria)	15° C.	37° C.	45°C.
3. Thermophilic bacteria	45° C.	55° C.	70°C.

The thermal death-point of an organism is another biological constant; and is that temperature which causes the death of the vegetative forms when the exposure is continued for a period of ten minutes (see pages 298-301).

3. *Light.*—Many organisms are indifferent to the presence of light. On the other hand, light frequently impedes growth, and alters to a greater or lesser extent the biochemical characters of the organisms—*e. g.*, chromogenicity or power of liquefaction. Pathogenic bacteria undergo a progressive loss of virulence when cultivated in the presence of light.

4. *Movements.*—Movements, if slight and simply of a flowing character, do not appear to injuriously affect the growth of bacteria; but violent agitation, such as shaking, absolutely kills them.

A condition of perfect rest would seem to be that most conducive to bacterial growth.

The Metabolic Products of Bacteria.—*Pigment Production.*—Many micro-organisms produce one or more vivid pigments—yellow, orange, red, violet, fluorescent, etc.—during the course of their life and growth. The colouring matter usually exists as an intercellular excrementitious substance. Occasionally, however, it appears to be stored actually within the bodies of the bacteria. The chromogenic bacteria are therefore classified, in accordance with the final destination of the colouring matter they elaborate, into—

Chromoparous Bacteria: in which the pigment is diffused out upon and into the surrounding medium.

Chromophorous Bacteria: in which the pigment is stored in the cell protoplasm of the organism.

Parachromophorous Bacteria: in which the pigment is stored in the cell wall of the organism.

Different species of chromogenic bacteria differ in their requirements as to environment, for the production of their characteristic pigments; *e. g.*, some need oxygen, light, or high temperature; others again favor the converse of these conditions.

Light Production.—Some bacteria, and usually those originally derived from water, whether fresh or salt, exhibit marked phosphorescence when cultivated under suitable conditions. These are classed as "photogenic."

Enzyme Production.—Many bacteria produce soluble ferments or enzymes during the course of their growth, as evidenced by the liquefaction of gelatine, the clotting of milk, etc. These ferments may belong to either of the following well-recognised classes: proteolytic, diastatic, invertin, rennet.

Toxin Production.—A large number, especially of the pathogenic bacteria, elaborate or secrete poisonous substances concerning which but little exact knowledge is available, although many would appear to be enzymic in their action.

These toxins are usually differentiated into—

Extracellular (or Soluble) Toxins: those which are diffused into, and held in solution by, the surrounding medium.

Intracellular (or Inseparate) Toxins: those which are so closely bound up with the cell protoplasm of the bacteria elaborating them that up to the present time no means has been devised for their separation or extraction.

End-products of Metabolism.—Under this heading are included—

Organic Acids (*e. g.*, lactic, butyric, etc.).

Alkalies (*e. g.*, ammonia).

Aromatic Compounds (*e. g.*, indol, phenol).

Reducing Substances (*e. g.*, those reducing nitrates to nitrites).

Gases (*e. g.*, sulphuretted hydrogen, carbon dioxide, etc.).

And while the discussion of their formation, etc., is beyond the scope of a laboratory handbook, the methods in use for their detection and separation come into the ordinary routine work and will therefore be described (*vide* page 276 *et seq.*).

X. NUTRIENT MEDIA.

In order that the life and growth of bacteria may be accurately observed in the laboratory, it is necessary—

1. To *isolate* individual members of the different varieties of micro-organisms.

2. To *cultivate* organisms, thus isolated, apart from other associated or contaminating bacteria—*i. e.*, in *pure culture*.

For the successful achievement of these objects it is necessary to provide nutriment in a form suited to the needs of the particular bacterium or bacteria under observation, and in a general way it may be said that the nutrient materials should approximate as closely as possible, in composition and character, to the natural pabulum of the organism.

The general requirements of bacteria as to their food-supply have already been indicated (page 142) and many combinations of proteid and of carbohydrate have been devised, from time to time, on those lines. These, together with various vegetable tissues, physiological or pathological fluid secretions, etc., are collectively spoken of as *nutrient media* or *culture media*.

The greater number of these media are primarily *fluid*, but, on account of the rapidity with which bacterial growth diffuses itself through a liquid, it is impossible to study therein the characteristics of individual organisms. Many such media are, therefore, subsequently rendered solid by the addition of substances like gelatine or agar, in varying proportions, the proportions of such added material being generally mentioned when referring to the media; *e. g.*, 10 per cent. gelatine, 2 per cent. agar. Gelatine is employed for the solidification of those media it is intended to use in the cultivation of bacteria at the room temperature or in the "cold" incubator. In the percentages usually employed, gelatine media become fluid at 25°C.; higher percentages remain solid at somewhat higher temperatures, but the difficulty of filtering strong solutions of gelatine militates against their general use.

Media, on the other hand which have been solidified by the addition of agar, only become liquid when exposed to 90° C. for about ten minutes, and again solidify when the temperature falls to 40°C.

When it becomes necessary to render these media fluid, heat is applied, upon the withdrawal of which they again assume their solid condition. Such media should be referred to as *liquefiable media*; in point of fact, however, they are usually grouped together with the solid media.

Note.—It must here be stated that the designation 10 per cent. gelatine or 2 per cent. agar refers only to the quantity of those substances actually added in the process of manufacture, and *not* to the percentage of gelatine or agar, as the case may be, present in the finished medium; the explanation being that the commercial products employed contain a large proportion of insoluble material which is separated off by filtration during the preparation of the liquefiable media.

Other media, again—*e. g.*, potato, coagulated blood-serum, etc.—cannot be again liquefied by physical means, and these are spoken of as *solid* media.

The following pages detail the method of preparing the various nutrient media, in ordinary use (see also Chapter XI), those which are only occasionally required for more highly specialised work are grouped together in Chapter XII. It must be premised that scrupulous cleanliness is to be observed with regard to all apparatus, vessels, funnels, etc., employed in the preparation of media; although in the preliminary stages of the preparation of most media absolute sterility of the apparatus used is not essential.

MEAT EXTRACT.

A watery solution of the extractives, etc., of lean meat (usually beef) forms the basis of several nutrient media. This solution is termed "meat extract" and it has been determined empirically that its preparation shall be carried out by extracting half a kilo of moist meat with one litre of water. For many purposes, however, it is more convenient to have a more concentrated extract; one kilo of meat should therefore be extracted with one litre of water, to form "Double Strength" meat extract.

It was customary at one time, and is even now in some laboratories to use either "shin of beef" or "beef-steak"—both contain muscle sugar which often needs to be removed before the nutrient medium can be completed. Heart muscle (bullock's heart or sheep's heart) is much to be preferred and from the point of economy, ease and cleanliness of manipulation, and extractive value, the imported frozen bullock's hearts provide the best extract.

Meat extract (Fleischwasser) is prepared as follows:

1. Measure 1000 c.c. of distilled water into a large flask (or glass beaker, or enamelled iron pot) and add 1000 grammes (roughly, 2-1/2 pounds) of fresh lean meat—*e. g.*, bullock's heart—finely minced in a mincing machine.

2. Heat the mixture gently in a water-bath, taking care that the temperature of the contents of the flask does not exceed 40° C. for the first twenty minutes. (This dissolves out the soluble proteids, extractives, salts, etc.)

3. Now raise the temperature of the mixture to the boiling-point, and maintain at this temperature for ten minutes. (This precipitates some of the albumins, the hæmoglobin, etc., from the solution.)

4. Strain the mixture through sterile butter muslin or a perforated porcelain funnel, then filter the liquid through Swedish filter paper into a sterile "normal" litre flask, and when cold make up to 1000 c.c. by the addition of distilled water—to replace the loss from evaporation.

5. If not needed at once, sterilise the meat extract in bulk in the steam steriliser for twenty minutes on each of three consecutive days.

Calf, sheep, or chicken flesh is occasionally substituted for the beef; or the meat extract may be prepared from animal viscera, such as brain, spleen, liver, or kidneys.

Note.—As an alternative method, 5 c.c. of Brand's meat juice or 3 grammes of Wyeth's beef juice, or 10 grammes Liebig's extract of meat (Lemco) may be dissolved in 1000 c.c. distilled water, and heated and filtered as above to form ordinary or single strength meat extract.

Media, prepared from such meat extracts are, however, eminently unsatisfactory when used for the cultivation of the more highly parasitic bacteria; although when working in tropical and subtropical regions their use is well-nigh compulsory.

Reaction of Meat Extract.—Meat extract thus prepared is acid in its reaction, owing to the presence of acid phosphates of potassium and sodium, weak acids of the glycolic series, and organic compounds in which the acid character predominates. Owing to the nature of the substances from which it derives its reaction, the total acidity of meat extract can only be estimated accurately when the solution is at the boiling-point.

Moreover, it has been observed that prolonged boiling (such as is involved in the preparation of nutrient media) causes it to undergo hydrolytic changes which increase its acidity, and **the meat extract only becomes stable in this respect after it has been maintained at the boiling-point for forty-five minutes**.

Although meat extract always reacts acid to phenolphthalein, it occasionally reacts neutral or even alkaline to litmus; and again, meat extract that has been rendered exactly neutral to litmus still reacts acid to phenolphthalein. This peculiar behaviour depends upon two factors:

1. Litmus is insensitive to many weak organic acids the presence of which is readily indicated by phenolphthalein.

2. Dibasic sodium phosphate which is formed during the process of neutralisation is a salt which reacts alkaline to litmus, but neutral to phenolphthalein. In order, therefore, to obtain an accurate estimation of the reaction of any given sample of meat extract, it is essential that—

1. The meat extract be previously exposed to a temperature of 100° C. for forty-five minutes.

2. The estimation be performed at the boiling-point.

3. Phenolphthalein be used as the indicator.

The estimation is carried out by means of titration experiments against standard solutions of caustic soda, in the following manner:

Method of Estimating the Reaction.—

Apparatus Required:

1. 25 c.c. burette graduated in tenths of a centimetre.

2. 1 c.c. pipette graduated in hundredths, and provided with rubber tube, pinch-cock, and delivery nozzle.

3. 25 c.c. measure (cylinder or pipette, calibrated for 98°C.—*not* 15°C).

Solutions Required:

1. 10N NaOH, accurately standardised.

2. n/1 NaOH, accurately standardised

3. n/10 NaOH, accurately standardised

4. Several 60 c.c. conical beakers or Erlenmeyer flasks.

4. 0.5 per cent. solution of phenolphthalein in 50 percent. alcohol.

5. White porcelain evaporating basin, filled with boiling water and arranged over a gas flame as a water-bath.

6. Bohemian glass flask, fitted as a wash-bottle, and filled with distilled water, which is kept boiling on a tripod stand.

Method.—Arrange the apparatus as indicated in figure 97.
(A) 1. Fill the burette with n/10 NaOH.
2. Fill the pipette with n/1 NaOH.

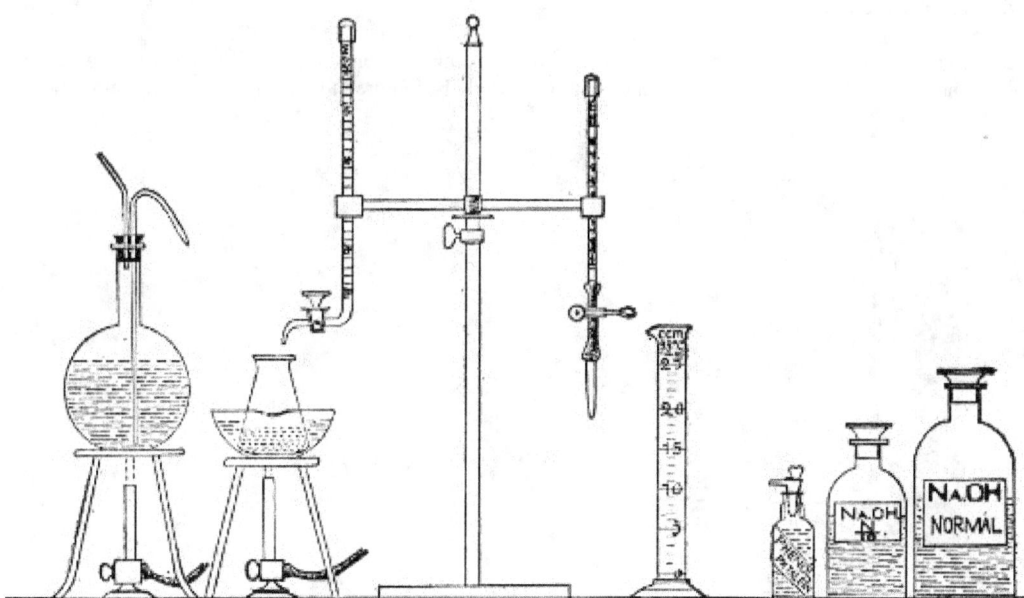

Fig. 97.—Arrangement of apparatus for titrating media.

3. Measure 25 c.c. of the meat extract (previously heated in the steamer at 100° C. for forty-five minutes) into one of the beakers by means of the measure; rinse out the measure with a very small quantity of boiling distilled water from the wash-bottle, and then add this rinse water to the meat extract already in the beaker.

4. Run in about 0.5 c.c. of the phenolphthalein solution and immerse the beaker in the water-bath, and raise to the boil.

5. To the medium in the beaker run in n/10 NaOH cautiously from the burette until the end-point is reached, as indicated by the development of a pinkish tinge, shown in figure 98 (*b*). Note the amount of decinormal soda solution used in the process.

Note.—Just before the end-point is reached, a very slight opalescence may be noted in the fluid, due to the precipitation of dibasic phosphates. After the true end-point is reached, the further addition of about 0.5 c.c. of the decinormal soda solution will produce a deep magenta colour (Fig. 98, *c*), which is the so-called "end-point" of the American Committee of Bacteriologists.

a, *b*, *c*

Fig. 98.—a, Sample of filtered meat extract or nutrient gelatine to which phenolphthalein has been added. The medium is acid, as evidenced by the unaltered colour of the sample. b, The same neutralised by the addition of n/10 NaOH. The production of this faint rose-pink colour indicates that the "end-point," or neutral point to phenolphthalein, has been reached. If such a sample is cooled down to say 30° or 20° C., the colour will be found to become more distinct and decidedly deeper and brighter, resembling that shown in c. c, Also if, after the end-point is reached, a further 0.5 c.c. or 1.0 c.c. n/10 NaOH be added to the sample, the marked alkalinity is evidenced by the deep colour here shown.

(B) Perform a "control" titration (occasionally two controls may be necessary), as follows:

1. Measure 25 c.c. of the meat extract into one of the beakers, wash out the measure with boiling water, and add the phenolphthalein as in the first estimation.

2. Run in n/1 NaOH from the pipette, just short of the equivalent of the amount of *deci*-normal soda solution required to neutralise the 25 c.c. of medium. (For example, if in the first estimation 5 c.c. of n/10 NaOH were required to render 25 c.c. of medium neutral to phenolphthalein, only add 0.48 c.c. of n/1 NaOH.) Immerse the beaker in the water-bath.

3. Complete the titration by the aid of the n/10 NaOH.

4. Note the amount of n/10 NaOH solution required to complete the titration, and add it to the equivalent of the n/1 NaOH solution previously run in. Take the total as the correct estimation.

Method of Expressing the Reaction.—

The reaction or *titre* of meat extract, medium, or any solution estimated in the foregoing manner, is most conveniently expressed by indicating the number of cubic centimetres of normal alkali (or normal acid) that would be required to render *one litre* of the solution exactly neutral to phenolphthalein.

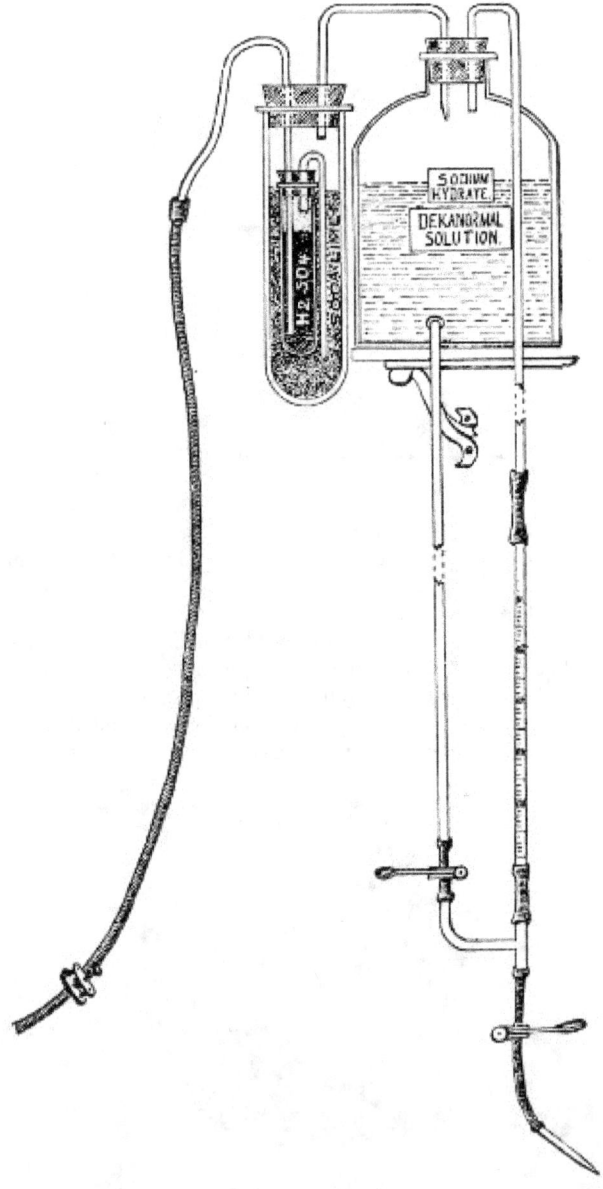

Fig. 99.—Stock bottle for dekanormal soda solution.

The sign + (plus) is prefixed to this number if the original solution reacts acid, and the sign - (minus) if it reacts alkaline.

For example, "meat extract + 10," indicates a sample of meat extract which reacts acid to phenolphthalein, and would require the addition of 10 c.c. of *normal* NaOH per litre, to neutralise it.

Note.—Such a solution would probably react alkaline to litmus.

Conversely, if as the result of our titration experiments we find that 25 c.c. of meat extract require the addition of 5 c.c. n/10 NaOH to neutralise, then 1000 c.c. of meat extract will require the addition of 200 c.c. n/10 NaOH = 20 c.c. n/1 NaOH.

And this last figure, 20, preceded by the sign + (*i. e.*, +20), to signify that it is acid, indicates the reaction of the meat extract.

Note.—The standard soda solutions should be prepared by accurate measuring operations, controlled by titrations, from a stock solution of 10N NaOH, which should be very carefully standardised. If a large supply is made or the consumption is small this stock solution must be kept in an aspirator bottle to which air can only gain access after it has been dried and rendered free from CO_2. This may be done by first leading it over H_2SO_4 and soda lime, or soda lime alone, by some such arrangement as is shown in figure 99, which also shows a constant burette arrangement for the delivery of small measured quantities of the dekanormal soda solution.

STANDARDISATION OF MEDIA.

Differences in the reaction of the medium in which it is grown will provoke not only differences in the rate of growth of any given bacterium, but also well-marked differences in its cultural and morphological characters; and nearly every organism will be found to affect a definite "optimum reaction"—a point to be carefully determined for each. For most bacteria, however, the "optimum" usually approximates fairly closely to +10; and as experiment has shown that this reaction is the most generally useful for routine laboratory work, it is the one which may be adopted as the standard for all nutrient media derived from meat extract.

Briefly, the method of standardising a litre of media to +10 consists in subtracting 10 from the initial *titre* of the medium mass; the remainder indicates the number of cubic centimetres of normal soda solution that must be added to the medium, per litre, to render the reaction +10.

Standardising Nutrient Bouillon.—For example, 1000 c.c. bouillon are prepared; at the first titration it is found

1. 25 c.c. require the addition of 5.50 c.c. n/10 NaOH to neutralise.

Two controls give the following results:

2. 25 c.c. require the addition of 5.70 c.c. n/10 NaOH to neutralise.

3. 25 c.c. require the addition of 5.60 c.c. n/10 NaOH to neutralise.

Averaging these two controls, 25 c.c. require the addition of 5.65 c.c. n/10 NaOH to neutralise, and therefore 1000 c.c. require the addition of 226 c.c. n/10 NaOH, or 22.60 c.c. n/1 NaOH, or 2.26 c.c. n/10 NaOH.

Initial *titre* of the bouillon = +22.6, and as such requires the addition of (22.6 c.c. - 10 c.c.) = 12.6 c.c. of n/1 NaOH per litre to leave its finished reaction +10.

But the three titrations, each on 25 c.c. of medium, have reduced the original bulk of bouillon to (1000 - 75 c.c.) = 925 c.c. The amount of n/1 NaOH required to render the reaction of this quantity of medium +10 may be deduced thus:

1000 c.c.:925 c.c.::12.6 c.c.:*x*.

Then *x* = 11.65 c.c. n/1 NaOH.

Whenever possible, however, the required reaction is produced by the addition of dekanormal soda solution, on account of the minute increase it causes in the bulk, and the consequent insignificant disturbance of the percentage composition of the medium. By means of a pipette graduated to 0.01 c.c. it is possible to deliver very small quantities; but if the calculated amount runs into thousandth parts of a cubic centimetre, these are replaced by corresponding quantities of normal or even decinormal soda.

In the above example it is necessary to add 11.65 c.c. normal NaOH or its equivalent, 1.165 c.c. dekanormal NaOH. The first being too bulky a quantity, and the second inconveniently small for exact measurement, the total weight of soda is obtained by substituting 1.16 c.c. dekanormal soda solution, and either 0.05 c.c. of normal soda solution or 0.5 c.c. of decinormal soda solution.

Standardising Nutrient Agar and Gelatine.—The method of standardising agar and gelatine is precisely similar to that described under bouillon.

THE FILTRATION OF MEDIA.

Fluid media are usually filtered through stout Swedish filter paper (occasionally through a porcelain filter candle), and in order to accelerate the rate of filtration the filter paper should be folded in that form which is known as the "physiological filter," not in the ordinary "quadrant" shape, as by this means a large surface is available for filtration and a smaller area in contact with the glass funnel supporting it.

To fold the filter proceed thus:

1. Take a circular piece of filter paper and fold it exactly through its centre to form a semicircle (Fig. 100, *a*).

2. Fold the semicircle exactly in half to form a quadrant; make the crease 2, distinct by running the thumbnail along it, then open the filter out to a semicircle again.

3. Fold each end of the semicircle in to the centre and so form another quadrant; smooth down the two new creases 3 and 3*a*, thus formed and again open out to a semicircle.

4. The semicircle now appears as in figure 100, *a*, the dark lines indicating the creases already formed.

5. Fold the point 1 over to the point 3, and 1*a* to 3*a*, to form the creases 4 and 4*a*, indicated in the diagram by the light lines. Fold point 1 over to 3*a*, and 1*a* to 3, to form the creases 5 and 5*a*.

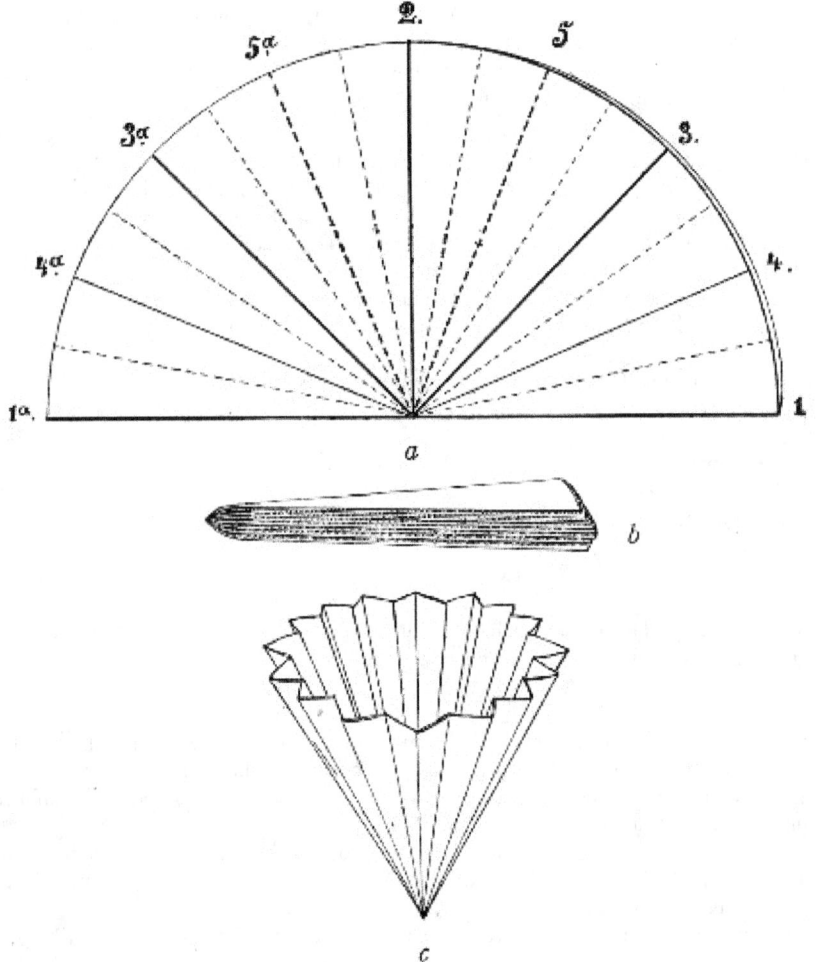

Fig. 100.—Filter folding: a, Filter folded in half, showing creases; b, appearance of filter on completion of folding; c, filter opened out ready for use.

6. Thus far the creases have all been made on the same side of the paper. Now subdivide each of the eight sectors by a crease through its centre on the opposite side of the paper, indicated by the faint broken lines in the diagram. Fold up the filter gradually as each crease is made, and when finished the filter has assumed the shape of a wedge, as in figure 100, b.

When opened out the filter assumes the shape represented in figure 100, c.

The folded filter is next placed inside a glass funnel supported on a retort stand, and moistened with hot distilled water before the filtration of the medium is commenced.

Liquefiable solid media are filtered through a specially made filter paper—"papier Chardin"—which is sold in boxes of twenty-five ready-folded filters.

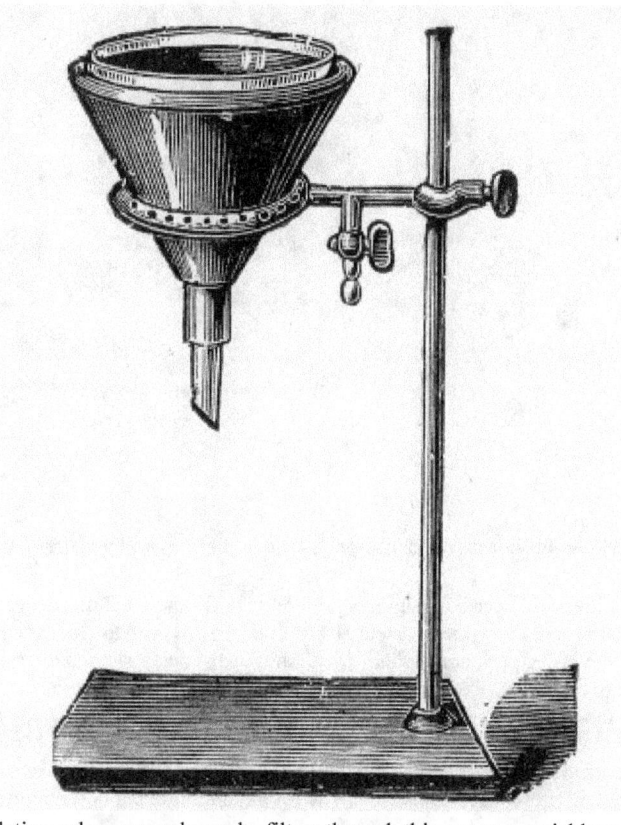

Fig. 101.—Hot-water filter funnel and ring burner.

Gelatine, when properly made, filters through this paper as quickly as bouillon does through the Swedish filter paper, and does *not* require the use of the hot-water funnel.

Agar, likewise, if properly made, filters readily, although not at so rapid a rate as gelatine. If badly "egged," and also during the winter months, it is necessary to surround the glass funnel, in which the filtration of the agar is carried on, by a hot-water jacket. This is done by placing the glass funnel inside a double-walled copper funnel—the space between the walls being filled with water at about 90° C.—and supporting the latter on a ring gas burner fixed to a retort stand (Fig. 101). The gas is lighted and the water jacket maintained at a high temperature until filtration is completed. If the steam steriliser of the laboratory is sufficiently large, it is sometimes more convenient to place the flask and filtering funnel bodily inside, close the steriliser and allow filtration to proceed in an atmosphere of live steam, than to use the gas ring and hot-water funnel.

STORING MEDIA IN BULK.

After filtration fill the medium into sterile litre flasks with cotton-wool plugs and sterilise in the steamer for twenty minutes on each of three consecutive days. After the third sterilisation, and when the flasks and contents are cool, cut off the top of the cotton-wool plug square with the mouth of the flask; push the plug a short distance down into the neck of the flask and fill in with melted paraffin wax to the level of the mouth. When the wax has set the flasks are stored in a cool dark cupboard for future use.

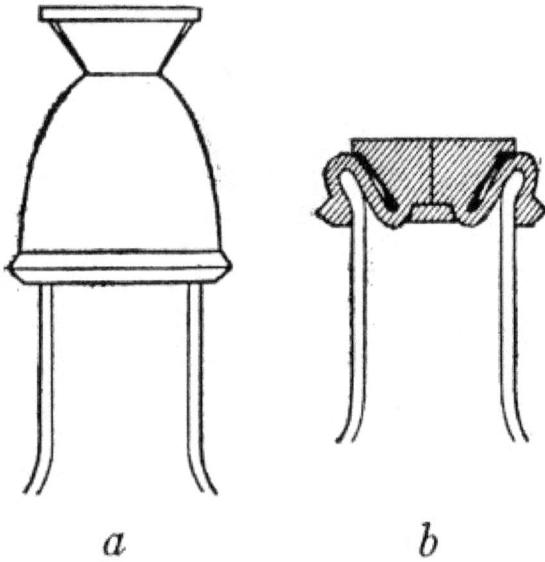

Fig. 102.—Rubber cap closing store bottle. a, before, and b, after sterilizing.

This plan is not absolutely satisfactory, although very generally employed on occasion, and it is preferable to fill the medium into long-necked flint glass bottles (the quart size, holding nearly 1000 c.c., such as those in which Pasteurised milk is retailed) and to close the neck of the bottle by a special rubber cap. This cap is made of soft rubber, the lower part, dome-shaped with thin walls, being slipped over the neck of the bottle (Fig. 102, a). The upper part is solid, but with a sharp clean-cut (made with a cataract or tenotomy knife) running completely through its axis from the centre of the disc to the top of the dome. During sterilisation the air in the neck of the bottle, expanded by the heat, is driven out through the valvular aperture in the solid portion of the stopper. On removing the bottle from the steam chamber, the liquid contracts as it cools, and the pressure of the external air drives the solid piece of rubber down into the neck of the bottle, and forces together the lips of the slit (Fig. 102, b). Thus sealed, the bottle will preserve its contents sterile for an indefinite period without loss from evaporation.

TUBING NUTRIENT MEDIA.

After the final filtration, the nutrient medium is usually "tubed"—*i. e.*, filled into sterile tubes in definite measured quantities, usually 10 c.c. This process is sometimes carried out by means of a large separator funnel fitted with a "three-way" tap which communicates with a small graduated tube (capacity 20 c.c. and graduated in cubic centimetres) attached to the side. The shape of this piece of apparatus, known as Treskow's funnel, renders it particularly liable to damage. It is better, therefore, to arrange a less expensive piece of apparatus which will serve the purpose equally well (Fig. 103).

A Geissler's three-way stop-cock has the tube on one side of the tap ground obliquely at its extremity, and the tube on the opposite side cut off within 3 cm. of the tap. The short tube is connected by means of a perforated rubber cork with a 10 cm. length of stout glass tubing (1.5 cm. bore). The third channel of the three-way tap is connected, by means of rubber tubing, with the nozzle of an ordinary separator funnel. Finally, the receiving cylinder above the three-way tap is graduated in cubic centimetres up to 20, by pouring into it measured quantities of water and marking the various levels on the outside with a writing diamond.

Fluid media containing carbohydrates are filled into fermentation tubes (*vide* Fig. 21); or into ordinary media tubes which already have smaller tubes, inverted, inside them (Fig. 104), to collect the products of growth of gas-forming bacteria. When first filled, the small tubes float on the surface of the medium after the first sterilisation nearly all the air is replaced by the medium, and after the final sterilisation the gas tubes will be submerged and completely filled with the medium.

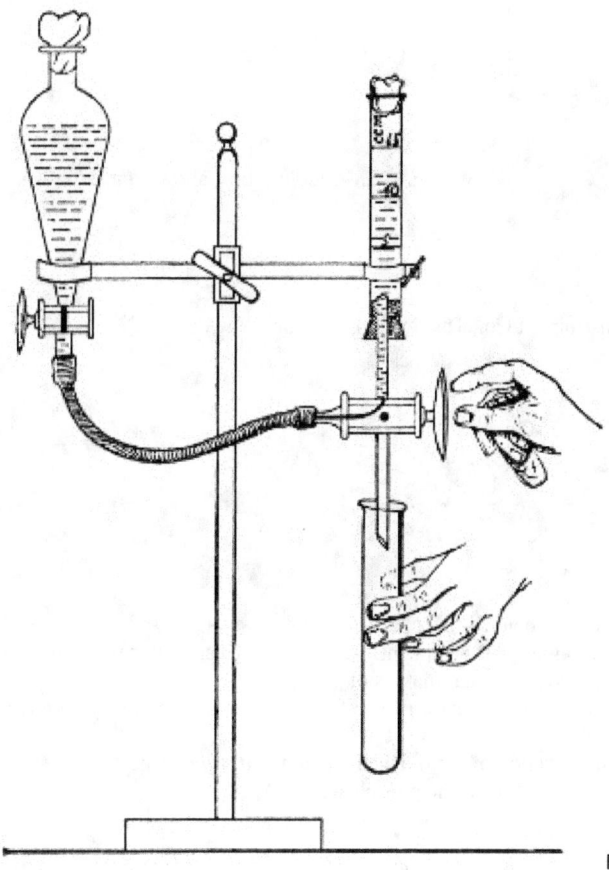

Fig. 103.—Separatory funnel and three-way tap arranged for tubing media.

Fig. 104.—Gas tube (Durham).

Storing "Tubed" Media.—Media after being tubed are best stored by packing, in the vertical position, in oblong boxes having an internal measurement of 37 cm. long by 12 cm. wide by 10 cm. deep. Each box (Fig. 105) has a movable partition formed by the vertical face of a weighted triangular block of wood, sliding free on the bottom (Fig. 105, A); or by a flat piece of wood sliding in a metal groove in the bottom of the box, which can be fixed at any spot by tightening the thumbscrew of a brass guide rod which transfixes the partition (Fig. 105, B). The front of the box is provided with a handle and a celluloid label for the name of the contained medium. These boxes are arranged upon shelves in a dark cupboard—or preferably an iron safe—which should be rendered as nearly air-tight as possible, and should have the words "media stores" painted on its doors.

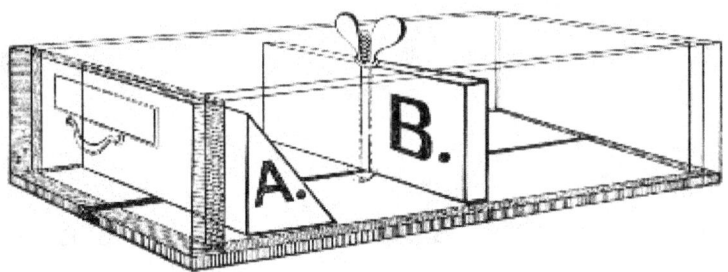

Fig. 105.—Medium box, showing alternative partitions A and B.

FOOTNOTES:

This rubber cap has been made for me by the Holborn Surgical Instrument Co., Thavies Inn, London, W. C.

XI. CULTURE MEDIA.

ORDINARY OR STOCK MEDIA.

Nutrient Bouillon.—
1. Measure out double strength meat extract, 500 c.c., into a litre flask and add 300 c.c. distilled water.
2. Weigh out Witté's peptone, 10 grammes (= 1 per cent.), salt, 5 grammes (= 0.5 per cent.), and mix into a smooth paste with 200 c.c. of distilled water previously heated to 60° C. (Be careful to leave no unbroken globular masses of peptone.)
3. Add the peptone emulsion to the meat extract in the flask and heat in the steamer for forty-five minutes (to completely dissolve the peptone, and to render the acidity of the meat extract stable).
4. Estimate the reaction of the medium; control the result; render the reaction of the finished medium +10 (*vide* page 155).
5. Heat for half an hour in the steamer at 100°C. (to complete the precipitation of the phosphates, etc.).
6. Filter through Swedish filter paper into a sterile flask.
7. Fill into sterile tubes (10 c.c. in each tube).
8. Sterilise in the steamer for twenty minutes on each of three consecutive days—*i. e.*, by the discontinuous method (*vide* page 35).

Note.—As an alternative method when neither fresh nor frozen meat is available nutrient bouillon may be prepared from a commercial meat extract, as follows:

Lemco Broth.—
1. Measure out 250 c.c. distilled water into a litre flask.
2. Weigh out 10 grammes Liebig's Lemco Meat Extract on a piece of clean filter paper and add to the water in the flask. Shake the flask well to make an even emulsion of the meat extract.
3. Weigh out Witté's peptone (10 grammes), salt (5 grammes). Mix into smooth paste with 100 c.c. distilled water previously heated to 60°C.
4. Add the peptone salt emulsion to the meat extract emulsion in the flask and add 650 c.c. distilled water. Heat in the steamer for forty-five minutes.
5. Standardise the medium and complete as for nutrient bouillon.

Nutrient Gelatine.—
1. Weigh a 2-litre flask on a trip balance (Fig. 106) and note the weight, or counterpoise carefully.

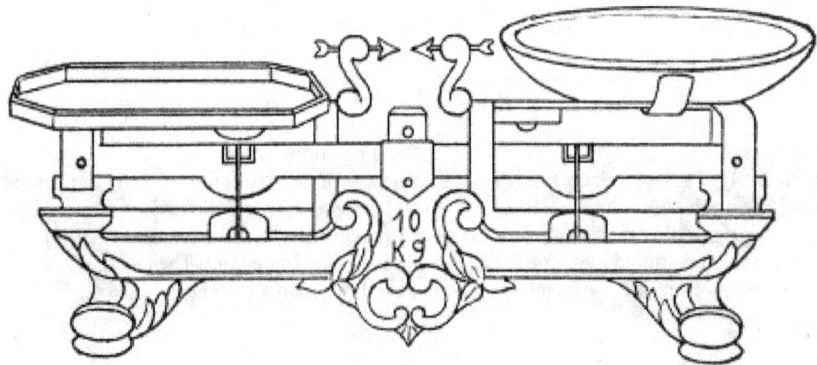

Fig. 106.—Trip balance.

An extremely useful counterpoise is a small sheet-brass cylinder about 38 mm. high and 38 mm. in diameter, with a funnel-shaped top and provided with a side tube by which its contents, fine "dust" shot, may be emptied out (Fig. 107).

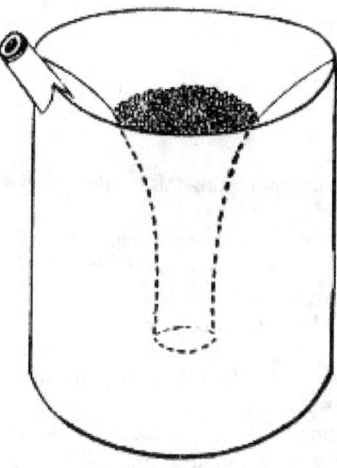

Fig. 107.—Counterpoise; weight when empty, 35 grammes; when full of dust shot, 200 grammes.

2. Measure out double strength meat extract, 500 c.c., into the "tared" flask.
3. Weigh out and mix 10 grammes of peptone, 5 grammes of salt, and make into a thick paste with 150 c.c. distilled water; then add the emulsion to the meat extract in the flask; also add 100 grammes sheet gelatine cut into small pieces; place the flask in the water-bath and raise to the boil.

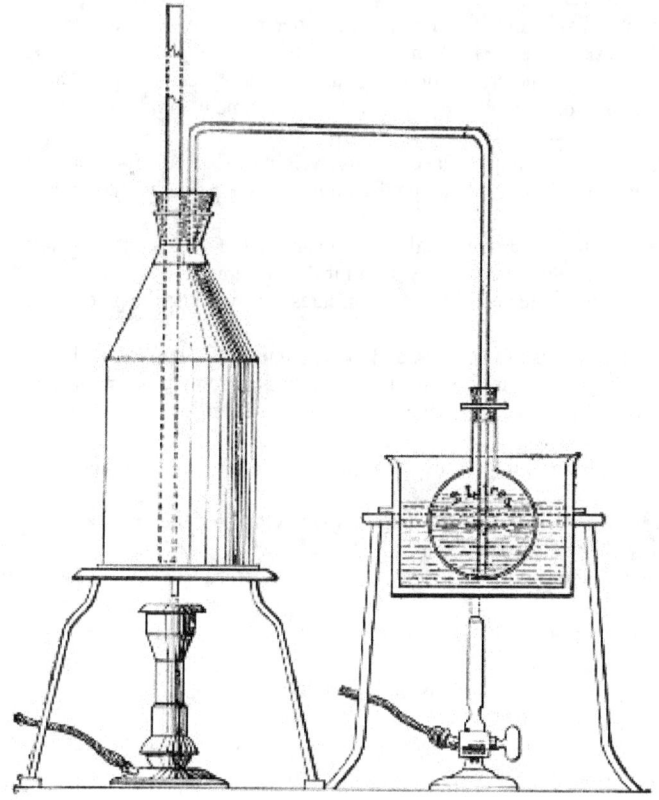

Fig. 108.—Arrangement of steam can and water-bath for the preparation of media.

4. Arrange a 5-litre tin can (with copper bottom, such as is used in the preparation of distilled water) by the side of the water bath, fill the can with boiling water and place a lighted Bunsen burner under it. Fit a long safety tube to the neck of the can and also a delivery tube, bent twice at right angles; adjust the tube to reach to the bottom of the interior of the flask containing the gelatine, etc. (Fig. 108).
5. Keep the water in the steam can vigourously boiling, and so steam at 100°C, bubbling through the medium mass, for ten minutes, by which time complete solution of the gelatine is effected. A certain amount of steam will condense as water in the medium flask during this

process—hence the necessity for the use of double strength meat extract—but if the water bath is kept boiling this condensation will not exceed 100 c.c.

6. Weigh the flask and its contents; then (1115 grammes + weight of the flask) minus (weight of the flask and its contents) equals the weight of water required to make up the bulk to 1 litre. The addition of the requisite quantity of water is carried out as follows:

In one pan of the trip balance place the counterpoise of the tared flask (or its equivalent in weights) together with the weights making up the *calculated medium weight*. In the opposite pan place the flask containing the medium mass. Now add boiling distilled water from a wash bottle until the two pans are exactly balanced.

7. Titrate and estimate the reaction of the medium mass; control the result. Calculate the amount of soda solution required to make the reaction of the medium mass +10 (*i. e.*, calculate for 1000 c.c., less the quantity used for the titrations).

8. Add the necessary amount of soda solution and heat in the steamer at 100° C. for twenty minutes, to precipitate the phosphates, etc.

9. Allow the medium mass to cool to 60° C. Well whip the whites of two eggs, add to the contents of the flask and replace in the steamer at 100° C. for about half an hour (until the egg-albumen has coagulated and formed large, firm masses floating on and in clear gelatine).

10. Filter through papier Chardin into a sterile flask.

11. Tube in quantities of 10 c.c.

12. Sterilise in the steamer at 100° C. for twenty minutes on each of three consecutive days—*i. e.*, by the discontinuous method.

Nutrient Agar-agar.—

1. Weigh a 2-litre flask and note the weight—or counterpoise exactly.

2. Measure out double strength meat extract, 500 c.c., into the "tared" flask.

3. Weigh out and mix 10 grammes of peptone, 5 grammes of salt, and 20 grammes of powdered agar, and make into a thick paste with 150 c.c. distilled water, and add to the meat extract in the flask; place the flask in a water-bath.

4. Arrange the steam can and water-bath as already directed (for the preparation of gelatine) and figured.

5. Bubble live steam (at 100° C.) through the medium mass, for twenty-five minutes, by which time complete solution of the agar is effected.

6. Now weigh the flask and its contents; then (1035 grammes + weight of flask) minus (weight of flask and its contents) equals the weight of water required to make up the bulk of the medium to 1 litre. Add the requisite amount (see preparation of gelatine, page 166, step 6).

7. Titrate, and estimate the reaction of the medium mass; control the result. Calculate the amount of soda solution required to make the reaction of the medium mass + 10 (*i. e.*, calculated for 1000 c.c., less the quantity used for the titrations).

8. Add the necessary amount of soda solution and replace in the steamer for twenty minutes (to complete the precipitation of the phosphates, etc.).

9. Allow the medium mass to cool to 60° C. Well whip the whites of two eggs, add to the contents of the flask, and replace in the steamer at 100° C. for about *one hour* (until the egg-albumen has coagulated and formed large, firm masses floating on and in clear agar.)

10. Filter through papier Chardin, by the aid of a hot-water funnel, if necessary (Fig. 101), into a sterile flask.

11. Tube in quantities of 10 c.c. or 15 c.c.

12. Sterilise in the steamer at 100° C. for thirty minutes on each of three consecutive days—*i. e.*, by the discontinuous method.

Blood-serum (Inspissated).—

1. Sterilise cylindrical glass jar (Fig. 109) and its cover by dry heat, or by washing first with ether and then with alcohol and drying.

2. Collect blood at the slaughter house from ox or sheep in the sterile cylinder.

3. Allow the vessel to stand for fifteen minutes for the blood to coagulate. (This must be done before leaving the slaughterhouse, otherwise the serum will be stained with hæmoglobin.)

4. Separate the clot from the sides of the vessel by means of a sterile glass rod (the yield of serum is much smaller when this is not done), and place the cylinder in the ice-chest for twenty-four hours.

5. Remove the serum with sterile pipettes, or syphon it off, and fill into sterile tubes (5 c.c. in each) or flasks.

6. Heat tubes containing serum to 56° C. in a water-bath for half an hour on each of two successive days.

7. On the third day, heat the tubes, in a sloping position, in a serum inspissator to about 72° C. (A coagulum is formed at this temperature which is fairly transparent; above 72° C., a thick turbid coagulum is formed.)

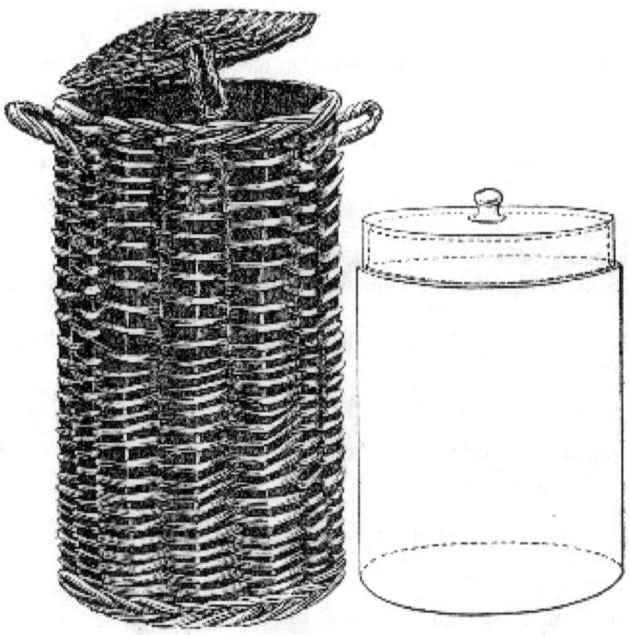

Fig. 109.—Blood-serum jar with wicker basket for transport.

The serum inspissator (Fig. 110) in its simplest form is a double-walled rectangular copper box, closed in by a loose glass lid, and cased in felt or asbestos—the space between the walls is filled with water. The inspissator is supported on adjustable legs so that the serum may be solidified at any desired "slant," and is heated from below by a Bunsen burner controlled by a thermo-regulator. The more elaborate forms resemble the hot-air oven (Fig. 26) in shape and are provided with adjustable shelves so that any desired obliquity of the serum slope can be obtained.

8. Place the tubes in the incubator at 37° C. for forty-eight hours in order to eliminate those that have been contaminated. Store the remainder in a cool place for future use.

Alternative Method.
Steps 1-5 as above.

6. Sterilise the serum by the fractional method—that is, by exposure in a water-bath to a temperature of 56° C. for half an hour on each of six consecutive days; store in the fluid condition.

7. Coagulate in the inspissator when needed.

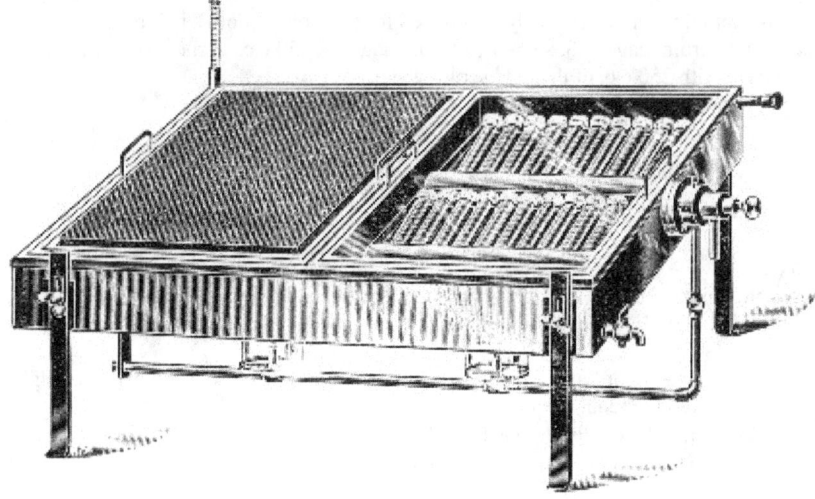

Fig. 110.—Serum inspissator.

Serum Water.—

This forms the basis of many useful media, and is prepared as follows:

1. Collect blood in the slaughterhouse (see page 168) and when firmly clotted collect all the expressed serum and measure in a graduated cylinder.

2. For every 100 c.c. of serum add 300 c.c. distilled water and mix in a flask.

3. Heat the mixture in the steamer at 100° C. for thirty minutes. (This destroys any diastatic ferment present in the serum and partially sterilises the fluid.)

4. Filter if turbid.

5. If not needed at once complete the sterilisation of the serum water by two subsequent steamings at 100° C. for twenty minutes at twenty-four hour intervals.

Citrated Blood Agar. Guy's.—

1. Kill a small rabbit with chloroform vapour, and nail it out on a board (as for a necropsy); moisten the hair thoroughly with 2 per cent. solution of lysol.
2. Sterilise several pairs of forceps, scissors, etc. by boiling.
3. Reflect the skin over the thorax with sterile instruments.
4. Open the thoracic cavity by the aid of a fresh set of sterile instruments.
5. Open the pericardium with another set of sterile instruments.
6. Sear the surface of the left ventricle with a red-hot iron.
7. Take a sterile capillary pipette (Fig. 13, c); break off the sealed extremity with a pair of sterile forceps.
8. Steady the heart in a pair of forceps and thrust the point of the pipette through the wall of the ventricle and through the seared area, apply suction to the plugged end of the pipette and fill it with blood.
9. Transfer the entire quantity of blood collected from the rabbit's heart to a small Erlenmeyer flask containing a number of sterile glass beads and 5 c.c. concentrated sod. citrate solution. (See page 378.)
10. Agitate thoroughly and set aside for a couple of hours.
11. Melt up several tubes of nutrient agar (see page 167) and cool to 42° C.
12. With a sterile 10 c.c. graduated pipette transfer 1 c.c. citrated blood from the Erlenmeyer flask to each tube of liquefied agar. Rotate the tube between the hands in order to diffuse the citrated blood evenly throughout the agar.
13. Place the tubes in a sloping position and allow the medium to set.
14. Place tubes of blood agar for forty-eight hours in the incubator at 37° C. and at the end of that time eliminate any contaminated tubes.
15. Store such tubes as remain sterile for future use.

Milk.—

1. Pour 1 litre of fresh cow's or goat's milk into a large separating funnel, and heat in the steamer at 100° C. for one hour.
2. Remove from the steamer and estimate the reaction of the milk (normal cows' milk averages +17). If of higher acidity than +20, or lower than +10, reject this sample of milk and proceed with another supply of milk from a different source.

Reject milk to which antiseptics have been added as preservatives.

3. Allow the milk to cool, when the fat or cream will rise to the surface and form a thick layer.
4. Draw off the subnatant fat-free milk into sterile tubes (10 c.c. in each).
5. Sterilise in the steamer at 100° C. for twenty minutes on each of five successive days.
6. Incubate at 37° C. for forty-eight hours and eliminate any contaminated tubes. Store the remainder for future use.

Litmus Milk.—

1. Prepare milk as described above, sections 1 to 3.
2. Draw off the subnatant fat-free milk into a flask.
3. Add sterile litmus solution, sufficient to colour the milk a deep lavender.
4. Tube, sterilise, etc., as for milk.

Nutrose Agar (Eyre).—

(This is a modification of the well known Drigalski-Conradi medium originally introduced for the isolation of B. typhosus).

1. Collect 250 c.c. perfectly fresh ox serum (*vide* Blood Serum, page 168, steps 1 to 5) and add to it 450 c.c. sterile distilled water.
2. Weigh out agar powder, 20 grammes, and emulsify it with 250 c.c. of the cold serum water.
3. Weigh out

Witté's peptone	10 grammes
Sodium chloride	5 grammes
Nutrose	10 grammes

and dissolve in 200 c.c. of serum water heated to 80° C.

4. Mix the agar emulsion and the peptone-nutrose solution in a "tared" flask of 2-litre capacity and add a further 100 c.c. serum water.
5. Complete the solution of the various ingredients by bubbling live steam through the flask as in making nutrient agar.
6. Add further 250 c.c. serum water.
7. Weigh the flask and its contents: then (1045 grammes + weight of flask) minus (weight of flask and its present contents) = weight of fluid required to make up the bulk of the medium to 1 litre. Add the requisite amount of sterile distilled water.
8. Titrate and estimate the reaction of the medium mass. Then standardise to reaction of +2.5.
9. Clarify with egg, and filter as for nutrient agar. (In clarifying, after the addition of the egg white the mixture should be in the steamer for full two hours.)
10. After filtration is complete measure the filtrate, and to every 150 c.c. of the medium add:

Litmus solution (Kahlbaum)	20 c.c.
Krystal violet aqueous solution (1:1000) (B. Hoechst)	1.5 c.c.
Lactose	1.5 grammes

11. Tube in quantities of 15 c.c.
12. Sterilise in the steamer at 100° C. for thirty minutes on each of three successive days—*i. e.*, by the discontinuous method for three days.

Egg Medium (Dorset).—

1. Prepare 1000 c.c. of a 0.85 per cent. solution of sodium chloride in a stout 2-litre flask.
2. Sterilise in the autoclave at 120° C. for twenty minutes. Cool to 20° C.
3. Take 12 fresh eggs; wash the shells first with water then with undiluted formalin: allow the shells to dry.
4. Break the eggs into a sterile graduated cylinder and measure the total volume of the mixed whites and yolks. Add one part sterile saline solution to three parts mixed eggs.
5. Transfer this mixture to a large wide-mouthed stoppered bottle previously sterilised. Add sterile glass beads and shake thoroughly in a mechanical shaker for about thirty minutes, or whip with an egg-whisk.
6. Filter through coarse butter muslin into a sterile flask.

Note.—A few drops of alcoholic solution of basic fuchsin (sufficient to give a definite pink colour), or a few drops of waterproof Chinese ink added to the medium at this stage facilitates the subsequent "fishing" of colonies.

7. Tube in quantities of 10 c.c.
8. Solidify in the sloping position in the inspissator at 75° C. for one hour.
9. Place the tubes for forty-eight hours in the incubator at 37° C., and eliminate any contaminated tubes.

To prevent drying, 0.5 c.c. glycerine bouillon (see page 209) may be added to each tube between steps 8 and 9.

10. Cap those tubes of media which remain sterile with india-rubber caps and store for future use.

Potato.—
1. Choose fairly large potatoes, wash them well, and scrub the peel with a stiff nail-brush.
2. Peel and take out the eyes.
3. Remove cylinders from the longest diameter of each potato by means of an apple-corer or a large cork-borer (*i. e.*, one of about 1.4 cm. diameter).

The reaction of the fresh potato is strongly acid to phenolphthalein. If, therefore, the potatoes are required to approximate +10, as for the cultivation of some of the vibrios, the cylinders should be soaked in a 1 per cent. solution of sodium carbonate for thirty minutes.

4. Cut each cylinder obliquely from end to end, forming two wedge-shaped portions.
5. Place a small piece of sterilised cotton-wool, moistened with sterile water, at the bottom of a sterile test-tube; insert the potato wedge into the tube so that its base rests upon the cotton-wool. Now plug the tube with cotton-wool (Fig. 111).
6. Sterilise in the steamer at 100° C. for twenty minutes on each of *five* consecutive days.

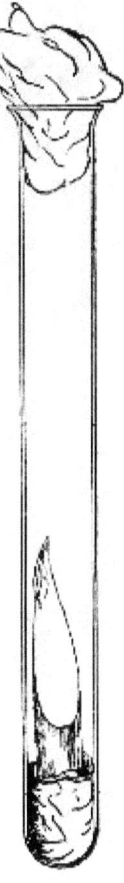

Fig. 111.—Potato tube.

Note.—The cork borer reserved for cutting the potato cylinders should be silver electro-plated both inside and out, and the knife used for dividing the cylinders should be of silver or silver plated. When these precautions are adopted the potato wedges will retain their white color and will not show the discoloration so often observed when steel instruments are employed.

Beer Wort.—Wort is chiefly used as a medium for the cultivation of yeasts, moulds, etc., both in its fluid form and also when made solid by the addition of gelatine or agar. The wort is prepared as follows:

1. Weigh out 250 grammes crushed malt and place in a 2-litre flask.
2. Add 1000 c.c. distilled water, heated to 70° C., and close the flask with a rubber stopper.
3. Place the flask in a water-bath regulated to 60°C. and allow the maceration to continue for one hour.
4. Strain through butter muslin into a clean flask and heat in the steamer for thirty minutes.
5. Filter through Swedish filter paper.
6. Tube in quantities of 10 c.c. or store in flasks.
7. Sterilise in the steamer at 100° C. for twenty minutes on each of three consecutive days.

The natural reaction of the wort should *not* be interfered with.

Note.—It is sometimes more convenient to obtain "*unhopped*" beer wort direct from the brewery. In this case it is diluted with an equal quantity of distilled water, steamed for an hour, filtered, filled into sterile flasks or tubes, and sterilised by the discontinuous method.

Wort Gelatine.—
1. Measure out wort (prepared as above), 900 c.c., into a sterile flask.
2. Weigh out gelatine, 100 grammes (= 10 per cent.), and add it to the wort in the flask.
3. Bubble live steam through the mixture for ten minutes, to dissolve the gelatine.
4. Cool to 60°C.; clarify with egg as for nutrient gelatine (*vide* page 164).
5. Filter through papier Chardin.
6. Tube, and sterilise as for nutrient gelatine.

Wort Agar.—
1. Measure out wort (as above), 700 c.c., into a sterile flask.
2. Weigh out powdered agar, 20 grammes; mix into a smooth paste with 200 c.c. of cold wort and add to the wort in the flask.
3. Bubble live steam through the mixture for twenty minutes, to dissolve the agar.
4. Cool to 60° C.; clarify with egg as for nutrient agar (*vide* page 167).
5. Filter through papier Chardin, using the hot-water funnel.
6. Tube, and sterilise as for nutrient agar.

Peptone Water (Dunham).—
1. Weigh out Witté's peptone, 10 grammes, and salt, 5 grammes, and emulsify with about 250 c.c. of distilled water previously heated to 60° C.
2. Pour the emulsion into a litre flask and make up to 1000 c.c. by the addition of distilled water.
3. Heat in the steamer at 100° C. for thirty minutes.
4. Filter through Swedish filter paper.
5. Tube in quantities of 10 c.c. each.
6. Sterilise in the steamer at 100° C. for twenty minutes on each of three consecutive days.

"Sugar" or "Carbohydrate" Media.—

Formerly the ability of bacteria to induce hydrolytic changes in carbohydrate substances was observed only in connection with a few well-defined sugars, but of recent years it has been shown that when using litmus as an indicator these so-called "fermentation reactions" facilitate the differentiation of closely allied species, and the list of substances employed in this connection has been considerably extended. The media prepared with them are now no longer regarded as special, but are comprised in the "stock media" of the laboratory. The chief of these substances are the following, arranged in accordance with their chemical constitution:

Monosaccharides	Dextrose (glucose), lævulose, galactose, mannose, arabinose, xylose.
Disaccharides	Maltose, lactose, saccharose.
Trisaccharides	Raffinose (mellitose).
Polysaccharides	Dextrin, inulin, starch, glycogen, amidon.
Glucosides	Amygdalin, coniferin, salicin, helicin, phlorrhizin.
Polyatomic alcohols	*Trihydric*, Glycerin.
	Tetrahydric, Erythrite.
	Pentahydric, Adonite.
	Hexahydric, Dulcite, (dulcitol or melampirite), isodulcite (rhamnose), mannite (mannitol), sorbite (sorbitol), inosite.

These substances should be obtained from Kahlbaum (of Berlin); in the pure form, and when possible as large crystals, and the method of preparing a medium containing either of them may be exemplified by describing Dextrose Solution.

Dextrose Solution.—
1. Weigh out

Peptone	20 grammes
Glucose	10 grammes

and grind together in a mortar; then emulsify in 100 c.c. of distilled water heated to 60° C.

2. Place in a flask and add

Distilled water   850 c.c.

3. Steam in the steamer at 100° C. for twenty minutes to dissolve the peptone and glucose.
4. Add

Kubel-Tiemann litmus solution (Kahlbaum)   50 c.c.

(The substances enumerated above react acid to phenolphthalein, but variously toward the neutral litmus solution. To such as react acid, add very cautiously n/1 sodium hydrate solution to the medium in bulk until the neutral tint has returned).

5. Fill into tubes in which have previously been placed the inverted Durham's gas tubes.
6. Sterilise in the steamer at 100° C. for *twenty minutes* on each of three successive days.

Note.—On no account should these media be sterilised in the autoclave, as temperatures above 100° C. themselves induce hydrolytic changes in the substances in question. It is equally important that the twenty minutes should not be exceeded in sterilisation, as neglect of this precaution may discolour the litmus or lead to the production of yellowish tints when the tubes are subsequently inoculated with acid-forming bacteria.

Neutral Litmus Solution.

The most satisfactory is the Kubel-Tiemann, prepared by Kahlbaum. It can however be made in the laboratory as follows:
1. Weigh out

Commercial litmus   50 grammes,

and place in a well stoppered 500 c.c. bottle; measure out and add 300 c.c. alcohol 95 per cent.
2. Shake well at least once a day for seven days—the alcohol acquires a green colour.
3. Decant off the green alcohol and fill a further 300 c.c. 95 per cent. alcohol into the bottle and repeat the shaking.
4. Repeat this process until on adding fresh alcohol the fluid only becomes tinged with violet.
5. Pour off the alcohol, leaving the litmus as dry as possible. Connect up the bottle to an air pump and evaporate off the last traces of alcohol.
6. Transfer the dry litmus to a litre flask, measure in 600 c.c. distilled water and allow to remain in contact 24 hours with frequent shakings.
7. Filter the solution into a clean flask and add one or two drops of pure concentrated sulphuric acid until the litmus solution is distinctly wine-red in colour.
8. Add excess of pure solid baryta and allow to stand until the reaction is again alkaline.
9. Filter.
10. Bubble CO_2 through the solution until reaction is definitely acid.
11. Sterilise in the steamer at 100° C. for thirty minutes on each of three consecutive days. This sterilises the solution and also drives off the carbon dioxide, leaving the solution neutral.

Media for anaerobic cultures. In addition to the foregoing media, all of which can be, and are employed in the cultivation of anaerobic bacteria, certain special media containing readily oxidised substances are commonly used for this purpose. The principal of these are as follows:

Bile Salt Broth (MacConkey).—
1. Weigh out Witté's peptone, 20 grammes (= 2 per cent.), and emulsify with 200 c.c. distilled water previously warmed to 60°C.
2. Weigh out sodium taurocholate (commercial), 5 grammes (= 0.5 per cent.), and glucose, 5 grammes (= 0.5 per cent.), and dissolve in the peptone emulsion.
3. Wash the peptone emulsion into a flask with 800 c.c. distilled water, and heat in the steamer at 100° C. for twenty minutes.
4. Filter through Swedish filter paper into a sterile flask.
5. Add sterile litmus solution sufficient to colour the medium to a deep purple, usually 13 per cent. required.
6. Fill, in quantities of 10 c.c., into tubes containing small gas tubes (*vide* Fig. 104, page 161). Sterilise in the steamer at 100° C. for twenty minutes on each of three consecutive days.

Glucose Formate Bouillon (Kitasato).—
1. Measure out nutrient bouillon, 1000 c.c. (*vide* page 163, sections 1 to 6).
2. Weigh out glucose, 20 grammes (= 2 per cent.), sodium formate, 4 grammes (= 0.4 per cent.), and dissolve in the fluid.
3. Tube, and sterilise as for bouillon.

Glucose Formate Gelatine (Kitasato).—
1. Prepare nutrient gelatine (*vide* page 164, sections 1 to 7) and measure out 1000 c.c.
2. Weigh out glucose, 20 grammes (= 2 per cent.), and sodium formate, 4 grammes (= 0.4 per cent.), and dissolve in the hot gelatine.
3. Filter through papier Chardin.
4. Tube, and sterilise as for nutrient gelatine.

Glucose Formate Agar (Kitasato).—
1. Prepare nutrient agar (*vide* page 167, sections 1 to 8). Measure out 1000 c.c.
2. Weigh out glucose, 20 grammes (= 2 per cent.), sodium formate, 4 grammes (= 0.4 per cent.), and dissolve in the agar.
3. Tube, and sterilise as for nutrient agar.

Sulphindigotate Bouillon (Weyl).—
1. Measure out nutrient bouillon (*vide* page 163, sections 1 to 6 1000 c.c.).
2. Weigh out glucose, 20 grammes (= 2 per cent.), sodium sulphindigotate, 1 gramme (= 0.1 per cent.), and dissolve in the fluid.
3. Tube, and sterilise as for bouillon.

Sulphindigotate Gelatine (Weyl).—

1. Prepare nutrient gelatine (*vide* page 164, sections 1 to 7). Measure out 1000 c.c.
2. Weigh out glucose, 20 grammes (= 2 per cent.), and sodium sulphindigotate, 1 gramme (= 0.1 per cent.), and dissolve in the hot gelatine.
3. Filter through papier Chardin.
4. Tube, and sterilise as for nutrient gelatine.

Sulphindigotate Agar.—
1. Prepare nutrient agar (*vide* page 167, sections 1 to 8). Measure out 1000 c.c.
2. Weigh out glucose, 20 grammes (= 2 per cent.), sodium sulphindigotate, 1 gramme (= 0.1 per cent.), and dissolve in the hot agar.
3. Tube, and sterilise as for nutrient agar.

Note.—The Sulphindigotate media are of a blue colour, which during the growth of anaerobic bacteria is oxidised and decolourised to a light yellow.

FOOTNOTES:

This figure is obtained by adding together 1 litre water, 1000 grammes; 10 per cent. gelatine, 100 grammes; 1 per cent. peptone, 10 grammes; 0.5 per cent. salt, 5 grammes; total, 1115 grammes. Modifications of the above process, as to quantities and percentages, will require corresponding alterations of the figures. The average weight of a measured litre of 10 per cent. nutrient gelatine when prepared in this way *after filtration* is 1080 grammes.

This figure is obtained by adding together 1 litre of water (meat extract), 1000 grammes; 2 per cent. agar, 20 grammes; 1 per cent. peptone, 10 grammes; 0.5 per cent. salt, 5 grammes—total 1035 grammes. Modifications of the process as to quantities or percentages will necessitate corresponding alterations in the calculated medium figure. The average weight of a measured litre of 2 per cent. agar when prepared in this way, *after filtration*, is 1010.5 grammes.

"Hopped" wort exerts a toxic effect upon many bacteria, including the lactic acid bacteria.

XII. SPECIAL MEDIA.

In this chapter are collected a number of media which have been elaborated by various workers for special purposes, grouped together under headings which indicate their chief utility. In many instances the name of the originator of the medium is given, but without reference to his original instructions, since these are in many cases inadequate to the requirements of the isolated worker, who would probably fail to reproduce the medium in a form giving the results attributed to it by its author. Such modifications have therefore been introduced as make for uniformity between the different batches of media.

A considerable number of coloured media, chiefly intended for work with intestinal bacteria, have been included; but beyond the fact that the author's modification of the Drigalski-Conradi medium has been included amongst the routine media of the laboratory, no comment has been made upon their relative values, since only by observation and practice can the skill necessary to utilise their full value be acquired.

The instructions as to sterilisation are rarely given in full; the routine method of exposure in the steam steriliser at 100° C. (without pressure) for twenty minutes on each of three successive days for all fluid media, and thirty minutes on each of three successive days for all liquefiable or solid media must be carried out; and only when these general rules are to be departed from are further details given.

Media for the Study of the Chemical Composition of Bacteria.

Asparagin Medium (Uschinsky).—
1. Weigh out and mix

Asparagin	3.4 grammes
Ammonium lactate	10.0 grammes
Sodium chloride	5.0 grammes
Magnesium sulphate	0.2 gramme
Calcium chloride	0.1 gramme
Acid potassium phosphate (KH_2PO_4)	1.0 gramme

2. Dissolve the mixture in distilled water 1000 c.c.
3. Add glycerine, 40 c.c.
4. Tube, and sterilise as for nutrient bouillon.

Asparagin Medium (Frankel and Voges).—
1. Weigh out and mix

Asparagin	4 grammes
Sodium phosphate, (Na_2HPO_4) 12OH	2 grammes
Ammonium lactate	6 grammes

Sodium chloride	5 grammes

and dissolve in

Distilled water	1000 c.c.

2. Tube, and sterilise as for nutrient bouillon.

Note.—Either of the above asparagin media, after the addition of 10 per cent. gelatine or 1.5 per cent. agar, may be advantageously employed in the solid condition.

Proteid Free Broth (Uschinsky).—

1. Weigh out and mix

Calcium chloride	0.1 gramme
Magnesium sulphate	0.2 gramme
Acid potassium phosphate (KH_2PO_4)	2.0 grammes
Potassium aspartate	3.0 grammes
Sodium chloride	5.0 grammes
Ammonium lactate	6.0 grammes

2. Dissolve the mixture in distilled water 1000 c.c.
3. Add glycerine 30 c.c.
4. Tube and sterilise as for nutrient broth.

Media for the Study of Biochemical Reaction.

Inosite-free Media—Bouillon (Durham).—

1. Prepare meat extract, 1000 c.c. (*vide* page 148), from bullock's heart which has been "hung" for a couple of days.
2. Prepare nutrient bouillon (+10), 1000 c.c. (*vide*, page 161), from the meat extract, and store in 1-litre flask.
3. Inoculate the bouillon from a pure cultivation of the B. lactis aerogenes, and incubate at 37° C. for forty-eight hours.
4. Heat in the steamer at 100° C. for twenty minutes to destroy the bacilli and some of their products.
5. Estimate the reaction of the medium and if necessary restore to +10.
6. Inoculate the bouillon from a pure cultivation of the B. coli communis and incubate at 37° C. for forty-eight hours.
7. Heat in the steamer at 100° C. for twenty minutes.

Now fill two fermentation tubes with the bouillon, tint with litmus solution, and sterilise; inoculate with B. lactis aerogenes. If no acid or gas is formed, the bouillon is in a sugar-free condition; but if acid or gas is present, again make the bouillon in the flask +10, reinoculate with one or other of the above-mentioned bacteria, and incubate; then test again. Repeat this till neither acid nor gas appears in the medium when used for the cultivation of either of the bacilli referred to above.

8. After the final heating, stand the flask in a cool place and allow the growth to sediment. Filter the supernatant broth through Swedish filter paper. If the filtrate is cloudy, filter through a porcelain filter candle.
9. Tube, and sterilise as for bouillon.

Bouillon prepared in the above-described manner will prove to be absolutely sugar-free; and from it may be prepared nutrient sugar-free gelatine or agar, by dissolving in it the required percentage of gelatine or agar respectively and completing the medium according to directions given on pages 166 and 167. The most important application of inosite-free bouillon is its use in the preparation of sugar bouillons, whether glucose, maltose, lactose, or saccharose, of exact percentage composition.

Sugar (Dextrose) Bouillon.—

1. Measure out nutrient bouillon, 1000 c.c. (*vide* page 163, sections 1 to 6) or sugar-free bouillon (*vide supra*).
2. Weigh out glucose (anhydrous), 20 grammes (= 2 per cent.), and dissolve in the fluid.
3. Tube, and sterilise as for bouillon.

Ordinary commercial glucose serves the purpose equally well, but is not recommended, as during the process of sterilisation it causes the medium to gradually deepen in colour.

Note.—In certain cases a corresponding percentage of lactose, maltose, or saccharose is substituted for glucose.

Sugar Gelatine.—

1. Prepare nutrient gelatine (*vide* page 164, sections 1 to 7). Measure out 1000 c.c.
2. Weigh out glucose, 20 grammes (= 2 per cent.), and dissolve in the hot gelatine.
3. Filter through papier Chardin.
4. Tube, and sterilise as for nutrient gelatine.

Sugar Agar.—

1. Prepare nutrient agar (*vide* page 167, sections 1 to 8). Measure out 1000 c.c.
2. Weigh out glucose, 20 grammes (= 2 per cent.), and dissolve in the clear agar.
3. Tube, and sterilise as for nutrient agar.

Note.—Other "sugar" media are prepared by substituting a corresponding percentage of lactose, maltose (or any other of the substances referred to under "Sugar Media," page 177) for the glucose.

Iron Bouillon.—

1. Measure out nutrient bouillon, 1000 c.c. (*vide* page 141, sections 1 to 6).
2. Weigh out ferric tartrate, 1 gramme (= 0.1 per cent.), and dissolve it in the bouillon.
3. Tube, and sterilise as for bouillon.

Note.—The lactate of iron may be substituted for the tartrate.

Lead Bouillon.—
1. Measure out nutrient bouillon, 1000 c.c. (*vide* page 163, sections 1 to 6).
2. Weigh out lead acetate, 1 gramme (= 0.1 per cent.), and dissolve it in the bouillon.
3. Tube, and sterilise as for bouillon.

Nitrate Bouillon.—
1. Measure out nutrient bouillon, 1000 c.c. (*vide* page 163, sections 1 to 6).
2. Weigh out potassium nitrate, 5 grammes (= 0.5 per cent.), and dissolve it in the bouillon.
3. Tube, and sterilise as for bouillon.

Note.—The nitrate of sodium or ammonium may be substituted for that of potassium, or the salt may be added in the proportion of from 0.1 to 1 per cent. to meet special requirements.

Iron Peptone Solution (Pakes).—
1. Weigh out peptone, 30 grammes, and emulsify it with 200 c.c. tap water, previously heated to about 60°C.
2. Wash the emulsion into a litre flask with 800 c.c. tap water.
3. Weigh out salt, 5 grammes, and sodium phosphate, 3 grammes, and dissolve in the mixture in the flask.
4. Heat the mixture in the steamer at 100° C. for thirty minutes, to complete the solution of the peptone, and filter into a clean flask.
5. Fill into tubes in quantities of 10 c.c. each.
6. Add to each tube 0.1 c.c. of a 2 per cent. neutral solution of ferric tartrate. (A yellowish-white precipitate forms.)
7. Sterilise as for nutrient bouillon.

Lead Peptone Solution.—
Prepare as for iron peptone solution but in step 6 substitute 0.1 c.c. of a 1 per cent. neutral aqueous solution of lead acetate.

Nitrate Peptone Solution (Pakes).—
1. Weigh out Witté's peptone, 10 grammes, and emulsify it with 200 c.c. ammonia-free distilled water previously heated to 60°C.
2. Wash the emulsion into a flask and make up to 1000 c.c., with ammonia-free distilled water.
3. Heat in the steamer at 100° C. for twenty minutes.
4. Weigh out sodium nitrate, 1 gramme, and dissolve in the contents of the flask.
5. Filter through Swedish filter paper.
6. Tube, and sterilise as for nutrient bouillon.

Litmus Bouillon.—
1. Measure out nutrient bouillon, 1000 c.c. (*vide* page 163, sections 1 to 6).
2. Add sufficient sterile litmus solution to tint the medium a dark lavender colour. (Media rendered +10 will usually react very faintly alkaline or occasionally neutral to litmus.)
3. Tube, and sterilise as for bouillon.

Rosolic Acid Peptone Solution.—
1. Weigh out rosolic acid (corallin), 0.5 gramme, and dissolve it in 80 per cent. alcohol, 100 c.c. Keep this as a stock solution.
2. Measure out peptone water (Dunham), 100 c.c., and rosolic acid solution, 2 c.c., and mix.
3. Heat in the steamer at 100° C. for thirty minutes.
4. Filter through Swedish filter paper.
5. Tube, and sterilise as for nutrient bouillon.

Capaldi-Proskauer Medium, No. I.—
1. Weigh out and mix

Sodium chloride	2.0 grammes
Magnesium sulphate	0.1 gramme
Calcium chloride	0.2 gramme
Monopotassium phosphate	2.0 grammes

2. Dissolve in water 1000 c.c. in a 2-litre flask
3. Weigh out and mix

Asparagin	2 grammes
Mannite	2 grammes

and add to contents of flask.

4. Measure out 25 c.c. of the solution and titrate it against decinormal sodic hydrate, using litmus as the indicator. Control the result and estimate the amount of sodic hydrate necessary to be added to render the remainder of the solution neutral to litmus. Add this quantity of sodic hydrate.
5. Filter.
6. Add litmus solution 47.5 c.c. (= 5 per cent.).
7. Tube, and sterilise as for nutrient bouillon.

Capaldi-Proskauer Medium No. II.—
1. Weigh out and mix

Peptone	20 grammes

Mannite   1 gramme

2. Dissolve in water 1000 c.c. in a 2-litre flask.
3. Neutralise to litmus as in No. I (*vide supra*, Step 4).
4. Filter.
5. Add litmus solution 47.5 c.c. (= 5 per cent.).
6. Tube, and sterilise as for nutrient bouillon.

Urine Media. Bouillon.—
1. Collect freshly passed urine in sterile flask.
2. Place the flask in the steamer at 100° C. for thirty minutes.
3. Filter through two thicknesses of Swedish filter paper.
4. Tube, and sterilise as for nutrient bouillon. (Leave the reaction unaltered.)

Urine Gelatine.—
1. Collect freshly passed urine in sterile flask.
2. Take the specific gravity, and, if above 1010, dilute with sterile water until that gravity is reached.
3. Estimate (with control) at the boiling-point, and note the reaction of the urine.
4. Weigh out gelatine, 10 per cent., and add to the urine in the flask.
5. Heat in the steamer at 100° C. for one hour to dissolve the gelatine.
6. Estimate the reaction and add sufficient caustic soda solution to restore the reaction of the medium mass to the equivalent of the original urine.
7. Cool to 60° C. and clarify with egg as for nutrient gelatine (*vide* page 166).
8. Filter through papier Chardin.
9. Tube, and sterilise as for nutrient gelatine.

Urine Gelatine (Heller).—
1. Collect freshly passed urine in sterile flask.
2. Filter through animal charcoal to remove part of the colouring matter.
3. Take the specific gravity, and if above 1010, dilute with sterile water till this gravity is reached.
4. Add Witté's peptone, 1 per cent.; salt, 0.5 per cent.; gelatine, 10 per cent.
5. Heat in the steamer at 100° C. for one hour, to dissolve the gelatine, etc.
6. Add normal caustic soda solution in successive small quantities, and test the reaction from time to time with litmus paper, until the fluid reacts faintly alkaline.
7. Cool to 60° C. and clarify with egg as for nutrient gelatine (*vide* page 166).
8. Filter through papier Chardin.
9. Tube, and sterilise as for nutrient gelatine.

Urine Agar.—
1. Collect freshly passed urine in sterile flask.
2. Take the specific gravity and if above 1010, dilute with sterile water till this gravity is reached.
3. Weigh out 1.5 per cent. or 2 per cent. powdered agar, and add it to the urine.
4. Heat in the steamer at 100° C. for ninety minutes to dissolve the agar.
5. Cool to 60° C. and clarify with egg as for nutrient agar (*vide* page 168).
6. Filter through papier Chardin, using the hot-water funnel.
7. Tube, and sterilise as for nutrient agar.
(Leave the reaction unaltered.)

Serum Sugar Media (Hiss).—
In these media the fermentation of carbohydrate substance by bacterial action is indicated by the coagulation of the serum proteids in addition to the production of an acid reaction.

Serum Dextrose Water (Hiss).—
1. Measure out into a litre flask

Serum water (See page 170)   1000 c.c.

2. Weigh out

Dextrose   10 grammes

and dissolve in the serum water.
3. Filter through Swedish filter paper.
4. Measure out and add to the medium

Litmus solution (Kahlbaum)   50 c.c.

5. Tube in quantities of 10 c.c. and sterilise in the steamer at 100° C. for twenty minutes on each of three successive days.
Lævulose, galactose, maltose, lactose, etc., can be substituted in similar amounts for dextrose and the medium completed as above.

Omeliansky's Nutrient Fluid (*For Cellulose Fermenters*).—
1. Weigh out and mix

Potassium phosphate   4.0 grammes

Magnesium sulphate	2.0 grammes
Ammonium sulphate	4.0 grammes
Sodium chloride	0.25 gramme

2. Dissolve in distilled water 4000 c.c.
3. Flask in quantities of 250 c.c.
4. Weigh out and add 5 grammes precipitated chalk to each flask.
5. Sterilise in the steamer at 100° C. for twenty minutes on each of three successive days.

Media for the Study of Chromogenic Bacteria.

Milk Rice (Eisenberg).—
1. Measure out nutrient bouillon, 70 c.c., and milk, 210 c.c., and mix thoroughly.
2. Weigh out rice powder, 100 grammes, and rub it up in a mortar with the milk and broth mixture.
3. Fill the paste into sterile capsules, spreading it out so as to form a layer about 0.5 cm. thick, over the bottom of each.
4. Heat over a water-bath at 100° C. until the mixture solidifies.
5. Replace the lids of the capsules. Sterilise in the steamer at 100° C. for thirty minutes on each of three consecutive days.
(A solid medium of the colour of *café au lait* is thus produced.)

Milk Rice (Soyka).—
1. Measure out nutrient bouillon, 50 c.c., and milk, 150 c.c., and mix thoroughly.
2. Weigh out rice powder, 100 grammes, and rub it up in a mortar with the milk and broth mixture.
3. Fill the paste into sterile capsules, to form a layer over the bottom of each.
4. Replace the lids of the capsules.
5. Sterilise in the steamer at 100° C. for thirty minutes on each of three consecutive days.
(A pure white, opaque medium is thus formed.)

Media for the Study of Phosphorescent and Photogenic Bacteria.

Fish Bouillon.—
1. Weigh out herring, mackerel, or cod, 500 grammes, and place in a large porcelain beaker (or enamelled iron pot).
2. Weigh out sodium chloride, 26.5 grammes; potassium chloride, 0.75 gramme; magnesium chloride, 3.25 grammes; and dissolve in 500 c.c. distilled water. Add the solution to the fish in the beaker.
3. Place the beaker in a water-bath and proceed as in preparing meat extract—*i. e.*, heat gently at 40° C. for twenty minutes, then rapidly raise the temperature to, and maintain at, the boiling-point for ten minutes.
4. Strain the mixture through butter muslin into a clean flask.
5. Weigh out peptone, 5 grammes, and emulsify with about 200 c.c. of the hot fish water; incorporate thoroughly with the remainder of the fish water in the flask.
6. Heat in the steamer at 100° C. for twenty minutes to complete the solution of the peptone.
7. Filter through Swedish filter paper.
8. When the fish bouillon is cold, if it is to be used as fluid medium, make up to 1000 c.c. by the addition of distilled water. If, however, it is to be used as the basis for agar or gelatine media store it in the "Double Strength" condition.
9. Tube and sterilise as for nutrient bouillon.
As an alternative method "Marvis" fish food (16 grammes) may be substituted for the 500 grammes of fresh fish.

Fish Gelatine.—
1. Measure out double strength fish bouillon, 500 c.c., into a "tared" 2-litre flask.
2. Add sheet gelatine, 100 grammes, cut into small pieces.
3. Bubble live steam through the mixture for fifteen minutes to dissolve the gelatine.
4. Weigh the flask and its contents; adjust the weight to the calculated figure for one litre of medium (1135.5 grammes) by the addition of distilled water at 100° C. (*vide* page 166).
5. Cool to below 60°C., and clarify with egg.
6. Filter through papier Chardin.
7. Tube, and sterilise as for nutrient gelatine.
Shake well after the final sterilisation, to aerate the medium.

Fish Gelatine-Agar.—
1. Weigh out powdered agar, 5 grammes, and emulsify it with 200 c.c. double strength fish bouillon.
2. Wash the emulsion into a "tared" 2-litre flask with 300 c.c. fish bouillon.
3. Weigh out sheet gelatine, 70 grammes, cut it into small pieces and add it to the contents of the flask.
4. Bubble live steam through the mixture to dissolve the gelatine and agar.
5. Weigh the flask and contents. Adjust the weight to the calculated figure for one litre of medium (1110.5 grammes) by the addition of distilled water at 100° C. (*vide* page 166).
6. Cool to below 60° C. and clarify with egg.
7. Filter through papier Chardin.
8. Tube, and sterilise as for nutrient gelatine.
Shake well after the final sterilisation, to aerate the medium.

Media for the Study of Yeasts and Moulds.

Pasteur's Solution.—
(Reaction alkaline).

1. Weigh out and mix the ash from 10 grammes of yeast; ammonium tartrate, 10 grammes; cane sugar, 100 grammes.
2. Dissolve the mixture in distilled water, 1000 c.c.
3. Tube or flask, and sterilise as for nutrient bouillon.

Yeast Water (Pasteur).—
1. Weigh out pressed yeast, 75 grammes; place in a 2-litre flask and add 1000 c.c. distilled water.
2. Heat in the steamer at 100° C. for thirty minutes.
3. Filter through papier Chardin.
4. Tube or flask, and sterilise as for nutrient bouillon.

Cohn's Solution.—
1. Weigh out and mix

Acid potassium phosphate (KH_2PO_4)	5.0 grammes
Calcium phosphate	0.5 gramme
Magnesium sulphate	5.0 grammes
Ammonium tartrate	10.0 grammes

and dissolve in

Distilled water   1000 c.c.

2. Tube, or flask and sterilise as for nutrient bouillon.

Naegeli's Solution.—
1. Weigh out and mix

Dibasic potassium phosphate (K_2HPO_4)	1.0 gramme
Magnesium sulphate	0.2 gramme
Calcium chloride	0.1 gramme
Ammonium tartrate	10.0 grammes

and dissolve in

Distilled water   1000 c.c.

2. Tube or flask; sterilise as for nutrient bouillon.

Plaster-of-Paris Discs.—
1. Take large corks, 2.5 cm. diameter, and roll a piece of stiff note-paper round each, so that about a centimetre projects as a ridge above the upper surface of the cork, and secure in position with a pin (Fig. 112).
2. Mix plaster-of-Paris into a stiff paste with distilled water, and fill each of the cork moulds with the paste.
3. When the plaster has set, remove the paper from the corks, and raise the plaster discs.
4. Place the plaster discs on a piece of asbestos board and sterilise by exposing in the hot-air oven to 150° C. for half an hour.

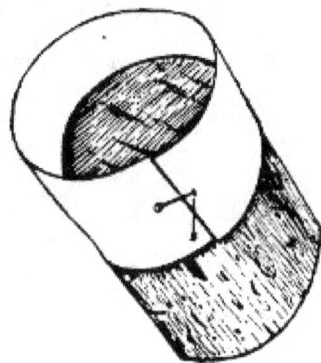

Fig. 112.—Cork and paper mould for plaster-of-Paris disc.

5. Remove the sterile discs from the oven by means of sterile forceps, place each inside a sterile capsule, and moisten with a little sterile water.
6. Sterilise in the steamer at 100° C. for thirty minutes on each of three consecutive days.

Gypsum Blocks (Engel and Hansen).—
These are in the form of truncated cones and for their preparation small tin moulds are required, each having a diameter of 5.5 cm. at the base and 4 cm. at the truncated apex. The height (or depth) of a mould is 4.5 to 5 cm.
1. Mix powdered calcined gypsum into a stiff paste with distilled water.
2. Fill the paste into the moulds and allow it to set and dry by exposure to air.
3. Remove the block from the mould and transfer it to a double glass dish of adequate size (7 cm. diameter × 7 cm. high).
4. Sterilise block in its dish for one hour in the hot-air oven at 115°C.
5. Carefully open the dish and add sterile distilled water to moisten the block and form a layer in the bottom of the dish 1 cm. deep.

Wine Must.—(Wine must is obtained from Sicily, in hermetically sealed tins, in a highly concentrated form—as a thick syrup—but not sterilised.)
1. Weigh out "wine must," 200 grammes, place in a 2-litre flask and add distilled water, 800 c.c.
2. Weigh out ammonium tartrate, 5 grammes, and add to the dilute must.
3. Place the flask in a water-bath regulated to 60° C. for one hour and incorporate the mixture thoroughly by frequent shaking.
4. Filter through papier Chardin.
5. Tube, and sterilise as for nutrient bouillon.

Wheat Bouillon (Gasperini).—
1. Weigh out and mix wheat flour, 150 grammes; magnesium sulphate, 0.5 gramme; potassium nitrate, 1 gramme; glucose, 15 grammes.
2. Dissolve the mixture in 1000 c.c. of water heated to 100°C.
3. Filter through papier Chardin.
4. Tube, and sterilise as for nutrient bouillon.

Bread Paste.—
1. Grate stale bread finely on a bread-grater.
2. Distribute the crumbs in sterile Erlenmeyer flasks, sufficient to form a layer about one centimetre thick over the bottom of each.
3. Add as much distilled water as the crumbs will soak up, but not enough to cover the bread.
4. Plug the flasks and sterilise in the steamer at 100° C. for thirty minutes on each of *four* consecutive days.

Media for the Study of Parasitic Moulds.

French Proof Agar (Sabouraud).—
1. Weigh out Chassaing's peptone, 10 grammes, and emulsify it with 200 c.c. distilled water previously heated to 60°C.
2. Weigh out powdered agar, 13 grammes, and emulsify with 200 c.c. cold distilled water.
3. Mix the two emulsions and wash into a tared 2-litre flask with 600 c.c. distilled water.
4. Bubble live steam through the mixture for twenty minutes, to dissolve the agar.
5. Cool to 60° C. and clarify with egg as for nutrient agar (*vide* page 168).
6. Filter through Papier Chardin, using the hot-water funnel.
7. Weigh out *French* maltose, 40 grammes, and dissolve in the agar.
8. Tube, and sterilise as for nutrient agar.

English Proof Agar (Blaxall).—Substitute Witté's peptone for that of Chassaing, and proceed as for French proof agar.

French Mannite Agar, Sabouraud.—(*For cultivation of Favus.*)
Proceed exactly as in preparing French Proof agar *vide supra* substituting Mannite (38 grammes) for maltose.

Media for the Study of Milk Bacteria.

Gelatine Agar.—This medium is prepared by adding to nutrient gelatine sufficient agar to ensure the solidity of the medium when incubated at temperatures above 22° C. If it is intended to employ an incubating temperature of 30°C., 10 per cent. gelatine and 0.5 per cent. agar must be dissolved in the meat extract before the addition of the peptone and salt; while for incubating at 37°C., 12 per cent. gelatine and 0.75 per cent. agar must be used. Avoid the addition of more agar than is absolutely necessary, otherwise the action upon the medium of such organisms as elaborate a liquefying ferment may be retarded or completely absent.
1. Measure out 400 c.c. double strength meat extract into a "tared" 2-litre flask, and add to it gelatine, 100 grammes.
2. Weigh out powdered agar, 5 grammes, emulsify with 100 c.c., cold distilled water and add to the contents of the flask.
3. Dissolve the agar and gelatine by bubbling live steam through the flask for twenty minutes.
4. Weigh out peptone, 10 grammes; salt, 5 grammes; emulsify with 100 c.c. double strength meat extract previously heated to 60°C., and add to the contents of the flask.
5. Replace in the steamer for fifteen minutes. Then adjust the weight to the calculated figure for one litre (in this instance 1120 grammes) by the addition of distilled water at 100°C.
6. Estimate the reaction; control the result. Then add sufficient caustic soda solution to render the reaction +10.
7. Replace in the steamer at 100° C. for twenty minutes.
8. Cool to 60° C. Clarify with egg as for nutrient agar.
9. Filter through papier Chardin, using the hot-water funnel.
10. Tube, and sterilise as for nutrient agar.

Agar Gelatine (Guarniari).—
1. Measure out double strength meat extract, 400 c.c., into a "tared" 2-litre flask, and add to it gelatine, 50 grammes.
2. Weigh out powdered agar, 3 grammes; emulsify with cold distilled water, 50 c.c., and add to the contents of the flask.
3. Dissolve the agar and gelatine by bubbling live steam through the flask for twenty minutes.
4. Weigh out Witté's peptone, 25 grammes; salt, 5 grammes, and emulsify with 100 c.c. double strength meat extract previously heated to 60°C., and add to the contents of the flask.
5. Replace in the steamer for fifteen minutes.
6. Weigh the flask and make up the medium mass to the calculated figure for one litre (1083 grammes) by the addition of distilled water at 100°C.
7. Neutralise carefully to litmus paper by the successive additions of small quantities of normal soda solution.
8. Replace in the steamer at 100° C. for twenty minutes.
9. Cool to 60° C. Clarify with egg as for nutrient agar.
10. Filter through papier Chardin, using the hot-water funnel.
11. Tube, and sterilise as for nutrient agar.

Whey Gelatine.—
1. Curdle fresh milk by warming to 60°C., and adding rennet; filter off the whey into a sterile "tared" flask.

2. Estimate and note the reaction of the whey.
3. Weigh out gelatine, 10 per cent., and add it to the whey in the flask.
4. Bubble live steam through the mixture fifteen minutes to dissolve the gelatine; and weigh.
5. Estimate the reaction of the medium mass; then add sufficient caustic soda solution to restore the reaction of the medium mass (*i. e.*, total weight minus weight of flask) to the equivalent of the original whey.
6. Cool to 60° C. and clarify with egg as for nutrient gelatine (*vide* page 166).
7. Filter through papier Chardin.
8. Tube, and sterilise as for nutrient gelatine.

Whey Agar.—
1. Curdle fresh milk by warming to 60°C., and adding rennet; filter off the whey into a sterile flask.
2. Weigh out agar, 1.5 or 2 per cent., and add it to the whey in the flask.
3. Bubble live steam through the mixture for twenty minutes, to dissolve the agar.
4. Cool to 60°C.; clarify with egg as for nutrient agar (*vide* page 168).
5. Filter through papier Chardin, using the hot-water funnel.
6. Tube, and sterilise as for nutrient agar.

Litmus Whey.—
1. Curdle fresh milk by warming to 60° C. and adding rennet.
2. Filter off the whey through butter muslin into a sterile flask.
3. Neutralise to litmus by the cautious addition of citric acid solution 4 per cent. (Do not neutralise with *mineral* acid.)
4. Heat in the steamer at 100° C. for one hour to coagulate all the proteid.
(If the whey is cloudy when removed from the steamer allow it to stand for forty-eight hours in the ice chest and then decant off the clear fluid—or filter through a Berkefeld filter candle.)
5. Filter into a sterile flask.
6. Tint the whey with litmus solution to a deep purple red.
7. Tube, and sterilise as for milk.

Litmus Whey (Petruschky).—
1. Measure out into a flask

Fresh milk   1000 c.c.

2. Add

Hydrochloric acid (or glacial acetic acid)   1.5 c.c.

and boil.
3. Filter off coagulated casein.
4. Neutralise to litmus by the addition of n/1 caustic soda solution and boil. Whey now cloudy and acid again.
5. Again neutralise to litmus by addition of n/10 caustic soda solution.
6. Filter.
7. Tint the whey with neutral litmus solution to a deep purple colour.
8. Tube and sterilise as for milk.

Litmus Whey Gelatine.—
1. Measure out milk 1000 c.c. into a tared 2-litre flask.
2. Add hydrochloric acid (or glacial acetic acid) 1.5 c.c. and boil for five minutes.
3. Filter off the casein, and make the whey faintly alkaline to litmus.
4. Weigh out

Peptone   10 grammes

and emulsify in a few cubic centimeters of the whey and return to the flask.
5. Weigh out

Gelatine   50 grammes

add it to the whey in the flask and incorporate the mixture by bubbling through live steam.
6. Clear with egg and filter.
7. Make the weight of the medium mass to the calculated figure for one litre (1060 grammes) by the addition of distilled water.
8. Weigh out

Dextrose   15 grammes

and dissolve in the fluid whey gelatine.
9. Add sterile litmus solution to the required tint.
10. Tube and sterilise for twenty minutes in steamer at 100°C. on each of five successive days.
This medium will remain semi-fluid at the room temperature, and may be used for cultures in the cool or hot incubator.

Litmus Whey Agar is prepared in a similar manner to Whey Gelatine, with the substitution of 15 grammes of agar for the gelatine.

Malt Extract Solution (Herschell).—
1. Measure into a flask distilled water 1000 c.c.
2. Weigh out

Extractum malti (malt extract)   25 grammes

and add to distilled water in flask.
3. Boil for five minutes, allow to stand, and decant off clear fluid from sediment.
4. Tube and sterilise as for nutrient bouillon.
Media for the Study of Earth Bacteria, Nitrogen Fixers.
Earthy Salts Agar (Lipman and Brown).—(*For the enumeration of soil organisms.*)
1. Measure out

Agar   20 grammes.

Emulsify in 200 c.c. distilled water.
2. Wash the agar emulsion into a tared 2-litre flask with 400 c.c. distilled water.
3. Weigh out

Peptone   0.5 gramme.

Emulsify in 50 c.c. distilled water and add to the contents of the flask.
4. Bubble live steam through the mixture for twenty minutes to dissolve the agar.
5. Weigh out and mix

Dextrose   10.0 grammes.

Potassium phosphate   0.5 gramme.

Magnesium sulphate   0.2 gramme.

Potassium nitrate   0.06 gramme.

and add to the contents of the flask.
6. Adjust the weight of the medium mass to the calculated figure for one litre (1025 grammes) by the addition of distilled water at 100°C.
7. Titrate the medium mass and adjust the reaction to +5.
8. Cool to 60° C. Clarify with egg and filter.
9. Tube in quantities of 10 c.c. and sterilise as for nutrient agar.
Beyrinck's Solution. I.—(*For the cultivation of nitrogen fixing organisms.*)
1. Weigh out and mix 1 gramme potassium hydrogen phosphate, 0.2 gramme magnesium sulphate, and 0.02 gramme sodium chloride.
2. Dissolve in water 1000 c.c., in a 2-litre flask.
3. Add 1 c.c. of a one per thousand aqueous solution of ferrous sulphate.
4. Add 1 c.c. of a one per thousand solution manganese sulphate.
5. Weigh out 20 grammes dextrose and add to the contents of the flask (dextrose up to 40 grammes may be used for the different organisms).
6. Steam for twenty minutes, filter.
7. Tube, and sterilise as for nutrient bouillon.
Beyrinck's Solution. II.—(*For growth of Azobacter.*)
Proceed as in preparing solution No. I, substituting mannite for dextrose in step 5.
Winogradsky's Solution (for Nitric Organisms).—
1. Weigh out and mix.

Potassium phosphate   1.0 gramme

Magnesium sulphate   0.5 gramme

Calcium chloride   0.01 gramme

Sodium chloride   2.0 grammes

and dissolve in

Distilled water   1000 c.c.

2. Fill into flasks, in quantities of 20 c.c. and add to each a small quantity of freshly washed magnesium carbonate.
3. Sterilise in the steamer at 100° C. for twenty minutes on each of three consecutive days.
4. Add to each flask containing 20 c.c. solution, 2 c.c. of a sterile 2 per cent. solution of ammonium sulphate.
5. Incubate at 37° C. for forty-eight hours and eliminate any contaminated culture flasks. Store the remainder for future use.
Winogradsky's Solution (for Nitrous Organisms).—
1. Weigh out and mix

Ammonium sulphate   1 gramme

Potassium sulphate   1 gramme

and dissolve in

Distilled water   1000 c.c.

2. Add 5 to 10 grammes basic magnesium carbonate, previously sterilised by boiling.

3. Fill into flasks and sterilise, etc., as for previous solution.
Silicate Jelly (Winogradsky).—
1. Weigh out and mix

Ammonium sulphate	0.40 gramme
Magnesium sulphate	0.05 gramme
Calcium chloride	0.01 gramme

and dissolve in

Distilled water   50 c.c.

Label—Solution A.
2. Weigh out and mix

Potassium phosphate	0.10 gramme
Sodium carbonate	0.60 gramme

and dissolve in

Distilled water   50 c.c.

Label—Solution B.
3. Weigh out

Silicic acid   3.4 grammes

and dissolve in

Distilled water   100 c.c.

4. Pour the silicic acid solution into a large porcelain basin.
5. Mix equal quantities of the solutions A and B; then add successive small quantities of the mixed salts to the silicic acid solution, stirring continuously with a glass rod, until a jelly of sufficiently firm consistence has been formed.
6. Spread a layer of this jelly over the bottom of each of several large capsules or "plates."
7. Sterilise in the steamer at 100° C. for thirty minutes on each of three consecutive days.
Media for the Study of Water Bacteria.

Naehrstoff Agar (Hesse and Niedner).—(*For enumeration of water organisms.*)
1. Weigh out: agar, 12.5 grammes and emulsify in 250 c.c. distilled water.
2. Wash the agar emulsion into a tared 2-litre flask with a further 250 c.c. distilled water.
3. Dissolve by bubbling live steam through the mixture.
4. Emulsify Naehrstoff-Heyden (albumose) 7.5 grammes in 200 c.c. cold distilled water and add to melted agar.
5. Adjust weight of medium mass to the calculated figure for one litre (1020 grammes) by addition of distilled water at 100° C.
6. Clarify with white of egg and filter.
7. Tube in quantities of 10 c.c. and sterilise in the steamer at 100° C. for twenty minutes on each of three successive days.

Bile Salt Broth—Double Strength.—
1. Weigh out Witté's peptone, 40 grammes, and emulsify with 300 c.c. distilled water previously warmed to 60° C.
2. Wash the peptone emulsion into a litre flask with 600 c.c. distilled water.
3. Weigh out sodium taurocholate, 10 grammes, and glucose, 10 grammes; dissolve in 100 c.c. distilled water and add to the peptone emulsion in the flask.
4. Heat in the steamer at 100° C. for twenty minutes.
5. Filter through Swedish filter paper into a sterile flask.
6. Add sterile neutral litmus solution sufficient to colour the medium to a deep purple.
7. Fill into small Erlenmeyer flasks in quantities of 25 c.c.
8. Sterilise as for nutrient bouillon.

Media for the Study of Plant Bacteria.

Beetroot.—   }

Carrot.—   } are prepared tubes and sterilised in a manner precisely

Turnip.—   } similar to that described for potato.

Parsnip.—   }

Hay Infusion.—
1. Weigh out dried hay, 10 grammes, chop it up into fine particles and place in a flask.
2. Add 1000 c.c. distilled water, heated to 70° C.; close the flask with a solid rubber stopper.
3. Macerate in a water-bath at 60° C. for three hours.
4. Replace the stopper by a cotton-wool plug, and heat in the steamer at 100° C. for one hour.
5. Filter through Swedish filter paper.
6. Tube, and sterilise as for nutrient bouillon.

Haricot Bouillon.—(*For cultivation of bacteria from tubercles of Legumes.*)

1. Measure out 1000 c.c. distilled water into a 2-litre flask.
2. Weigh out 250 grammes haricot beans and add to the water in the flask.
3. Weigh out 10 grammes sodium chloride and add to the contents of the flask.
4. Add 1 c.c. of a 1 per cent. solution of sodium bicarbonate.
5. Place in the steamer at 100° C. for thirty minutes.
6. Filter.
7. Weigh out 20 grammes saccharose and add to the filtrate.
8. Tube, and sterilise as for nutrient bouillon.

Haricot Agar.—
1. Measure out 400 c.c. distilled water into a "tared" 2-litre flask.
2. Weigh out 15 grammes agar and mix into a thick paste with 100 c.c. cold distilled water, and add to the flask.
3. Dissolve the agar by bubbling live steam through the mixture as in making nutrient agar.
4. Weigh out 250 grammes haricot beans, place in the flask with the agar mixture.
5. Add 1 c.c. of 1 per cent. aqueous solution sodium bicarbonate.
6. Weigh out 10 grammes sodium chloride and add to the contents of the flask.
7. Place in the steamer at 100° C. for thirty minutes.
8. Adjust the weight of the medium mass to 1030 grammes (the figure per litre obtained experimentally) by the addition of distilled water at 100° C.
9. Cool to 60°C., clarify with egg and filter.
10. Weigh out 20 grammes saccharose and add to the contents of the flask.
11. Tube, and sterilise as for nutrient agar.

Wood Ash Agar.—
1. Measure 400 c.c. distilled water into a tared 2-litre flask.
2. Weigh out 10 grammes agar and make into a thick paste with 100 c.c. cold distilled water.
3. Add this agar paste to the distilled water in the flask.
4. Dissolve the agar by passing live steam through it, as in preparing nutrient agar.
5. Weigh out 5 grammes clean wood ash and place in a second flask containing 200 c.c. distilled water with some sterile glass beads: shake thoroughly in a mechanical shaker for ten minutes.
6. Heat in steamer at 100°C., for thirty minutes.
7. After removal from the steamer dry the outside of the flask thoroughly, place it over a Bunsen flame and boil for one minute.
8. Filter directly into the flask containing the melted agar mixture.
9. Weigh out 4 grammes maltose. Add to the contents of the flask.
10. Adjust the weight of the medium mass to the calculated figure for one litre (1019 grammes) by the addition of distilled water at 100°C.
11. Replace the flask in the steamer for twenty minutes, cool to 60°C., and clarify with egg and filter.
12. Tube, and sterilise as for nutrient agar.

Media for the Study of Special Bacilli.
B. Acnes.

Oleic Acid Agar (Fleming).—
1. Measure out into a sterile stout glass bottle which already contains about 10 sterile glass beads

Ascitic fluid   250 c.c.

2. Weigh out

Oleic acid   25 grammes

and add it to the ascitic fluid in the bottle.
3. Emulsify evenly by shaking (either by hand or in a shaking machine) for ten minutes.
4. Liquefy and measure out into a flask

Nutrient agar   750 c.c.

then cool to 55°C.
5. Mix the oleic acid emulsion with the agar.
6. Add 10 c.c. sterile neutral red, 1 per cent. aqueous solution.
7. Tube in quantities of 10 c.c., slant, and allow to set.
8. Incubate for forty-eight hours at 37° C. and reject any contaminated tubes. Store the sterile tubes for future use.

Coli-typhoid Group.

Parietti's Bouillon.—
1. Measure out pure hydrochloric acid, 4 c.c., and add to it carbolic acid solution (5 per cent.), 100 c.c. Allow the solution to stand at least a few days before use.
2. This solution is added in quantities of 0.1, 0.2. and 0.3 c.c. (delivered by means of a sterile graduated pipette) to tubes each containing 10 c.c. of previously sterilised nutrient bouillon (*vide* page 163).
3. Incubate at 37° C. for forty-eight hours to eliminate contaminated tubes. Store the remainder for future use.

Carbolised Bouillon.—
1. Prepare nutrient bouillon (*vide* page 163, sections 1 to 6). Measure out 1000 c.c.
2. Weigh out carbolic acid, 1 gramme (2.5 or 5 grammes may be needed for special purposes), and dissolve it in the medium.

3. Tube, and sterilise as for bouillon.

Carbolised Gelatine.—
1. Prepare nutrient gelatine (*vide* page 164, sections 1 to 7). Measure out 1000 c.c.
2. Weigh out carbolic acid, 5 grammes (= 0.5 per cent.), and dissolve it in the gelatine.
3. Filter if necessary through papier Chardin.
4. Tube, and sterilise as for nutrient gelatine.

One or 2.5 grammes of carbolic acid (= 0.1 per cent. or 0.25 per cent.) are occasionally used in place of the 5 grammes to meet special requirements.

Carbolised Agar.—
1. Prepare nutrient agar (*vide* page 167, sections 1 to 8). Measure out 1000 c.c.
2. Weigh out 1 gramme pure phenol and dissolve in the medium.
3. Filter if necessary through papier Chardin.
4. Tube, and sterilise as for nutrient agar.

Litmus Gelatine.—
1. Prepare nutrient gelatine (*vide* page 164, sections 1 to 8).
2. Add sterile litmus solution, sufficient to tint the medium a deep lavender colour.
3. Tube, and sterilise as for nutrient gelatine.

Lactose Litmus Bouillon (Lakmus Molke).—
1. Weigh out peptone, 4 grammes, and emulsify it with 200 c.c. meat extract (*vide* page 148), previously heated to 60°C.
2. Weigh out salt, 2 grammes, and lactose, 20 grammes, and mix with the emulsion.
3. Wash the mixture into a sterile litre flask with 200 c.c. meat extract and add 600 c.c. distilled water.
4. Heat in the steamer at 100° C. for thirty minutes, to completely dissolve the peptone, etc.
5. *Neutralise carefully to litmus paper* by the successive additions of small quantities of decinormal soda solution.
6. Replace in the steamer for twenty minutes to precipitate phosphates, etc.
7. Filter through two thicknesses of Swedish filter paper.
8. Add sterile litmus solution, sufficient to colour the medium a deep purple.
9. Tube, and sterilise as for bouillon.

Lactose Litmus Gelatine (Wurtz).—
1. Prepare nutrient gelatine (*vide* page 164, sections 1 to 4).
2. Render the reaction of the medium mass -5.
3. Replace in the steamer at 100° C. for twenty minutes.
4. Clarify with egg as for gelatine.
5. Weigh out lactose, 20 grammes (= 2 per cent.), and dissolve it in the medium.
6. Filter through papier Chardin.
7. Add sufficient sterile litmus solution to colour the medium pale lavender.
8. Tube, and sterilise as for nutrient gelatine.

Lactose Litmus Agar (Wurtz).—
1. Prepare nutrient agar (*vide* page 167, sections 1 to 4).
2. Render the reaction of the medium mass -5.
3. Replace in the steamer at 100° C. for twenty minutes.
4. Cool to 60° C. and clarify with egg as for nutrient agar.
5. Weigh out lactose, 20 grammes (= 2 per cent.), and dissolve it in the medium.
6. Filter through papier Chardin, using the hot-water funnel.
7. Add sterile litmus solution, sufficient to colour the medium a pale lavender.
8. Tube, and sterilise as for nutrient agar.

Glycerine Potato Bouillon.—
1. Take 1 kilo of potatoes, wash thoroughly in water, peel, and grate finely on a bread-grater.
2. Weigh the potato gratings, place them in a 2-litre flask, and add distilled water in the proportion of 1 c.c. for every gramme weight of potato. Allow the flask to stand in the ice-chest for twelve hours.
3. Strain the mixture through butter muslin and filter through Swedish filter paper into a graduated cylinder. Note the amount of the filtrate.
4. Place the filtrate in a flask, add an equal quantity of distilled water, and heat in the steam steriliser for sixty minutes.
5. Add glycerine, 4 per cent., mix thoroughly, and again filter.
6. Tube and sterilise as for nutrient bouillon.

Potato Gelatine (Elsner).—
1. Take 1 kilo of potatoes, wash thoroughly in water, peel, and finally grate finely on a bread-grater.
2. Weigh the potato gratings, place them in a 2-litre flask, and add distilled water in the proportion of 1 c.c. for every gramme weight of potato. Allow the flask to stand in the ice-chest for twelve hours.
3. Strain the mixture through butter muslin, and filter through Swedish filter paper into a graduated cylinder.
4. Add 15 per cent. gelatine to the potato decoction and bubble live steam through the mixture for ten minutes.
5. Estimate the reaction; adjust the reaction of the medium mass to +25.
6. Cool the medium to below 60°C.; clarify with egg as for nutrient gelatine (*vide* page 166).
7. Add 1 per cent. potassium iodide (powdered) to the medium.
8. Filter through papier Chardin.

9. Tube and sterilise as for nutrient gelatine.

Aesculin Agar.—(B. coli and allied organisms give black colonies surrounded by black halo.)

1. Measure out 400 c.c. distilled water into a tared 2-litre flask.
2. Weigh out

Agar	15 grammes
Peptone	10 grammes
Sodium taurocholate	5 grammes

and make into a thick paste with 150 c.c. distilled water.

3. Add this paste to the distilled water in the flask.
4. Dissolve the ingredients by bubbling live steam through the mixture.
5. Weigh out

Aesculin	1.0 gramme
Ferric citrate	0.5 gramme

and dissolve in a second flask containing 100 c.c. distilled water.

6. Mix the contents of the two flasks—adjust the weight to the calculated medium figure (in this case 1031.5 grammes) by the addition of distilled water at 100°C.
7. Clarify with egg and filter.
8. Tube and sterilise as for nutrient agar.

Bile Salt Agar (MacConkey).—
1. Weigh out powdered agar, 15 grammes (= 1.5. per cent.), and emulsify with 200 c.c. *cold tap* water.
2. Weigh out peptone, 20 grammes (= 2 per cent.), and emulsify with 200 c.c. *tap* water previously warmed to 60°C.
3. Mix the peptone and agar emulsions thoroughly.
4. Weigh out sodium taurocholate, 5 grammes (= 0.5 per cent.), dissolve it in 300 c.c. *tap* water, and use the solution to wash the agar-peptone emulsion into a tared 2-litre flask.
5. Bubble live steam through the mixture for twenty minutes.
6. Adjust the weight of the medium mass to the calculated figure for one litre (1040 grammes).
7. Cool to 60° C. and clarify with egg as for nutrient agar (*vide* page 168).
8. Filter through papier Chardin, using the hot-water funnel.
9. Weigh out lactose, 10 grammes (= 1 per cent.), and dissolve it in the agar.

If desired, add 5 c.c. of a 1 per cent. (= 0.5 per cent.) aqueous solution of neutral red.

10. Tube, and sterilise as for nutrient agar.

Litmus Nutrose Agar (Drigalski-Conradi).—

This medium should be prepared in precisely the same manner as the Nutrose agar described on page 172 substituting meat extract for serum water, and increasing the percentage of agar added per litre to 3 per cent.

Fuchsin Agar (Braun).—
1. Liquefy and measure out into a sterile flask:

Nutrient agar	1000 c.c.

2. Weigh out: lactose 10 grammes and dissolve in the fluid agar.
3. Adjust the reaction to -5 and filter.
4. Measure out and mix thoroughly with agar:

Fuchsin, alcoholic solution	5 c.c.

The fuchsin solution is prepared by mixing:

Fuchsin (basic)	3 grammes.
Absolute alcohol	60 c.c.

Allow to stand twenty-four hours, then centrifugalise thoroughly and decant the supernatant fluid into a well-stoppered bottle.

5. Measure out and add to the nutrient agar, sodium sulphite, 10 per cent. aqueous solution, freshly prepared 25 c.c.
6. Tube and sterilise as for nutrient agar.
7. Store in a dark cupboard.

Fuchsin Sulphite Agar (Endo).—
1. Liquefy and measure out into a sterile flask:

Nutrient agar	1000 c.c.

2. Weigh out

Lactose	10 grammes.

and dissolve in the fluid agar.

3. Adjust the reaction to +3 and filter.
4. Measure out and mix thoroughly with the fluid agar.

Fuchsin, alcoholic solution (*vide supra*)   5 c.c.

5. Measure out and add to the medium

Sodium sulphite, 10 per cent. aqueous solution   25 c.c.

6. Tube and sterilise as for nutrient agar.

Brilliant Green Agar (Conradi).—

1. Liquefy and measure out into a sterile flask

Nutrient agar   1000 c.c.

2. Adjust reaction to +30 by the addition of normal phosphoric acid; and filter.
3. Measure out and mix thoroughly with the fluid medium

Brilliant green (Hoechst) 1 per thousand aqueous solution   6.5 c.c.

4. Measure out and add to the medium

Picric acid (Gruebler), 1 per cent. aqueous solution   6.5 c.c.

5. Tube and sterilise as for nutrient agar.

Brilliant Green Bile Salt Agar (Fawcus).—

1. Weigh out agar 20 grammes and emulsify in 100 c.c. cold distilled water.
2. Wash the emulsion into a "tared" 2-litre flask with 500 c.c. distilled water.
3. Dissolve the agar by bubbling live steam through the flask.
4. Cool, clarify with egg and filter.
5. Weigh out

Sodium taurocholate   5 grammes

Peptone   20 grammes

and add to the medium in the flask.

6. Weigh out

Lactose   5 grammes

and add to the medium in the flask.

7. Adjust reaction to +15 and filter if necessary.
8. Measure out

Brilliant green, 1 per thousand aqueous solution   20 c.c.

and mix thoroughly with the fluid agar.

9. Measure out and add to the medium

Picric acid, 1 per cent. aqueous solution   20 c.c.

10. Tube and sterilise as for nutrient agar.

China Green Agar (Werbitski).—

1. Liquefy and measure out into a sterile flask

Nutrient agar   1000 c.c.

2. Adjust the reaction accurately to +13 and filter.
3. Measure out and mix thoroughly with the fluid agar

China green 0.2 per cent. aqueous solution   15 c.c.

4. Tube and sterilise as for nutrient agar.

Malachite Green Agar (Loeffler).—

1. Liquefy and measure out into a sterile flask

Nutrient agar   1000 c.c.

2. Weigh out

Dextrose   10 grammes.

and dissolve in nutrient agar.

3. Adjust the reaction to +3, and filter.
4. Measure out and mix thoroughly in the fluid agar

Malachite green, 0.1 per cent. aqueous solution   16 c.c.

for **weak** medium.

4a. To the filtered agar add

Malachite green, 2 per cent. aqueous solution   25 c.c.

for **strong** medium.

5. Tube and sterilise as for nutrient agar.

Double Sugar Agar (Russell).—
1. Liquefy and measure out into a sterile flask

Nutrient agar   1000 c.c.

2. Add 100 c.c. litmus solution to the fluid agar.
3. Weigh out and dissolve in the fluid agar.

Lactose   10 grammes

Dextrose   10 grammes.

4. Render the reaction of the medium neutral to litmus paper by the cautious addition of normal caustic soda.
5. Tube in quantities of 10 c.c. and sterilise in the steamer at 100° C. for twenty minutes on each of three successive days.
6. Store for use in a cool dark place.

B. Diphtheriæ.

Glycerine Blood-serum.—
1. Prepare blood-serum as described, page 168, sections 1 to 4.
2. Add 5 per cent. pure glycerine.
3. Complete as described above for ordinary blood-serum, sections 5 to 7.

Note.—Different percentages of glycerine—from 4 per cent. to 8 per cent.—are used for special purposes. Five per cent. is that usually employed.

Blood-serum (Loeffler).—
1. Prepare nutrient bouillon (*vide* page 163), using meat extract made from veal instead of beef.
2. Add 1 per cent. glucose to the bouillon, and allow it to dissolve completely.
3. Now add 300 c.c. clear blood-serum (*vide* page 168, sections 1 to 4) to every 100 c.c. of this bouillon.
4. Fill into sterile tubes and complete as for ordinary blood-serum.

Blood-serum (Lorrain Smith).—
1. Collect blood-serum (*vide* page 168, sections 1 to 4), as free from hæmoglobin as possible.
2. Weigh out 0.15 per cent. sodium hydrate and dissolve it in the fluid (or add 0.375 c.c. of dekanormal soda solution for every 100 c.c. of serum).
3. Tube, and stiffen at 100° C. in the serum inspissator.
4. Incubate at 37° C. for forty-eight hours to eliminate any contaminated tubes. Store the remainder for future use.

Blood Serum (Councilman and Mallory).—
1. Collect blood serum in slaughterhouse, coagulate, remove serum and tube (*vide* page 168).

Great care must be taken to avoid the inclusion of air bubbles—indeed if only a few tubes are filled at one time, it is a good plan to stand them upright in the receiver of an air pump and to exhaust as completely as possible before transferring to the serum inspissator.

2. Heat the tubes in a slanting position in hot-air steriliser at 90° C. till firmly coagulated, say half an hour.
3. Sterilise in steam steriliser at 100° C. for 20 minutes on each of three successive days.

Resulting medium not translucent, but opaque and firm.

B. Tuberculosis.

Egg Medium (Lubenau).—
This modification of Dorset's egg medium (*quod vide* page 174) is preferred by some for the growth of the tubercle bacillus of the human type. It consists in the addition of one part of 6 per cent. glycerine in normal saline solution, to the egg mixture between steps 4 and 5.

Glycerine Bouillon.—
1. Measure out nutrient bouillon, 1000 c.c. (*vide* page 163, sections 1 to 6).
2. Measure out glycerine, 60 c.c. (= 6 per cent.), and add to the bouillon.
3. Tube, and sterilise as for bouillon.

Glycerine Agar.—
1. Prepare nutrient agar (*vide* page 167, sections 1 to 8). Measure out 1000 c.c.
2. Measure out pure glycerine, 60 c.c. (= 6 per cent.), and add to the agar.
3. Tube, and sterilise as for nutrient agar.

Glycerine Blood-serum.—
1. Prepare blood-serum as described, page 168, sections 1 to 4.
2. Add 5 per cent. pure glycerine.
3. Complete as described above for ordinary blood-serum, sections 5 to 7.

Note.—Different percentages of glycerine—from 4 per cent. to 8 per cent.—are used for special purposes. Five per cent. is that usually employed.

Glycerinated Potato.—
1. Prepare ordinary potato wedges (*vide* page 174, sections 1 to 4).
2. Soak the wedges in 25 per cent. solution of glycerine for fifteen minutes.
3. Moisten the cotton-wool pads at the bottom of the potato tubes with a 25 per cent. solution of glycerine.
4. Insert a wedge of potato in each tube and replug the tubes.
5. Sterilise in the steamer at 100° C. for twenty minutes on each of *five* consecutive days.

Animal Tissue Media (Frugoni).—

1. Take a number of sterile test-tubes 16 × 3 or 4 cm., plugged with cotton wool, and into each insert a 2 cm. length of stout glass tubing (about 1 cm. diameter); fill in glycerine (6 per cent.) bouillon to the upper level of the piece of glass tubing. Sterilise in the steamer at 100° C. for twenty minutes on each of three successive days.

2. Kill a small rabbit by means of chloroform vapour.

3. Under strictly aseptic precautions remove the lungs, liver and other solid organs and transfer them to a sterile double glass dish.

4. With the help of sterile scissors and forceps divide the organs into roughly rectangular blocks 3 × 1.5 × 1 cm.

5. Pour into the dish a sufficient quantity of sterile glycerine solution (6 per cent. in normal saline), cover, and allow to stand for one hour.

6. Introduce a block of tissue into each tube so that it rests upon the upper end of the piece of glass tubing. (The surface of the tissue will now be kept moist by capillary attraction and condensation).

7. Sterilise in the autoclave at 120° C. for thirty minutes.

8. Cap the tubes and store them in the ice chest for future use.

Tissues obtained at postmortems can also be used after preliminary sterilisation by boiling or autoclaving.

Media for the Study of Special Cocci.

Diplococcus Gonorrhϣ.

Ascitic Bouillon (Serum Bouillon).—

1. Collect ascitic fluid (pleuritic fluid, hydrocele fluid, etc.), by aspiration directly into sterile flasks, under strictly aseptic precautions.

2. Mix the serum with twice its bulk of sterile nutrient bouillon (*vide* page 163).

3. If considered necessary (on account of the presence of blood, crystals, etc.), filter the serum bouillon through porcelain filter candle.

4. Tube, and sterilise in the water bath at 56° C. for half an hour on each of five consecutive days.

5. Incubate at 37° C. for forty-eight hours and eliminate contaminated tubes. Store the remainder for future use.

Serum Agar (Heiman).—

1. Prepare nutrient agar (*vide* page 167), to following formula:

Agar	2.0 per cent.
Peptone	1.5 per cent.
Salt	0.5 per cent.
Meat extract	*quantum sufficit.*

2. Make reaction of medium + 10.

3. Filter; tube in quantities of 6 c.c.

4. Sterilise as for nutrient agar.

5. After the third sterilisation cool the tubes to 42°C., and add to each 3 c.c. of sterile hydrocele fluid, ascitic fluid, or pleuritic effusion (previously sterilised, if necessary, by the fractional method); allow the tubes to solidify in a sloping position.

6. When solid, incubate at 37° C. for forty-eight hours, and eliminate any contaminated tubes. Store the remainder for future use.

Serum Agar (Wertheimer).—

1. Prepare nutrient agar (*vide* page 167), to the following formula:

Agar	2.0 per cent.
Peptone	2.0 per cent.
Salt	0.5 per cent.
Meat extract	*quantum sufficit.*

2. Make reaction of medium +10.

3. Filter; tube in quantities of 5 c.c.

4. Sterilise as for nutrient agar.

5. After the last sterilisation cool to 42°C., then add 5 c.c. sterile blood-serum from human placenta (sterilised, if necessary, by the fractional method) to each tube; slope the tubes.

6. When solid, incubate at 37° C. for forty-eight hours, and eliminate any contaminated tubes. Store the remainder for future use.

Serum Agar (Kanthack and Stevens).—

1. Collect ascitic, pleuritic, or hydrocele fluid in sterile flasks and allow to stand in the ice-chest for twelve hours to sediment.

2. Decant 1000 c.c. of the clear fluid into a measuring cylinder and transfer to sterile litre flask.

3. Add 0.5 c.c. dekanormal NaOH solution for every 100 c.c. serum (*i. e.*, 5.0 c.c.), and mix thoroughly.

4. Heat in the steamer for twenty minutes.

5. Weigh out 15 grammes agar, emulsify in a separate vessel with 200 c.c. of the alkaline fluid previously cooled to about 20°C., and then add to the remainder of the fluid in the flask.

6. Bubble live steam through the mixture for twenty minutes to dissolve the agar.

7. Filter through papier Chardin, using a hot-water funnel.

8. Weigh out glucose 10 grammes (= 1 per cent.), and dissolve it in the clear agar.

8*a.* If desired, add glycerine, 5 per cent., to the clear agar.

9. Tube, and sterilise as for nutrient agar.

Serum Agar (Libman).—

1. Prepare nutrient agar (*vide*, page 167) using, however, 1.5 per cent. peptone (that is 15 grammes per litre instead of 10 grammes).
2. Adjust the reaction to 0 (*i. e.*, neutral to phenolphthalein).
3. Filter and transfer 1000 c.c. liquefied medium to a sterile flask.
4. Weigh out dextrose 20 grammes and dissolve in the fluid agar.
5. Tube in quantities of 6 c.c.; and sterilise in the steamer at 100° C. for thirty minutes on each of three consecutive days.
6. After the third sterilisation cool to 42° C. and add to each tube 3 c.c. of sterile hydrocele fluid, ascitic fluid or pleuritic effusion (previously sterilised, if necessary, by the fractional method); allow the tubes to solidify in a sloping position.
7. When solid, incubate at 37° C. for forty-eight hours, and eliminate any contaminated tubes. Store the remainder for future use.

Egg-albumen, Inspissated.—
1. Break several fresh eggs (hens', ducks', or turkeys' eggs), and collect the "whites" in a graduated cylinder, taking care to avoid admixture with the yolks.
2. Add 40 per cent. distilled water, and incorporate the mixture thoroughly by the aid of an egg-whisk.
3. Weigh out 0.15 per cent. sodium hydrate and dissolve it in the fluid (or add the amount of dekanormal caustic soda solution calculated to yield the required percentage of soda in the total bulk of the fluid—*i. e.*, 0.375 c.c. of dekanormal NaOH solution per 100 c.c. of the mixture).

3a. Glucose to the extent of 1 to 2 per cent. may now be added, if desired.
4. Strain the mixture through butter muslin and filter through a porcelain filter candle into a sterile filter flask.
5. Tube, and stiffen at 100° C. in the serum inspissator.
6. Incubate at 37° C. for forty-eight hours and eliminate any contaminated tubes; store the remainder for future use.

Egg-albumen (Tarchanoff and Kolesnikoff).—
1. Place unbroken hens' eggs in dekanormal caustic soda solution for ten days. (After this time the white becomes firm like gelatine.)
2. Carefully remove the shell and cut the egg into fine slices.
3. Wash for two hours in running water.
4. Place the egg slices in a large beaker and sterilise in the steamer at 100° C. for one hour.
5. Transfer each slice of egg by means of a pair of sterilised forceps to a Petri dish or large capsule.
6. Sterilise in the steamer at 100° C. for twenty minutes on each of three consecutive days.

Egg Albumin Broth (Lipschuetz).—
1. Weigh out

Egg albumin (extra fine powder, Merck).   4 grammes

and place in a 2-litre flask with a number of sterile glass beads.
2. Measure out distilled water 200 c.c. into a half-litre flask and warm to 37° C. in the incubator.
3. Add the water to the flask containing the albumin and beads and dissolve by shaking.
4. Add n/10-NaOH, 40 c.c. Allow the mixture to stand for thirty minutes with frequent shaking.
5. Filter through Swedish filter paper.
6. Sterilise by boiling two or three times at intervals of two hours.
7. Add ordinary nutrient bouillon 600 c.c.
8. Fill into small Erlenmeyer flasks in quantities of 50 c.c.
9. Incubate for forty-eight hours at 37°C.—discard any contaminated flasks and store the remainder for future use.

Egg Albumin Agar.—
1. Prepare egg albumin solution as above 1-6.
2. Liquefy and measure out ordinary nutrient agar 600 c.c. and add to the egg albumin solution (in place of the nutrient broth).
3. Complete as above 8-9.

Diplococcus Meningitidis Intracellularis.

Ascitic Fluid Agar (Wassermann) *Synonym* **N-as-gar (Mervyn Gordon).**
1. Liquefy and measure out into a sterile flask:

Nutrient agar   600 c.c.

2. Measure out into a half litre flask

Distilled water   210 c.c.

and add to it

Ascitic fluid   90 c.c.

Nutrose   6 grammes

3. Heat over a bunsen flame, shaking constantly until the fluid boils, and the nutrose is dissolved.
4. Add the nutrose ascitic solution to the fluid agar.
5. Heat in the steamer for thirty minutes, then filter.
6. Tube and sterilise as for nutrient agar.

Note.—The finished medium in this case measures 900 c.c. only since inconvenient fractions would be introduced in making up to one litre exactly.

Diplococcus Pneumoniæ.

Blood Agar (Washbourn).—
1. Melt up several tubes of nutrient agar (*vide* page 167) and allow them to solidify in the oblique position.

2. Place the tubes, in the horizontal position, in the "hot" incubator for forty-eight hours, to evaporate off some of the condensation water.

3. Kill a small rabbit with chloroform and nail it out on a board (as for a necropsy). Moisten the hair thoroughly with 2 per cent. solution of lysol.

4. Sterilise several pairs of forceps, scissors, etc., by boiling.

5. Reflect the skin over the thorax with sterile instruments.

6. Open the thoracic cavity by the aid of a fresh set of sterile instruments.

7. Open the pericardium with another set of sterile instruments.

8. Sear the surface of the left ventricle with a red-hot iron and remove fluid blood from the heart by means of sterile pipettes (*e. g.*, those shown in Fig. 13, *c*).

9. Deliver a small quantity of the blood on the slanted surface of the agar in each of the tubes, and allow it to run over the entire surface of the medium.

10. Place the tubes in the slanting position and allow the blood to coagulate.

11. Return the "blood agar" to the hot incubator for forty-eight hours and eliminate any contaminated tubes. Store the remainder for future use.

Media for the Study of Mouth Bacteria Generally.

Potato Gelatine (Goadby).—

1. Prepare glycerine potato broth (see page 203, sections 1 to 5).

2. Add 10 per cent. gelatine to the potato decoction and bubble live steam through the mixture for ten minutes.

3. Estimate the reaction; adjust the reaction of the medium to +5.

4. Cool the medium to below 60°C., clarify with egg as for nutrient gelatine.

5. Filter through papier Chardin.

6. Tube, and sterilise as for nutrient gelatine.

Media for the Study of Protozoa.

Tissue Medium (Noguchi).—*For spirochætes (cultivations must be grown anaerobically).*

1. Plug and sterilise test-tubes 20 × 2 cm.

2. Kill a small rabbit with chloroform vapour. Open the abdomen with all aseptic precautions, remove kidneys and testicles and transfer to a sterile glass dish. Cut up the organs with sterile scissors into small pieces—say 4 millimetre cubes. The four organs should yield from 25 to 30 pieces of tissue.

3. Drop a small piece of sterile tissue into the bottom of each sterilised tube.

4. Take a flask containing about 400 c.c. nutrient agar (+10 reaction), liquefy the medium by heat and cool in a water bath to 50°C.

5. Add 200 c.c. ascitic or hydrocele fluid (horse or sheep serum may be employed, but is not so good) to the liquid agar and mix carefully to avoid formation of air bubbles.

6. Fill about 20 c.c. of the ascitic agar into each of the sterilised tubes which already contains a piece of sterile rabbit's tissue, stand all the tubes upright in racks or a jar, and allow agar to set.

7. After solidification pour sterile paraffin oil on the surface of the medium in each tube to the depth of 3 centimetres.

8. Incubate tubes at 37° C. for several days and discard any which prove to be contaminated.

9. Store such tubes as are sterile for future use.

XIII. INCUBATORS.

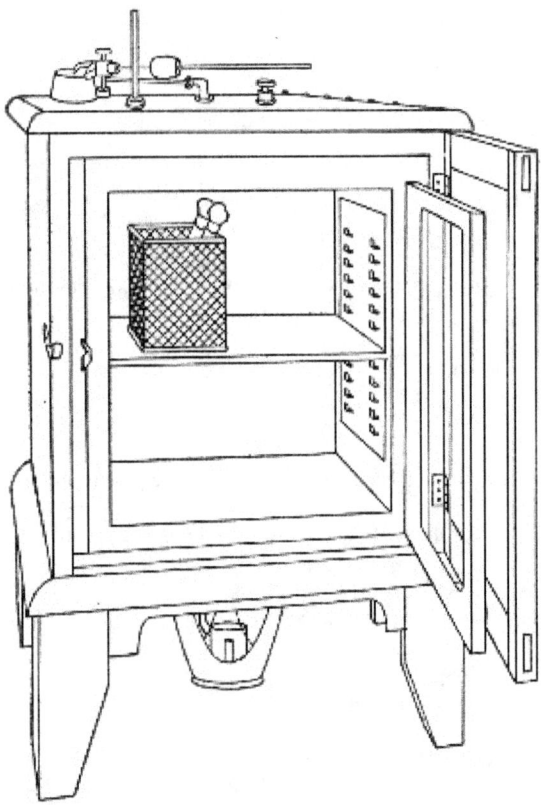

Fig. 113.—Incubator.

An incubator (Fig. 113) consists essentially of a chamber for the reception of cultivations, etc., surrounded by a water jacket, the walls of which are of metal, usually copper, and outside all an asbestos or felt jacket, or wooden casing. The water in the jacket is heated by gas or electricity and maintained at some constant temperature by a thermo-regulator. The cellular incubator (Fig. 114) which was made for me some years ago is of the greatest practical utility. Here the central cavity is subdivided by five double-walled partitions (in which water circulates in connection with the water tanks at the top and base of the incubator) and again by iron shelves to form twenty-four pigeon holes. Into each of these slides an iron drawer 35 cm. long × 12 cm. wide × 22 cm. high forming a self-contained incubator. The drawer is fitted with a wooden form to which is fixed a handle and a numbered label. The thermo-regulating apparatus is the well-known Hearson capsule.

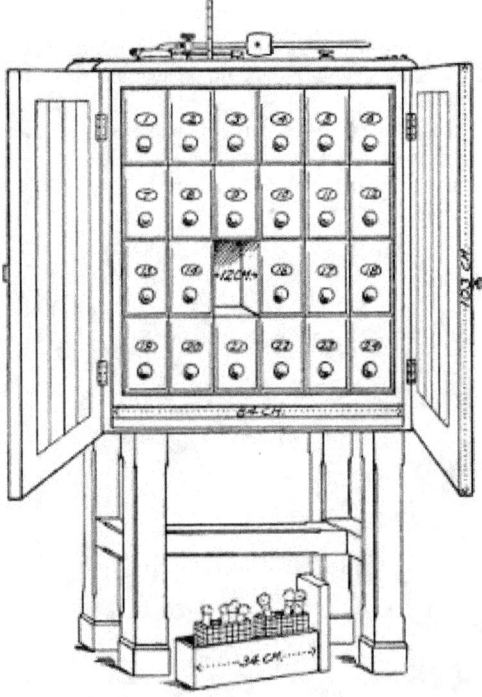

Fig. 114.—Cellular incubator.

Two incubators at least are required in the laboratory, for the cultivation of bacteria the one regulated to maintain a temperature of 37°C., and known as the "hot" incubator; the other, 20° C. to 22°C., and known as the "cool" or "cold" incubator.

Two other incubators, regulated to 42° C. and 60°C. respectively, whilst not absolutely, necessary very soon justify their purchase.

Thermo-regulators.—The thermo-regulator is the most essential portion of the incubator, as upon its efficient working depends the maintenance of a constant temperature in the cultivation chamber. It is also used in the fitting up of water and paraffin baths, and for many other purposes.

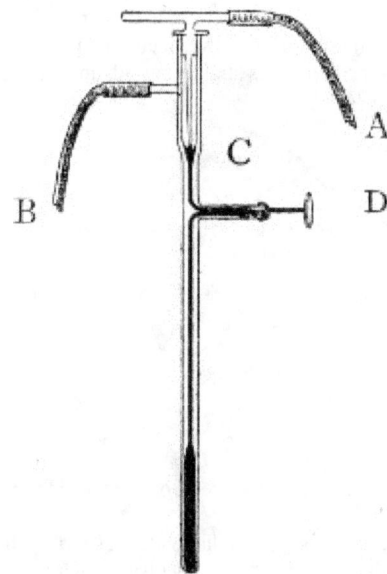

Fig. 115.—Reichert's thermo-regulator.

Of the many forms and varieties of thermo-regulator (other than electrical), two only are of sufficiently general use to need mention. In one of these the flow of gas to the gas-jet is controlled by the expansion or contraction of mercury within a glass bulb; in the other, by alterations in the position of the walls of a metallic capsule containing a fluid, the boiling-point of which corresponds to the temperature at which the incubator is intended to act. They are:

(*a*) *Reichert's* (Fig. 115), consists of a bulb containing mercury which is to be suspended in the medium, whether air or water, the temperature of which it is desired to regulate. Gas enters at A, and passes out to the jet by B. As the temperature rises the mercury expands and cuts off the main gas supply. As the temperature falls the mercury contracts and reopens the narrow tube C. By means of a thumbscrew D (which mechanically raises or lowers the column of mercury irrespective of the temperature) and the aid of a thermometer the apparatus can be set to keep the incubator at any desired temperature. With this form a special gas burner is required, with separate supply of gas to a pilot jet at the side.

(*b*) *Hearson's capsule regulator* consists of a metal capsule hermetically sealed and filled with a liquid which boils at the required temperature, this is adjusted in the interior of the incubator. Soldered to the upper side of the capsule is a thick piece of metal having a central cup to receive the lower end of a rigid rod, through which the movements of the walls of the capsule are transmitted to the gas valve fixed outside the incubator.

The gas valve or governor is shown in figure 116. A is the inlet for gas, C the outlet to burner heating the water jacket, B D a lever pivoted to standards at G, and acted upon by the capsule, through the rigid rod which enters the socket below the screw P.

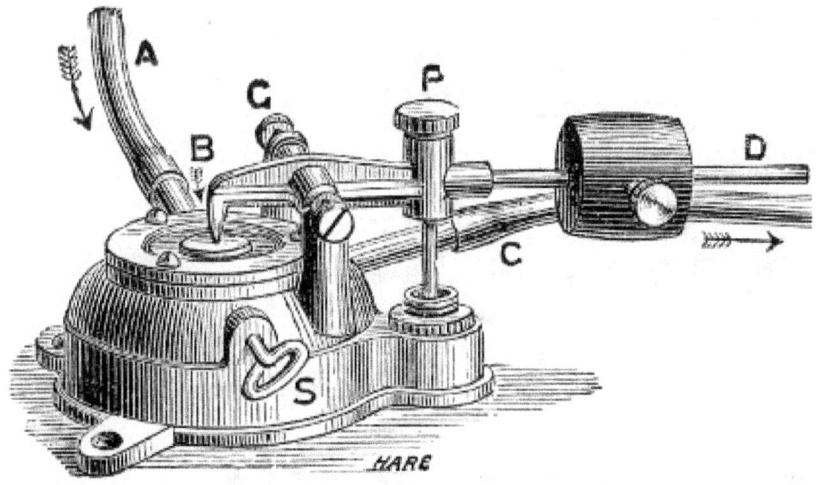

Fig. 116.—Capsule thermo-regulator.

The construction of the valve is such that, whenever the short arm of the lever B D presses on the disc below the end B, the main supply of gas is entirely cut off. At such times, however, a very small portion of gas passes from A to C, through an aperture inside the valve, the size of which aperture can be adjusted by the screw needle S, hence the gas flame below the incubator is never extinguished.

The expansion of the metal walls of the capsule, which takes place upon the boiling of its contents, provides the motive force, transmitted through the rigid rod to raise the long arm of the lever B D, and as this expansion only takes place at a predetermined temperature, the lever will only be acted upon when the critical temperature is reached, no sensible effect being produced at even 1° C. below that at which the capsule is destined to act.

W is a weight sliding on the lever rod D; by increasing the distance between the weight and the fulcrum of the lower increased pressure is brought to bear upon the walls of the capsule with the result that the boiling-point of the liquid in the capsule is slightly raised, and a range of about two degrees can thus be obtained with any particular capsule.

FOOTNOTES:

Made by the firm of Chas. Hearson & Co., 235 Regent St., London, W.

XIV. METHODS OF CULTIVATION.

Cultivations of micro-organisms are usually prepared in the laboratory in one of three ways:
Tube cultures. Plate cultures. Hanging-drop cultures.

These may be incubated either **aerobically** (*i. e.,* in the presence of oxygen) or **anaerobically** (*i. e.,* in the absence of oxygen, or in the presence of an indifferent gas, such as hydrogen, nitrogen, or carbon dioxide).

With regard to the temperature at which the cultivations are grown, it may be stated as a general rule that all media rendered solid by the addition of gelatine are incubated at 20°C., or at any rate at a temperature not exceeding 22° C. (that is, in the "cold" incubator); whilst fluid media and all other solid media are incubated at 37° C. (that is, in the "hot" incubator). Exceptions to this rule are numerous. For instance, in studying the growth of the psychrophylic bacteria, the yeasts and the moulds, the cold incubator is employed for all media.

Tube cultivations are usually packed in the incubator in small tin cylinders, such as those in which American cigarettes are sold, or in square tin boxes. Beakers or tumblers may be used for the same purpose, but being fragile are not so convenient. Metal test-tube racks, long enough to just fit into the interior of the incubator and each accommodating two rows of tubes, are also exceedingly useful.

AEROBIC.

The Preparation of Tube Cultivations.

The preparation of a tube cultivation consists in:

(*a*) Inoculating a tube of sterile nutrient medium with a portion of the material to be examined.

(*b*) Incubating the inoculated tube at a suitable temperature.

The details of the first of these processes must be varied somewhat according to whether the tubes of nutrient media are inoculated or "planted" from—

1. Pre-existing cultivations.
2. Morbid material previously collected (*vide* page 373).
3. Fluids, tissues, etc., or from the animal body direct.

The method of preparing tube cultivations from pre-existing cultivations is as follows:

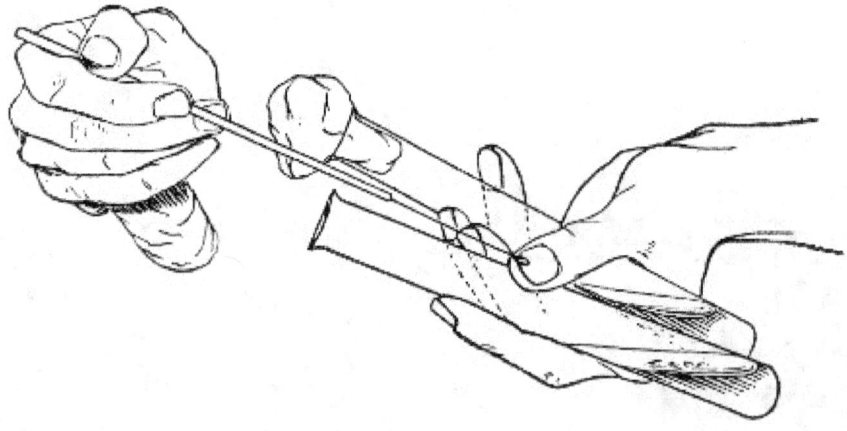

Fig. 117.—Inoculating tubes, seen from the front.

1. Fluid Media (*e. g.,* Nutrient Bouillon).—

1. Flame the cotton-wool plug of the tube containing the cultivation and also that of the tube of sterile bouillon.

2. Hold the two tubes, side by side, between the left thumb and the first and third fingers, allowing the sealed ends to rest on the dorsum of the hand, and separating the mouths of the tubes (which are pointed to the right) by the tip of the second finger. Keep the tubes as nearly horizontal as is possible without allowing the fluid in the bouillon tube to reach the cotton-wool plug (Fig. 117).

3. Sterilise the platinum loop and allow it to cool.

4. Grasp the plug of the tube containing the cultivation between the little finger and palm of the hand and remove it from the tube.

5. Grasp the plug of the bouillon tube between the fourth finger and the ball of the thumb and remove it from the tube.

6. Pass the platinum loop into the tube containing the culture—do not allow the loop to touch the sides of the tube, or the handle to touch the medium—and remove a small portion of the growth; withdraw the loop from the tube, keeping the infected side of the loop downward.

7. Pass the loop into the bouillon tube almost down to the level of the fluid, reverse the loop so that the infected side faces upward, emulsify the portion of the growth in the moisture adhering to the side of the tube which is uppermost. Withdraw the loop.

8. Replug both tubes.

9. Sterilise the platinum loop.

10. Label the bouillon tube with (*a*) the name of the organism and (*b*) the date of inoculation.

11. Incubate.

2. Solid Media.—Solid media are stored in tubes in one of two ways:

1. Oblique tube or slanted tube (Fig. 118), in which the medium has been allowed to solidify whilst the tube was retained in an inclined position, so forming an extensive surface of medium extending from the bottom of the tube almost to its mouth.

This is employed for "streak" or "smear" cultivations (*Strichcultur*).

2. Straight tube (Fig. 119), in which the medium forms a cylindrical mass in the lower portion of the tube and presents an upper surface which is at right angles to the long axis of the tube.

This is employed for "stab" or "stick" cultivations (*Stichcultur*), or by inoculating the medium whilst fluid, and allowing to solidify in this position, for "shake" cultivations.

Streak Culture.—

1. Flame the plugs, sterilise the platinum loop (or spatula). Open the tubes and charge the loop as in previous inoculation.

2. Pass the infected loop to the bottom of the tube to be inoculated and draw it, as lightly as possible, along the centre of the surface of the medium, terminating the "streak" over the thin layer of medium near the mouth of the tube.

3. Replug the tubes, sterilise the platinum loop.

4. Label the newly inoculated tube and incubate.

Smear Culture.—Proceed generally as in streak culture, but rub the infected loop all over the surface of the medium, instead of restricting the inoculation to a narrow line.

Note.—Gelatine and agar oblique tubes should be freshly "slanted" before use.

Stab Culture.—

1. Flame the plugs, open the tubes, sterilise the platinum needle and charge it with the inoculum as in the previous cultivations.

2. Pass the platinum needle into the tube to be inoculated until it touches the centre of the surface of the medium. Now thrust it deeply into the substance of the medium, keeping the needle as nearly as possible in the axis of the cylinder of medium. Then withdraw the needle.

3. Replug the tubes. Sterilise the platinum needle.

4. Label the newly planted tube and incubate.

Note.—When gelatine is stored for some time the upper surface of the cylinder becomes concave owing to evaporation. Tubes showing this appearance should be liquefied and again allowed to set before use for stab culture, otherwise when the needle enters the medium, the surface tension will cause the gelatine cylinder to split.

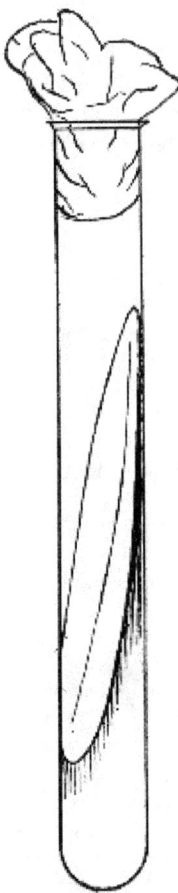

Fig. 118.—Sloped or slanted medium for streak or smear culture.

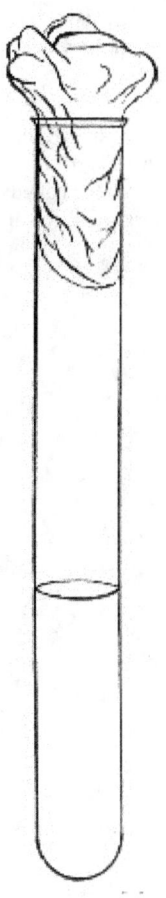

Fig. 119.—Straight tube.

Shake Culture.—
1. Liquefy a tube of nutrient gelatine (or agar, or other similar medium), by heating in a water-bath (Fig. 121).
2. Inoculate the liquefied medium and label it, etc., precisely as if dealing with a tube of bouillon.
3. Place the newly planted tube in the upright position (*e. g.*, in a test-tube rack) and allow it to solidify.
4. Label the tube; when solid, incubate.

Esmarch's Roll Cultivation.—
1. Liquefy three tubes of gelatine by heat.
2. Prepare three dilutions of the inoculum (as described for plate cultivations, page 228, steps 4 to 7).
3. Roll the tubes, held almost horizontally, in a groove made in a block of ice, until the gelatine has set in a thin film on the inner surface of tube (Fig. 120); or under the cold-water tap.

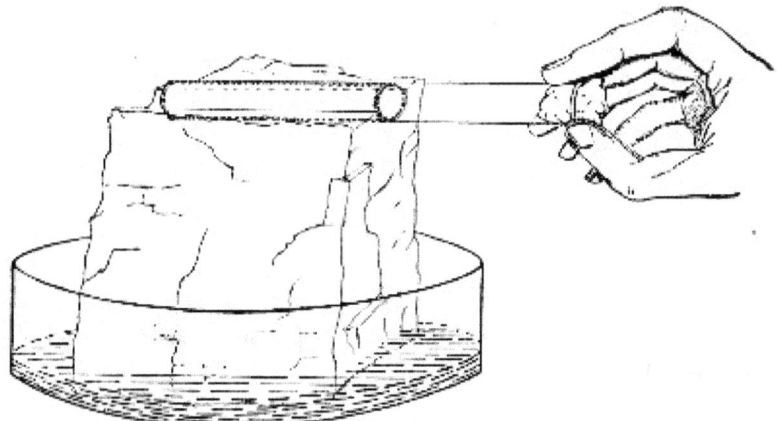

Fig. 120. Esmarch's roll culture on block of ice.

In order that the medium may adhere firmly to the glass, the agar used for roll cultivation should have 1 per cent. gelatine or 1 per cent. gum arabic added to it before sterilisation.

Roll cultivations, which served a most important purpose in the days before the introduction of Petri dishes for plate cultivations, are now obsolete in modern laboratories and are merely mentioned for the benefit of students, since examiners who are interested in the academic and historical aspects of bacteriology sometimes expect candidates to be acquainted with the method of preparing them.

The Preparation of Plate Cultures.

If a small number of bacteria are suspended in liquefied gelatine, agar, or other similar medium, and the infected medium spread out in an even layer over a flat surface and allowed to solidify, each individual micro-organism becomes fixed to a certain spot and its further development is restricted to the vicinity of this spot. After a variable interval the growth of this organism becomes visible to the naked eye as a "colony." This is the principle upon which the method of plate cultivation is based and its practice enables the bacteriologist to study the particular manner of development affected by each species of microbe when growing (*a*) unrestricted upon the surface of the medium, (*b*) in the depths of the medium. The method itself is as follows:

Apparatus Required.—

1. Tripod levelling stand.
2. Large shallow glass dish, with a square sheet of plate glass to cover it.
3. Spirit level.
4. Case of sterile Petri dishes.
5. Tubes of sterile nutrient media, gelatine (or agar) previously liquefied by heating in the water-bath and cooled to 42°C., otherwise the heat of the medium would destroy many, if not all, of the bacteria introduced.
6. Tube of cultivation to be planted from.
7. Platinum loop.
8. Bunsen burner.
9. Grease pencil.

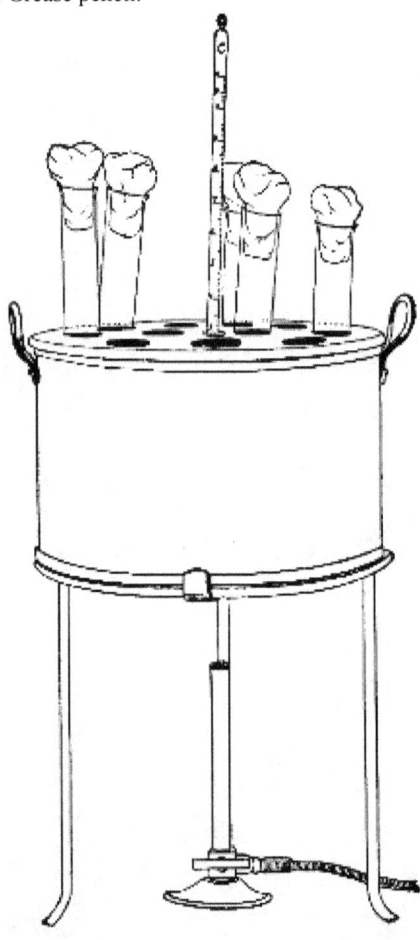

Fig. 121.—Handy form of water-bath for melting tubes of agar and gelatine previous to slanting them; or to making shake cultures or pouring plates.

Method of "Pouring" Plates.—

1. Place the glass dish on the levelling tripod (Figs. 122, 123); if gelatine plates are to be poured fill the dish with ice water—gelatine solidifies so slowly that it is necessary to hasten the process; if agar is to be used fill with water at 50°C.—agar sets almost immediately at the room temperature and by slightly retarding the process lumpiness is avoided; cover the dish with the square sheet of glass.

2. Place the spirit level on the sheet of glass and by means of the levelling screws adjust the surface of the glass to the horizontal.

This leveling is an important matter since the development of a colony is to some extent proportionate to the supply of medium available for its nutrition. Thus in a "smear" on sloped tube culture, the colonies at the upper part of the medium are stunted and small but increase in size and luxuriance of growth the nearer they approach to the bottom of the tube, where there is the greatest depth of medium.

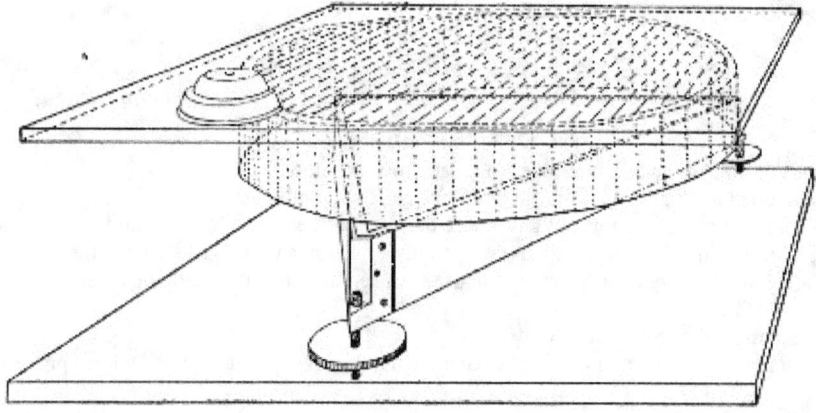

Fig. 122.—Plate-levelling stand.

3. Place three sterile Petri dishes in a row on the surface of the glass plate and number them 1, 2, and 3, from left to right.

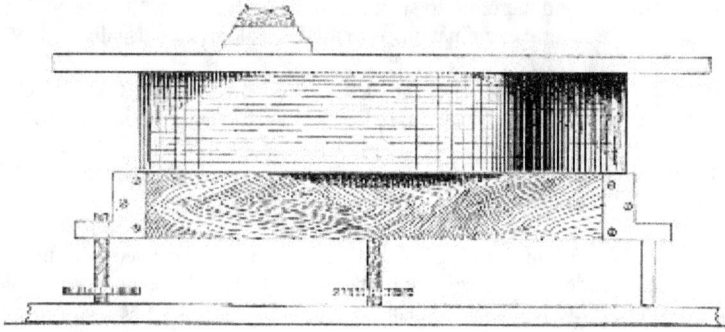

Fig. 123.—Plate-levelling stand, side view.

4. Number the previously liquefied tubes of nutrient media 1, 2, and 3. Flame the plugs and see that each plug can be readily removed from the mouth of its tube.

5. Add one loopful of the inoculum to tube No. 1, treating the liquefied medium as bouillon. After replugging, grasp the tube near its mouth by the thumb and first finger of the right hand, and with an even circular movement of the whole arm, diffuse the inoculum throughout the medium; avoid jerky movements, as these cause bubbles of air to form in the medium.

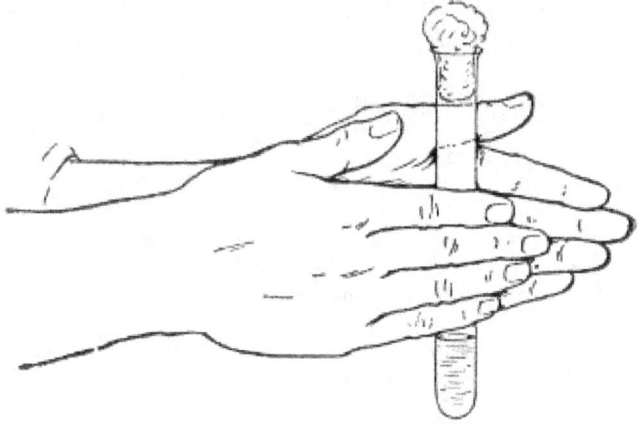

Fig. 124.—Mixing emulsion for plates.

The knack of mixing evenly without producing air bubbles, is not always easily acquired, by this method. An alternative plan is to hold the inoculated tube vertically upright between the opposed palms and to rotate it between them by rapid backward and forward movements of the two hands (Fig. 124).

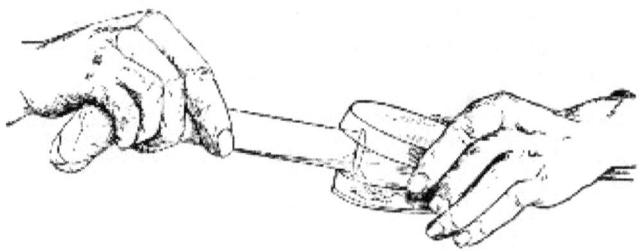

Fig. 125.—Pouring plates.

6. Sterilise the platinum loop, and add two loopfuls of diluted inoculum to tube No. 2, and mix as before.

7. In a similar manner transfer three loopfuls of liquefied medium from tube No. 2 to tube No. 3, and mix thoroughly.

8. Flame the plug of tube No. 1, remove it, then flame the lips of the tube; slightly raise the cover of Petri dish No. 1, introduce the mouth of the tube; then, elevating the bottom of the tube, pour the liquefied medium into the Petri dish, to form a thin layer. Remove the mouth of the tube and close the "plate." If the medium has failed to flow evenly over the bottom of the plate, raise the plate from the levelling platform and by tilting in different directions rectify the fault.

9. Pour plates No. 2 and No. 3, in a similar manner, from tubes Nos. 2 and 3.

10. Label the plates with the distinctive name or number of the inoculum, also the date; the number of the dilution having been previously indicated (step 3).

11. Place in the cool incubator for three or more days, as may be necessary.

In this way colonies may be obtained quite pure and separate from each other.

In plate No. 1, probably, the colonies will be so numerous and crowded, and therefore so small, as to render it useless. In plate No. 2 they will be more widely separated, but usually No. 3 is the plate reserved for careful examination, as in this the colonies are usually widely separated, few in number, and large in size.

Agar plates are poured in a similar manner, but the agar must be melted in boiling water and then allowed to cool to 45° C. or 42° C. in a carefully regulated water-bath before being inoculated, and the entire process must be carried out very rapidly, otherwise the agar will have solidified before the operation is completed.

Note.—In pouring plates, since tube No. 1 (for the first dilution) rarely gives a plate that is of any practical value it is frequently replaced by a tube of bouillon or sterile salt solution, and in such case plate No. 1 is not poured.

Surface Plates.—

This method of pouring what may be termed "whole" plates (since colonies may appear both on the surface and in the depths of the medium) is essential to the accurate study of the formation of colonies under various conditions, but when the main object of the separation of the bacteria is to obtain subcultivations from a number of individual bacteria, "surface" plates must be prepared, since here colony formation is restricted to the surface of the medium. The method adopted varies slightly according to whether the medium employed is gelatine or agar, or one of the derivatives or variants of the latter.

(a) Gelatine Surface Plates.—

1. Liquefy three tubes of nutrient gelatine.

2. Pour each tube into a separate Petri dish and allow it to solidify. Then turn each plate and its cover upside down.

Fig. 126.—Surface plate spreader.

3. When quite cold raise the bottom of plate 1, revert it and deposit a drop of the inoculum (whether a fluid culture or an emulsion from solid culture) upon the surface of the gelatine with a platinum loop—close to one side of the plate; replace the bottom half of the Petri dish in its cover.

4. Take a piece of thin glass rod, stout platinum wire or best of all a piece of aluminium wire (say 2 mm. diameter) about 28 cm. long. Bend the terminal 4 cm. at right angles to the remainder, making an L-shaped rod (Fig. 126). Sterilise the short arm and adjacent portion of the long arm, in the Bunsen flame, and allow it to cool.

5. Now raise the bottom of the Petri dish in the left hand, leaving the cover on the laboratory bench, and holding it vertically, smear the drop of inoculum all over the surface of the gelatine with the short arm of the spreader by a rotatory motion, (Fig. 127). Replace the dish in its cover.

6. Raise the bottom of plate 2 and rub the infected spreader all over the surface of the gelatine—then go on in like manner to the third plate in the series.

7. Sterilise the spreader.

8. Label and incubate the plates.

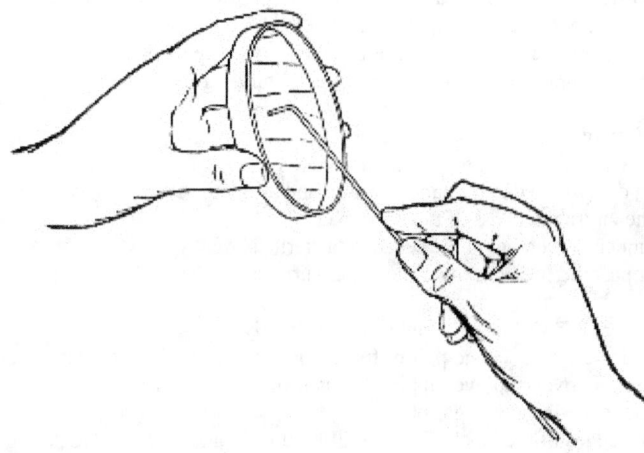
Fig. 127.—Spreading surface plate.

After incubation, plate No. 1 will probably yield an enormous number of colonies; plate 2 will show fewer colonies, since only those bacteria adhering to the rod after rubbing over plate 1 would be deposited on its surface, and by the time the rod reached plate 3 but very few organisms should remain upon it. So that the third plate as a rule will only show a very few scattered colonies, eminently suitable for detailed study.

(b) Agar Surface Plates.—
1. Liquefy three tubes of nutrient agar—nutrose agar or the like.
2. Pour each tube into a separate Petri dish and allow it to solidify.
3. When quite solid invert each dish, raise the bottom half and rest it obliquely on its inverted cover (Fig. 128) and place it in this position in an incubator at 60° C. for forty-five minutes (or in an incubator at 42° C. for two hours). This evaporates the water of condensation and gives the medium a firm, dry surface.
4. On removing the plates from the incubator close each dish and place it—still upside down—on the laboratory bench.

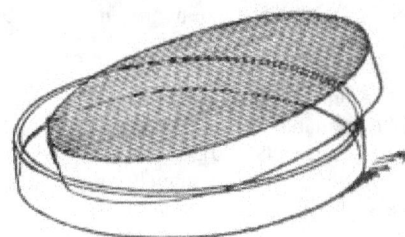
Fig. 128.—Drying surface plate of agar.

5. Inoculate the plates in series of three, as described for gelatine surface plates 3-8.
Hanging-drop Cultivation.

Apparatus Required.—
Hanging-drop slides. Cover-slips. Section rack (Fig. 75). Blotting paper. Bell glass to cover slides. Original culture. Tubes of broth, or liquefied gelatine or agar. Forceps. Platinum loop. Bunsen burner. Grease pencil. Sterile vaseline. Lysol.

(a) Fluid Media.—
1. Prepare first and second dilutions of the inoculum as directed for plate cultivations (*vide* pages 228-229, sections 4 to 6), substituting tubes of nutrient broth for the liquefied gelatine.
2. Sterilise a hanging-drop slide by washing thoroughly in water and drying, then plunging it into a beaker of absolute alcohol, draining off the greater part of the spirit, grasping the slide in a pair of forceps, and burning off the remainder of the alcohol in the flame.
3. Place the hanging-drop slide on a piece of blotting paper moistened with 2 per cent. lysol solution and cover it with a small bell glass that has been rinsed out with the same solution and *not dried*.
4. Raise the bell glass slightly and smear sterile vaseline around the rim of the metal cell by means of a sterile spatula of stout platinum wire.
5. Remove a clean cover-slip from the alcohol pot with sterile forceps and burn off the alcohol; again raise the bell glass and place the sterile cover-slip on the blotting paper by the side of the hanging-drop slide.
6. Remove a drop of the broth from the second dilution tube with a large platinum loop; raise the bell glass and deposit the drop on the centre of the cover-slip. Sterilise the loop.
7. Raise the bell glass sufficiently to allow of the cover-slip being grasped with forceps, inverted, and adjusted over the cell of the hanging-drop slide. Remove the bell glass altogether and press the cover-slip firmly on to the cell.
8. Either incubate and examine at definite intervals, or observe continuously with the microscope, using a warm stage if necessary (Fig. 53).

(b) Solid Media.—Observing precisely similar technique, a few drops of liquefied gelatine or agar from the second dilution tube may be run over the surface of the sterile cover-slip and a hanging-drop plate cultivation thereby prepared.

This method is extremely useful in connection with the study of yeasts, when the circular cell on the hanging-drop slide should be replaced by a rectangular cell some 38 by 19 mm., and the gelatine spread over a cover-slip of similar size. After sealing down the preparation, the upper surface of the cover-slip may be ruled into squares by the aid of the grease pencil or a writing diamond and numbered to facilitate the subsequent identification of the colonies which are observed to develop from solitary germs.

Hanging-block Culture (Hill).—

Apparatus required: As for hanging-drop cultivation with the addition of a scalpel.

Carry out the method as far as possible under cover of a bell glass, to avoid aerial contamination.

1. Liquefy a tube of nutrient agar (or gelatine) and pour into a Petri dish to the depth of about 4 mm. and allow to set.
2. With a sharp scalpel cut out a block some 8 mm. square, from the entire thickness of the agar layer.
3. Raise the agar block on the blade of the scalpel and transfer it, under side down, to the centre of a sterile slide.
4. Spread a drop of fluid cultivation (or an emulsion of growth from a solid medium) over the upper surface of the agar block as if making a cover-slip film.
5. Place the slide and block covered by the bell glass in the incubator at 37° C. for ten minutes to dry slightly.
6. Take a clean dry sterile cover-slip in a pair of forceps, and with the help of a second pair of forceps lower it carefully on the inoculated surface of the agar (avoiding air bubbles), so as to leave a clear margin of cover-slip overlapping the agar block.
7. Invert the preparation and with the blade of the scalpel remove the slide from the agar block.
8. With a platinum loop run a drop or two of melted agar around the edges of the block. This solidifies at once and seals the block to the cover-slip.
9. Prepare a sterile hanging-drop slide, and smear hard vaseline or melted white wax on the rim of the metal cell.
10. Invert the cover-slip with the block attached on to the hanging-drop slide, and seal the cover-slip firmly in place.
11. Observe as for hanging-drop cultivations.

ANAEROBIC CULTIVATIONS.

Numerous methods have been devised for the cultivation of anaerobic bacteria, the majority requiring the employment of special apparatus. The principle upon which any method is based and upon which it depends for its success falls under one or another of the following headings:

(a) **Exclusion of air** from the cultivation.

(b) **Exhaustion of air** from the vessel containing the cultivation by means of an air pump—*i. e.*, cultivation *in vacuo*.

(c) **Absorption of oxygen** from the air in contact with the cultivation by means of pyrogallic acid rendered alkaline with caustic soda—*i. e.*, cultivation in an atmosphere of nitrogen.

(d) **Displacement of air** by an indifferent gas, such as hydrogen or coal gas—*i. e.*, cultivation in an atmosphere of hydrogen.

(e) A combination of two or more of the above methods.

A selection of the simplest and most generally useful methods is given here.

Whenever possible, the nutrient media that are employed in any of the processes should contain some easily oxidisable substance, such as sodium formate (0.4 per cent.) or sodium sulphindigotate (0.1 per cent.), which will absorb all the available oxygen held in solution by the medium. The further addition of glucose, 2 per cent., favors the growth of anaerobic bacteria (*vide*, pages 189-190).

Further, it is advisable to seal all joints between india-rubber stoppers and tubulures or the mouths of the tubes with melted paraffin; glass stoppers and taps should be lubricated with resin ointment or a mixture of beeswax 1 part, olive oil 4 parts.

(A) **Method I** (Hesse's Method).—

1. Make a stab culture in gelatine or agar, choosing for the purpose a straight tube containing a deep column of medium, and thrusting the inoculating needle to the bottom of the tube.
2. Pour a layer of sterilised oil (olive oil, vaseline, or petroleum), 1 or 2 cm. deep, upon the surface of the medium.
3. Incubate.

Method II.—This method is only available when dealing with pure cultivations.

1. Liquefy a tube of gelatine (or agar) by heat, pour it into a Petri dish, and allow it to solidify.
2. Inoculate the surface of the medium in one spot only.
3. Remove a cover-slip from the pot of absolute alcohol with sterile forceps; burn off the alcohol in the gas flame.
4. Lower the now sterile cover-slip carefully on to the inoculated surface of the medium, carefully excluding air bubbles, and press it down firmly with the points of the forceps. (A sterile disc of mica may be substituted for the cover-slip.)
5. Incubate.

Method III (Roux's Physical Method).—

1. Prepare tube cultures of fluid media (or solid media rendered fluid by heat) in the usual way.
2. Aspirate some of the inoculated media into capillary pipettes.
3. Seal both ends of each pipette in the blowpipe flame.
4. Incubate.

Method IV (Roux's Biological Method).—

1. Plant a deep stab, as in method I.
2. Pour a layer, 1 or 2 cm. deep, of broth cultivation of a vigourous aerobe—*e. g.*, B. aquatilis sulcatus or B. prodigiosus—upon the surface of the medium; or an equal depth of liquefied gelatine, which is then inoculated with the aerobic organism.
3. Incubate.

The growth of the aerobe will use up all the oxygen that reaches it and will not allow any to pass through to the medium below, which will consequently remain in an anaerobic condition.

(B) **Method V.**—

1. Prepare tube or flask cultivations in the usual way.

2. Replace the cotton-wool plug by an india-rubber stopper perforated with one hole and fitted with a length of glass tubing which has a constriction about 3 cm. above the stopper and is then bent at right angles (Fig. 129). The stopper and glass tubing are sterilised by being boiled in a beaker of water for five minutes.

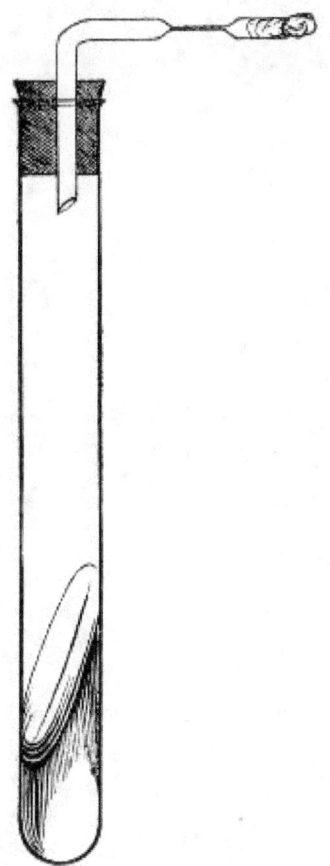

Fig. 129.—Vacuum culture.

3. Connect the tube leading from the culture vessel with a water or air pump, interposing a Wulff's bottle fitted as a wash-bottle and containing sulphuric acid.
4. Exhaust the air from the culture vessel.
5. Before disconnecting the apparatus, seal the glass tube from the culture vessel at the constriction, using the blowpipe flame.
6. Incubate.

(C) **Method VI** (Buchner's Method).

Apparatus and Solutions Required.—
Buchner's tube (a stout glass test-tube 23 cm. long and 4 cm. in diameter, fitted with india-rubber stopper, Fig. 130).
Pyrogallic acid in compressed tablets each containing 1 gram.
Dekanormal solution of caustic soda.

Method.—
1. Prepare the tube cultivation in the usual way.
2. Moisten the india-rubber stopper of the Buchner's tube with water and see that it fits the mouth of the tube accurately.
3. Remove the stopper from the caustic soda bottle.
4. Drop one of the pyrogallic acid tablets into the Buchner's tube (roughly, use 1 gramme pyrogallic acid for every 100 c.c. air capacity of the receiving vessel).
5. Add about 1 c.c. of the soda solution.
6. Place the inoculated tube inside the Buchner's tube. The pyrogallic tablet acts as a buffer and prevents damage to either the inoculated tube or the Buchner's tube even should it be slipped in hurriedly.
7. Fit the india-rubber stopper tightly into the mouth of the Buchner's tube.

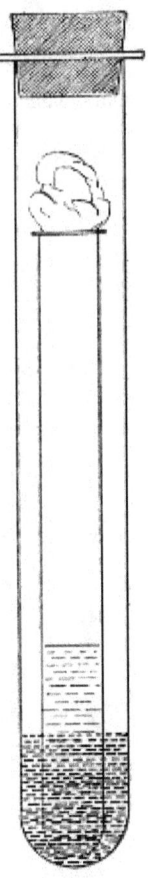

Fig. 130.—Buchner's tube.

The pyrogallic acid tablet dissolves slowly in the soda solution and its oxidation proceeds very slowly at first so that ample time is available when this method is adopted.

8. Restopper the caustic soda bottle.
9. Place Buchner's tube in a wire support, and incubate.

Method VII (Wright's Method).—

1. Prepare tube cultivation in the usual way.
2. Cut off that portion of the cotton-wool plug projecting above the mouth of the tube with scissors, then push the plug into the tube for a distance of 2 or 3 cm.
3. By means of a pipette drop about 1 c.c. of pyrogallic acid 10 per cent. aqueous solution on to the plug. It will immediately be absorbed by the cotton-wool.
4. With another pipette run in an equal quantity of the caustic soda solution.
5. Quickly close the mouth of the tube with a tightly fitting india-rubber stopper.
6. Incubate.

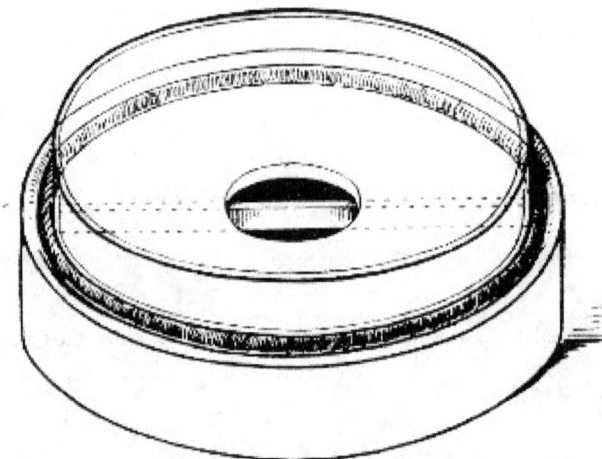

Fig. 131.—McLeod's anaerobic plate base with half petri dish inverted in situ

Method VIII (McLeod's Method).—
Apparatus and Solutions Required.—
McLeod's plate base (a hollow glazed earthenware disc 9 cm. in diameter and 2 cm. deep: the upper surface is pierced by a central hole, 2 cm. in diameter, giving access to the interior, the lower part of which is divided into two by a low partition. A shallow groove encircles the upper surface near to the edge).

Plasticine. Pyrogallic acid (1 gramme) compressed tablets. Sodic hydroxide (0.4 gramme) compressed tablets. Wash bottle of distilled water. Surface plates of one or other agar medium (in petri dishes of 8 cm. diameter). Surface plate spreader.

Method.—
1. Roll out a long cylinder of plasticine and fit it into the groove on the upper surface of the earthenware base.
2. Place a tablet of pyrogallic acid in one division of the interior of the plate base, and two tablets of sodic hydroxide in the other.
3. Prepare surface plate culture of the organism to be cultivated.
4. Run a few cubic centimetres of distilled water into that division of the plate base containing the sodic hydroxide.
5. Invert the bottom half of the surface plate over the plate base and press its edges firmly down into the plasticine filling the groove.
6. Label and incubate.

(D) **Method IX.**—
Apparatus Required.—
Small Ruffer's or Woodhead's flask (Fig. 33). Sterile india-rubber stopper. India-rubber tubing. Glass tubing. Metal screw clips. Cylinder of compressed hydrogen; or hydrogen gas apparatus

Method.—
1. Sterilise a glass vessel, shaped as in a Ruffer's or Woodhead's flask, in the hot-air oven. (The tubulure and the side tubes are plugged with cotton-wool.) After sterilisation, fix a short piece of rubber tubing occluded by a metal clip to each side tube.
2. Inoculate a large quantity (*e. g.*, 200 c.c.) of the medium. Where solid media are employed they must first be liquefied by heat.
3. Remove the cotton-wool plug from the tubulure and pour the inoculated medium into the glass vessel.
4. Close the tubulure by means of an india-rubber stopper previously sterilised by boiling in a beaker of water.

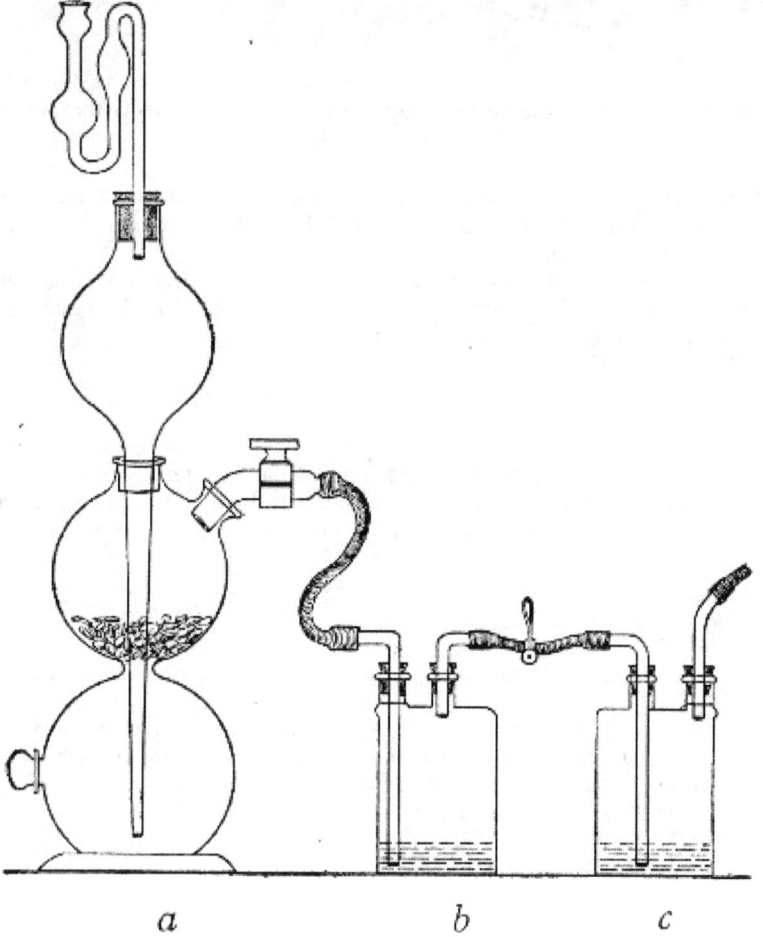

Fig. 132.—Kipp's hydrogen apparatus, (a) connected up to two washing bottles containing (b) lead acetate 10 per cent. solution, to remove H2S and (c) silver nitrate solution to remove AsH3. A third washing bottle containing pyrogallic acid 10 per cent. solution, rendered alkaline, to remove any trace of oxygen, is sometimes introduced.

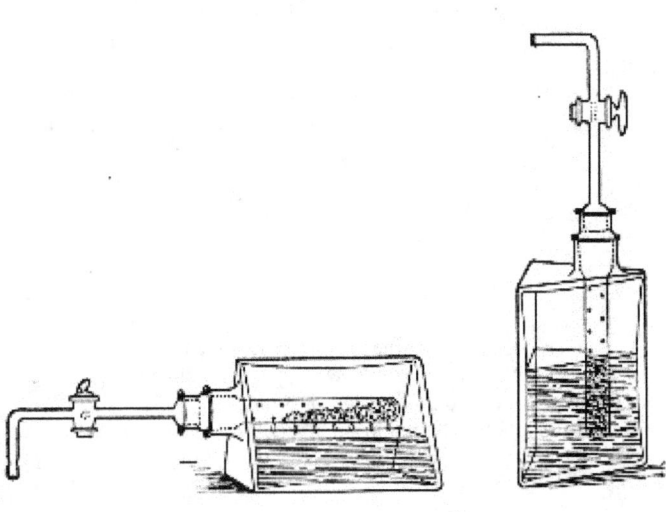

Fig. 133.—Improved gas apparatus; the metal is contained in a perforated glass tube which is submerged in acid when the triangular bottle is upright (a), but is above the level of the liquid when the bottle is turned on its side (b).

5. Connect up the india-rubber tubing on one of the side tubes with a cylinder of compressed hydrogen (or the delivery tube of a Kipp's Fig. 132 or other hydrogen apparatus, Fig. 133), interposing a short piece of glass tubing; and in like manner connect a long piece of rubber tubing which should be led into a basin of water, to the opposite side tube.

6. Open both metal clips and pass hydrogen through the vessel until the atmospheric air is replaced by hydrogen. This is determined by collecting some of the gas which bubbles through the water in the basin in a test-tube and testing it by means of a lighted taper.

7. Close the metal clip on the tube through which the gas is entering; close the clip on the exit tube.

8. Disconnect the gas apparatus.

9. Incubate.

Method X (Botkin's Method).—

Apparatus Required.—

Large glass dish 20 cm. diameter and 8 cm. deep. Flat leaden cross slightly shorter than the internal diameter of the glass dish. Bell glass about 15 cm. diameter and 20 to 25 cm. high. Metal frame for plate cultivations. *Or*, glass battery jar for tube cultivations. Cylinder of compressed hydrogen. Rubber tubing. Two pieces of **U**-shaped glass tubing (each arm 8 cm. in length). Half a litre of glycerine (or metallic mercury).

Method.—

1. Place the leaden cross inside the glass dish, resting on the bottom.
2. Prepare the cultivations in the usual way.
3. Place the tube cultivations in a glass battery jar (or the plate cultivations on a metal frame), resting on the centre of the leaden cross.
4. Cover the cultivations with the bell jar.
5. Adjust the U-shaped pieces of glass tubing in a vertical position on opposite sides of the bell jar, one arm of each inside the jar, the other outside. These tubes are best held in position by embedding the U-shaped bends in two lumps of plasterine stuck on the bottom of the glass dish. Fix a short length of rubber tubing clamped with a metal clip to each of the outside arms (Fig. 134).
6. Fill the glass dish with glycerine or metallic mercury to a depth of about 5 cm.

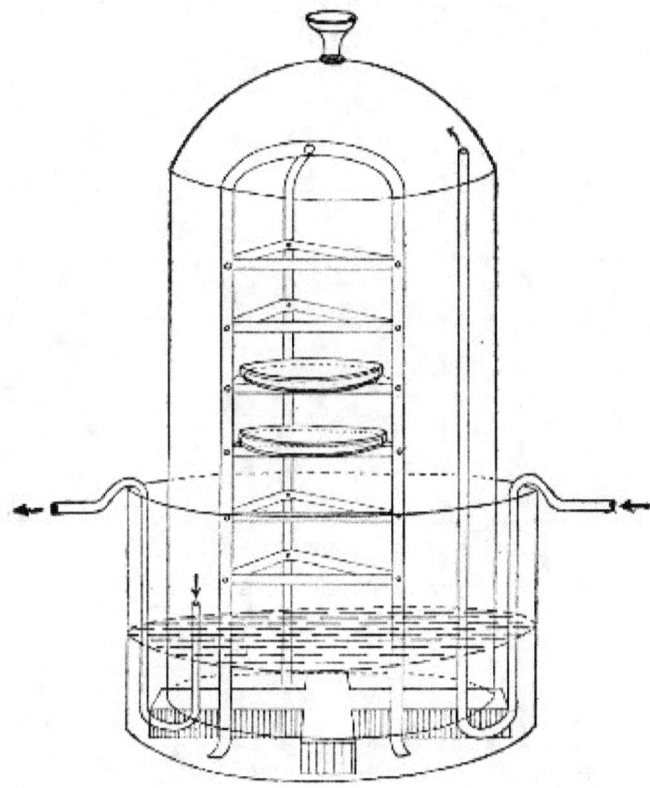

Fig. 134.—Botkin's apparatus.

7. Connect up one U-shaped tube with the hydrogen cylinder (or gas apparatus) by means of rubber tubing. Replace the atmospheric air by hydrogen, as in method IX.

8. Clamp the tubes and disconnect the gas apparatus.

9. Incubate.

Method XI (Novy's Method).—

Apparatus Required.—

Jar for plate cultivations (Fig. 135). *Or*, jar for tube cultivations (Fig. 136). Lubricant for stopper of jar. Rubber tubing. Cylinder of compressed hydrogen.

Method.—

1. Prepare cultivations in the usual way.
2. Place these inside the jar.
3. Lubricate the stopper and insert it in the mouth of the jar, with the handle in a line with the two side tubes.
4. Connect up the delivery tube *a* with the hydrogen gas supply by means of rubber tubing.

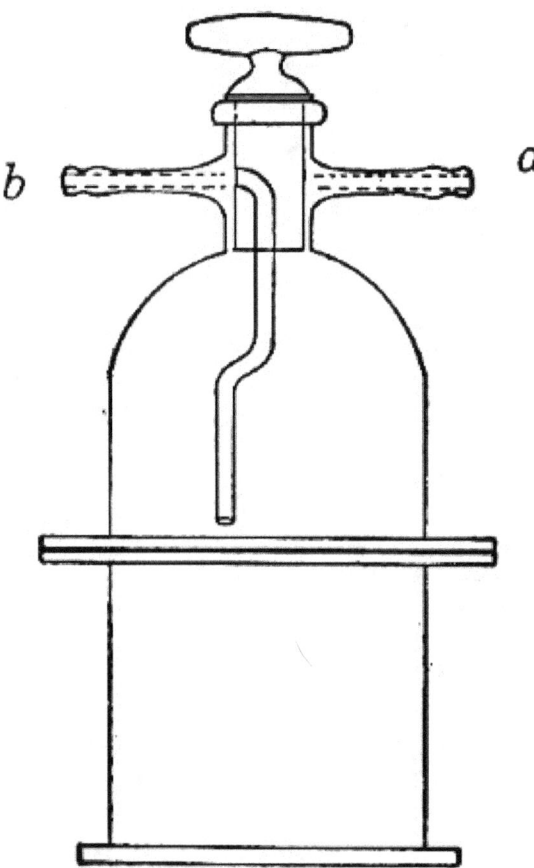

Fig. 135.—Novy's jar for plate cultivations.

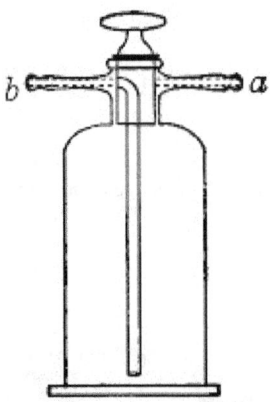

Fig. 136.—Novy's jar for tube cultivations.

5. Attach a piece of rubber tubing to the exit tube *b* and collect samples of the issuing gas (over water) and test from time to time.

6. When the air is completely displaced by hydrogen, turn the handle of the stopper at right angles to the line of entry and exit tubes; this seals the orifice of both tubes.

7. Disconnect the gas apparatus and incubate.

(E) **Method XII** (Bulloch's Method).—

Apparatus Required.—

Bulloch's jar. Pot of resin ointment. Small glass dish 14 cm. diameter by 5 cm. deep. Vessel for tube cultures or metal rack for plate cultures. Pyrogallic acid tablets. Cylinder of compressed hydrogen. Geryk or other air pump. Rubber pressure tubing. 10 c.c. pipette. Glass tubing. Dry granulated caustic soda or compressed tablets each, containing 0.4 grammes sodic hydroxide. Small beaker of water.

Method.—

1. Prepare the cultivations in the usual way.

2. Place the glass dish in the centre of the glass slab, and stand the cultivations inside this.

3. Place a sufficient number of pyrogallic acid tablets at one side of the glass dish (*i. e.*, 1 tablet for each 100 cubic centimeters air capacity of the bell jar). Place a small heap of dry granulated soda (or half a dozen tablets of sodic hydroxide) by the side of the pyro tablets.

4. Smear the flange of the bell jar with resin ointment and apply the jar firmly to the glass slab, covering the cultivations—so arranged that the long tube passes with its lower end into the glass dish at a point directly opposite to the pyrogallic acid tablets. Lubricate the two stop-cocks with resin ointment (Fig. 137).

5. Connect up the short tube *a* with the gas-supply by means of rubber pressure tubing and open both stop-cocks.

6. Connect a long, straight piece of glass tubing to the long tube *b* by means of a piece of rubber tubing interposing a screw clamp: and collect samples of the issuing gas from time to time and test.

7. When the air is displaced, shut off the stop-cock of the entry tube, then that of the exit tube *b*. Screw down the clamp and remove the glass tube from the rubber connection and connect up the short tube *a* to the air pump by means of pressure tubing.

8. Open the stop-cock of tube *a* and with two or three strokes of the air pump, aspirate a small quantity of gas, so creating a slight vacuum. Then shut off the stop-cock and disconnect the air pump.

9. Fill the 10 c.c. bulb pipette with water; insert its point into the rubber tubing on the long tube *b* as far as the screw clamp. Open the screw clamp and run in water until stopped by the internal pressure. Shut off stop-cock. (The water dissolves the soda and pyrogallic acid converting the latter into alkaline pyro. and so bringing its latent capacity for oxygen into action).

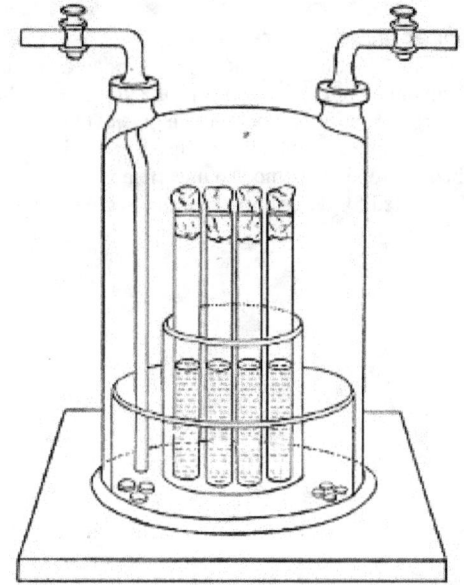

Fig. 137.—Bulloch's jar.

10. Reverse the tubes from the tubulures so that they meet, out of harm's way, over the top of the bell glass; again see that all joints are tight and transfer the apparatus to the incubator.

This last method is the most satisfactory for anaerobic cultivations, as by its means complete anaerobiosis can be obtained with the least expenditure of time and trouble.

FOOTNOTES:

See also method of opening and closing culture tubes, pages 74-76.
If compressed tablets of pyrogallic acid cannot be obtained make up a stock "acid pyro" solution
Pyrogallic acid, 10 grammes Hydrochloric acid, 1.5 c.c. Distilled water, 100 c.c.
and at step 4, run in 10 c.c. of the solution.

XV. METHODS OF ISOLATION.

The work in the preceding sections, arranged to demonstrate the chief biological characters of bacteria in general, is intended to be carried out by means of cultivations of various organisms previously isolated and identified and supplied to the student in a state of purity. A cultivation which comprises the progeny of a single cell is termed a "pure culture"; one which contains representatives of two or more species of bacteria is spoken of as an "impure," or "mixed" "cultivation," and it now becomes necessary to indicate the chief methods by which one or more organisms may be isolated in a state of purity from a mixture; whether that mixture exists as an impure laboratory cultivation, or is contained in pus and other morbid exudations, infected tissues, or water or food-stuffs.

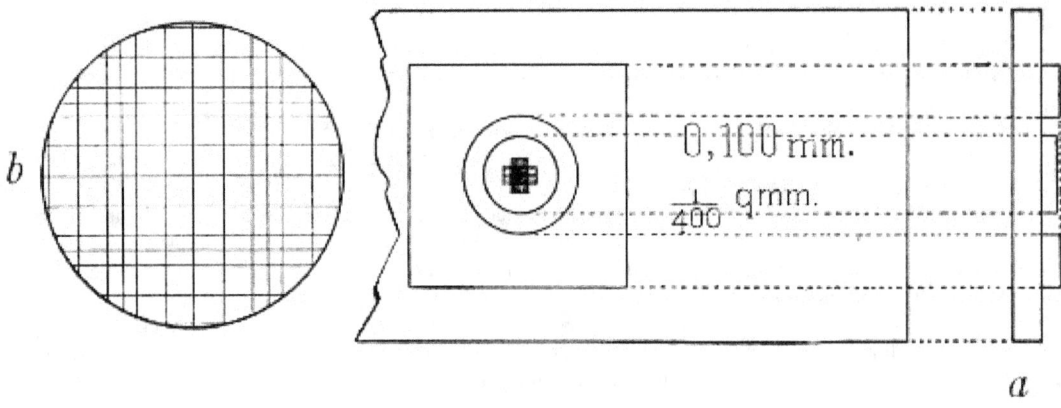

Fig. 138.—Hæmatocytometer cell, showing, a, section through the centre of the cell, and b, a magnified image of the cell rulings.

Before the introduction of solid media the only method of obtaining pure cultivations was by "dilution"—by no means a reliable method. "Dilution" consisted in estimating approximately the number of bacteria present in a given volume of fluid (by means of a graduated-celled slide resembling a hæmatocytometer, Fig. 138), and diluting the fluid by the addition of sterile water or bouillon until a given volume (usually 1 c.c.) of the dilution contained but one organism. By planting this volume of the fluid into several tubes or flasks of nutrient media, it occasionally happened that the resulting growth was the product of one individual microbe. A method so uncertain is now fortunately replaced by many others, more reliable and convenient, and in those methods selected for description here, the segregation and isolation of the required bacteria may be effected—

A. By Mechanical Separation.

1. By surface plate cultivation:
(*a*) Gelatine. (*b*) Agar. (*c*) Serum agar. (*d*) Blood agar. (*e*) Hanging-drop or block.

3. By serial cultivation.

B. By Biological Differentiation.

4. By differential media.
(*a*) Selective. (*b*) Deterrent.
5. By differential incubation.
6. By differential sterilisation.
7. By differential atmosphere cultivation.
8. By animal inoculation.

The selection of the method to be employed in any specific instance will depend upon a variety of circumstances, and often a combination of two or more will ensure a quicker and more reliable result than a rigid adherence to any one method. Experience is the only reliable guide, but as a general rule the use of either the first or the third method will be found most convenient, affording as each of them does an opportunity for the simultaneous isolation of several or all of the varieties of bacteria present in a mixture.

1. Surface Plate Cultivations.—

(*a*) *Gelatine* (*vide* page 164).

(*b*) *Agar* (*vide* page 167).

(*c*) *Alkaline serum agar* (*vide* page 211).

These plates are prepared in a manner precisely similar to that adopted for nutrient gelatine and agar surface plates (*vide* pages 231-233).

(*d*) *Serum Agar.—*

1. Melt three tubes of nutrient agar, label them 1, 2, and 3, and place them, with three tubes of sterile fluid serum, also labelled 1*a*, 2*a*, and 3*a*, in a water-bath regulated at 45° C.; allow sufficient time to elapse for the temperature of the contents of each tube to reach that of the water-bath.

2. Take serum tube No. 1*a* and agar tube No. 1. Flame the plugs and remove them from the tubes (retaining the plug of the agar tube in the hand); flame the mouths of the tubes, pour the serum into the tube of liquefied agar and replace the plug of the agar tube.

3. Mix thoroughly and pour plate No. 1 *secundum artem*.

4. Treat the remaining tube of agar and serum in a similar fashion, and pour plates Nos. 2 and 3.

5. Dry the serum agar plates in the incubator running at 60° C. for one hour (see page 232).

6. Inoculate the plates in series as described for gelatine surface plates (page 231).

(*e*) *Blood Agar, Human.—*

1. Melt a tube of sterile agar and pour it into a sterile plate; let it set.

2. Collect a few drops of human blood, under all aseptic conditions, in a sterile capillary teat pipette.

3. Raise the cover of the Petri dish very slightly, insert the extremity of the capillary pipette, and deposit the blood on the centre of the agar surface. Close the dish.

4. Charge a platinum loop with a small quantity of the inoculum. Raise the cover of the plate, introduce the loop, mix its contents with the drop of blood, remove the loop, close the dish and sterilise the loop.

5. Finally smear the mixture over the surface of the agar with a sterilised L-shaped rod.

6. Label and incubate.

(If considered necessary, two, three, or more similar plates may be inoculated in series.)

(*f*) *Blood Agar, Animal.*—

When preparing citrated blood agar (page 171) it is always advisable to pour several blood agar tubes into plates, which can be stored in the ice chest ready for use at any moment for surface plate cultures.

(*g*) Hanging-drop or block culture, (*vide* page 233).

3. Serial Cultivations.—These are usually made upon agar or blood-serum, although gelatine may also be used. The method is as follows:

1. Take at least four "slanted" tubes of media and number them consecutively.

2. Flame all the plugs and see that each can be readily removed.

3. Charge the platinum loop with a small quantity of the inoculum, observing the usual routine, and plant tube No. 1, smearing thoroughly all over the surface. If any water of condensation has collected at the bottom of the tube, use this as a diluent before smearing the contents of the loop over the surface of the medium.

4. Without sterilising or recharging the loop, inoculate tube No. 2, by making three parallel streaks from end to end of the slanted surface.

5. Plant the remainder of the tubes in the series as "smears" like tube No. 1.

6. Label with distinctive name or number, and date; incubate.

The growth that ensues in the first two or three tubes of the series will probably be so crowded as to be useless. Toward the end of the series, however, discrete colonies will be found, each of which can be transferred to a fresh tube of nutrient medium without risk of contamination from the neighbouring colonies.

"Working" up Plates.—

Having succeeded in obtaining a plate (or tube cultivation) in which the colonies are well grown and sufficiently separated from each other, the process of "working up," "pricking out," or "fishing" the colonies in order to obtain subcultures in a state of purity from each of the different bacteria present must now be proceeded with.

Occasionally it happens that this is quite a simple matter. For example, the original mixed cultivation when examined microscopically was found to contain a Gram positive micrococcus, a Gram positive straight bacillus and a Gram negative short bacillus. The third gelatine plate prepared from this mixture, on inspection after four day's incubation, showed twenty-five colonies—seven moist yellow colonies, each sinking into a shallow pit of liquefied gelatine, fourteen flat irridescent filmy colonies, and four raised white slimy colonies. A film preparation (stained Gram) from each variety examined microscopically showed that the yellow liquefying colony was composed of Gram positive micrococci; the flat colony of Gram positive bacilli and the white colony of gram negative bacilli. One of each of these varieties of colonies would be transferred by means of the sterilised loop to a fresh gelatine culture tube, and after incubation the growth in each subculture would correspond culturally and microscopically with that of the plate colony from which it was derived,—the object aimed at would therefore be achieved.

Usually, however, the colonies cannot be thus readily differentiated, and unless they are "worked up" in an orderly and systematic manner much labour will be vainly expended and valuable time wasted. The following method minimises the difficulties involved.

(A) Inspection.

a. Without opening the plate carefully study the various colonies with the naked eye, with the assistance of a watchmaker's lens or by inverting the plate on the stage of the microscope and viewing with the 1-inch objective through the bottom of the plate and the layer of medium.

b. If gross differences can be detected mark a small circle on the bottom of the plate around the site of each of the selected colonies, with the grease pencil.

c. If no obvious differences can be made out choose nine colonies haphazard and indicate their positions by pencil marks on the bottom of the plate.

(B) Fishing Colonies.—

a. Take a sterile Petri dish and invert it upon the laboratory bench. Rule two parallel lines on the bottom of the dish with a grease pencil, and two more parallel lines at right angles to the first pair—so dividing the area of the dish into nine portions. Number the top right-hand portion 1, and the central bottom portion 8 (Fig. 139). Revert the dish. The numbers 1 and 8 can be readily recognised through the glass and by their positions enable any of the other divisions to be localised by number. This is the stock dish.

b. Slightly raise the cover of the dish, and with a sterile teat-pipette deposit a small drop of sterile water in the centre of each of the nine divisions.

c. With the sterilised platinum spatula raise one of the marked colonies from the "plate 3" and transfer it to the first division in the ruled plate and emulsify it in the drop of water awaiting it. Repeat this process with the remaining colonies, emulsifying a separate colony in each drop of water.

(C) Preliminary Differentiation of Bacteria.—

a. Prepare a cover-slip film preparation from each drop of emulsion in the "stock dish" and number to correspond to the division from which it was taken. Stain by Gram's method.

b. Examine microscopically, using the oil immersion lens and note the numbers of those cover-slips which morphologically and by Gram results appear to be composed of different species of bacteria.

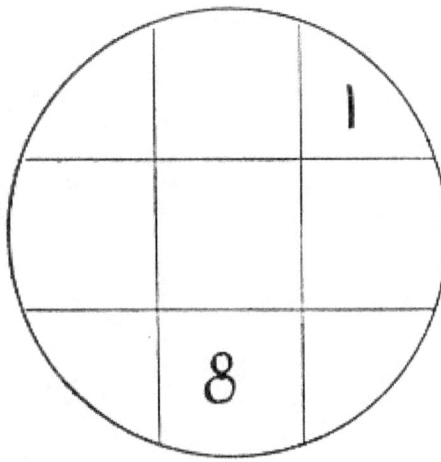

Fig. 139.—Diagram for stock plate.

(D) Preparing Isolation Subcultures.—

a. Inoculate an agar slope and a broth tube from the emulsion in the stock dish corresponding to each of these specially selected numbers.

b. Ascertain whether the cover-slips from the nine emulsions in the stock dish include all the varieties represented in the cover-slip film preparation made from the original mixture before plating.

c. If some varieties are missing prepare a second stock dish from other colonies on plate 3, and repeat the process until each morphological form or tinctorial variety has been secured in subculture.

d. Place the stock dishes in the ice chest to await the results of incubation. (If any of the subcultures fail, further material can be obtained from the corresponding emulsion; or if it has dried, by moistening it with a further drop of sterile distilled water.)

e. Incubate all the subcultures and identify the organisms picked out.

4. Differential Media.—

(*a*) *Selective.*—Some varieties of media are specially suitable for certain species of bacteria and enable them to overgrow and finally choke out other varieties; *e. g.*, wort is the most suitable medium-base for the growth of torulæ and yeasts and should be employed when pouring plates for the isolation of these organisms. To obtain a pure cultivation of yeast from a mixture containing bacteria as well, it is often sufficient to inoculate wort from the mixture and incubate at 37° C. for twenty-four hours. Plant a fresh tube of wort from the resulting growth and incubate. Repeat the process once more, and from the growth in this third tube plant a streak on wort gelatine, and incubate at 20°C. The resulting growth will almost certainly be a pure culture of the yeast.

(*b*) *Deterrent.*—The converse of the above also obtains. Certain media possess the power of inhibiting the growth of a greater or less number of species. For instance, media containing carbolic acid to the amount of 1 per cent. will inhibit the growth of practically everything but the Bacillus coli communis.

5. Differential Incubation.—

In isolating certain bacteria, advantage is taken of the fact that different species vary in their optimum temperature. A mixture containing the Bacillus typhosus and the Bacillus aquatilis sulcatus, for example, may be planted on two slanted agar tubes, the one incubated at 40°C., and the other at 12° C. After twenty-four hours' incubation the first will show a pure cultivation of the Bacillus typhosus, whilst the second will be an almost pure culture of the Bacillus aquatilis.

6. Differential Sterilisation.—

(*a*) *Non-sporing Bacteria.*—Similarly, advantage may be taken of the varying thermal death-points of bacteria. From a mixture of two organisms whose thermal death-points differ by, say, 4°C.—*e. g.*, Bacillus pyocyaneus, thermal death-point 55°C., and Bacillus mesentericus vulgatus, thermal death-point 60°C.—a pure cultivation of the latter may be obtained by heating the mixture in a water-bath to 58° C. and keeping it at that point for ten minutes. The mixture is then planted on to fresh media and incubated, when the resulting growth will be found to consist entirely of the B. mesentericus.

(*b*) *Sporing Bacteria.*—This method finds its chief practical application in the differentiation of a spore-bearing organism from one which does not form spores. In this case the mixture is heated in a water-bath at 80° C. for fifteen to twenty minutes. At the end of this time the non-sporing bacteria are dead, and cultivations made from the mixture will yield a growth resulting from the germination of the spores only.

Differential sterilisation at 80° C. is most conveniently carried out in a water-bath of special construction, designed by Balfour Stewart (Fig. 140). It consists of a double-walled copper vessel mounted on legs, and provided with a tubulure communicating with the space between the walls. This space is nearly filled with benzole (boiling-point 80°C.; pure benzole, free from thiophene must be employed for the purpose, otherwise the boiling-point gradually and perceptibly rises in the course of time), and to the tubulure is fitted a long glass tube, some 2 metres long and about 0.75 cm. diameter, serving as a condensing tube (a tube half this length if provided with a condensing bulb at the centre will be equally efficient). The interior of the vessel is partly filled with water and covered with a lid which is perforated for a thermometer. This latter dips into the water and records its temperature. A very small Bunsen flame under the apparatus suffices to keep the benzole boiling and the water within at a constant temperature of 80° C. The bath is thus always ready for use.

Method.—To use the apparatus.

1. Place some of the mixture itself, if fluid, containing the spores, or an emulsion of the same if derived from solid material, in a test-tube.

2. Immerse the test-tube in the water contained in the benzole bath, taking care that the upper level of the liquid in the tube is at least 2 cm. beneath the surface of the water in the copper vessel.

3. The temperature of the water, of course, falls a few degrees after opening the bath and introducing a tube of colder liquid, but after a few minutes the temperature will have again reached 80°C.

4. When the thermometer again records 80°C., note the time, and fifteen minutes later remove the tube containing the mixture from the bath.

5. Make cultures upon suitable media; incubate.

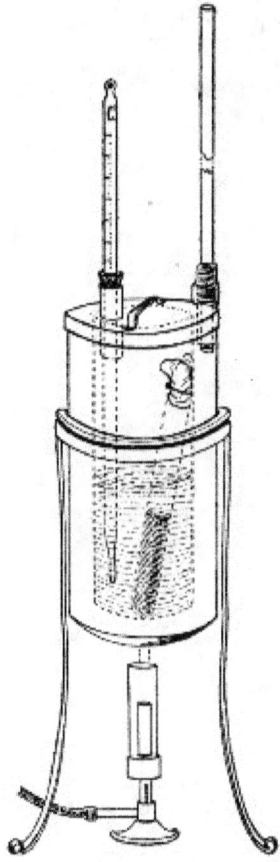

Fig. 140.—Benzole bath.

7. Differential Atmosphere Cultivation.—

(*a*) By adapting the atmospheric conditions to the particular organism it is desired to isolate, it is comparatively easy to separate a strict aerobe from a strict anaerobe, and *vice versa*. In the first case, however, it is important that the cultivations should be made upon solid media, for if carried out in fluid media the aerobes multiplying in the upper layers of fluid render the depths completely anaerobic, and under these conditions the growth of the anaerobes will continue unchecked.

(*b*) When it is desired to separate a facultative anaerobe from a strict anaerobe, it is generally sufficient to plant the mixture upon the sloped surface agar, incubate aerobically at 37°C., and examine carefully at frequent intervals. At the first sign of growth, subcultivations must be prepared and treated in a similar manner. As a result of these rapid subcultures, the facultative anaerobe will be secured in pure culture at about the third or fourth generation.

(*c*) If, on the other hand, the strict anaerobe is the organism required from a mixture of facultative and strict anaerobes, pour plates of glucose formate agar (or gelatine) in the usual manner, place them in a Bulloch's or Novy's jar, and incubate at a suitable temperature. Pick off the colonies of the required organism when the growth appears, and transfer to tubes of the various media.

Incubate under suitable conditions as to temperature and atmosphere.

8. Animal Inoculation.—

Finally, when dealing with pathogenic organisms, it is often advisable to inoculate some of the impure culture (or even some of the original *materies morbi*) into an animal specially chosen on account of its susceptibility to the particular pathogenic organism it is desired to inoculate. Indeed, with some of the more sensitive and strictly parasitic bacteria this method of animal inoculation is practically the only method that will yield a satisfactory result.

XVI. METHODS OF IDENTIFICATION AND STUDY.

The Elements of Bacteriological Technique

In order to identify an organism after isolation, tube, plate, and other cultivations must be prepared, incubated under suitable conditions as to temperature and environment, and examined from time to time (**a**) **macroscopically**, (**b**) by **microscopical methods**, (**c**) by **chemical methods**, (**d**) by **physical methods**, (**e**) by **inoculation methods**, and the results of these examinations duly recorded.

It must be stated definitely that no micro-organism can be identified by any *one* character or property, whether microscopical, biological or chemical, but that on the contrary its entire life history must be carefully studied and then its identity established from a consideration of the sum total of these observations.

In order to give to the recorded results their maximum value it is essential that they should be exact and systematic, therefore some such scheme as the following should be adhered to; and especially is this necessary in describing an organism not previously isolated and studied.

SCHEME OF STUDY.

Designation:
Originally isolated by (*observer's name*) in (*date*), from (*source of organism*).

1. Cultural Characters.—(*Vide* Macroscopical Examination of Cultivation, page 261.)

Gelatine plates,	}
Gelatine streak,	} at 20°C.
Gelatine stab,	}
Gelatine shake,	}
Agar plates,	}
Agar streak or smear,	}
Agar stab,	}
Inspissated blood-serum,	} at 20° C. and 37°C.
Bouillon,	}
Litmus milk,	}
Potato,	}

Special media for the purpose of demonstrating characteristic appearances.

2. Morphology.—(*Vide* Microscopical Examination of Cultivations, page 272.)
Vegetative forms: Shape. Size. Motility. Flagella (if present). Capsule (if present). Involution forms. Pleomorphism (if observed). Sporing forms (if observed). Of which class? Staining reactions.

3. Chemical Products of Growth.—(*Vide* Chemical Examination of Cultivations, page 276.)
Chromogenesis. Photogenesis. Enzyme formation. Fermentation of carbohydrates: Acid formation. Alkali formation. Indol formation. Phenol formation. Reducing and oxidising substances. Gas formation.

4. Biology.—(*Vide* Physical Examination of Cultures, page 295.)
Atmosphere. Temperature.
Reaction of nutrient media. Resistance to lethal agents: Physical: Desiccation. Light. Colours. Chemical germicides. Vitality.

5. Pathogenicity:
Susceptible animals, subsequently arranged in order of susceptibility. Immune animals. Experimental inoculation, symptoms of disease. Post-mortem appearances. Virulence: Length of time maintained. Optimum medium? Minimal lethal dose. Exaltation and attenuation of virulence? Toxin formation.

MACROSCOPICAL EXAMINATION OF CULTIVATIONS.

In describing the naked-eye and low-power appearances of the bacterial growth the descriptive terms introduced by Chester (and included in the following scheme) should be employed.

Solid Media.
Plate Cultures.—

Gelatine.—Note the presence or absence of liquefaction of the surrounding medium. If liquefaction is present, note shape and character (*vide* page 269, "stab" cultures).

Agar.—No liquefaction takes place in this medium. The liquid found on the surface of the agar (or at the bottom of the tube in agar tube cultures) is merely water which has been expressed during the rapid solidification of the medium and has subsequently condensed.

Gelatine and Agar.—Examine the colonies at intervals of twenty-four hours.
(a) With the naked eye.
(b) With a hand lens or watchmaker's glass.
(c) Under a low power (1 inch) of the microscope, or by means of a small dissecting microscope.
Distinguish superficial from deep colonies and note the characters of the individual colonies.
(*A*) **Size.**—The diameter in millimetres, at the various ages.
(*B*) **Shape.**—

Punctiform: Dimensions too slight for defining form by naked eye; minute, raised, hemispherical.
Round: Of a more or less circular outline.
Elliptical: Of a more or less oval outline.
Irregular: Outlines not conforming to any recognised shape.
Fusiform: Spindle-shaped, tapering at each end.
Cochleate: Spiral or twisted like a snail shell (Fig. 141, *a*).

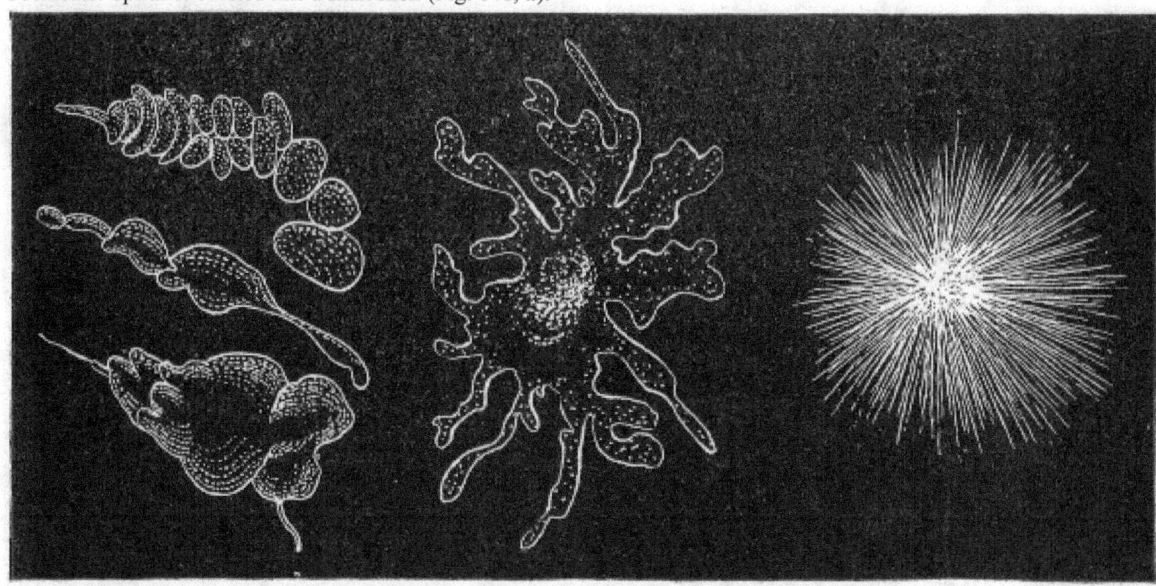

Fig. 141.—Types of colonies: a, Cochleate; b, amœboid; c, mycelioid.

Amœboid: Very irregular, streaming (Fig. 141, *b*).
Mycelioid: A filamentous colony, with the radiate character of a mould (Fig. 141, *c*).
Filamentous: An irregular mass of loosely woven filaments (Fig. 142, *a*).
Floccose: Of a dense woolly structure.
Rhizoid: Of an irregular, branched, root-like character (Fig. 142, *b*).
Conglomerate: An aggregate of colonies of similar size and form (Fig. 142, *c*).
Toruloid: An aggregate of colonies, like the budding of the yeast plant (Fig. 142, *d*).
Rosulate: Shaped like a rosette.

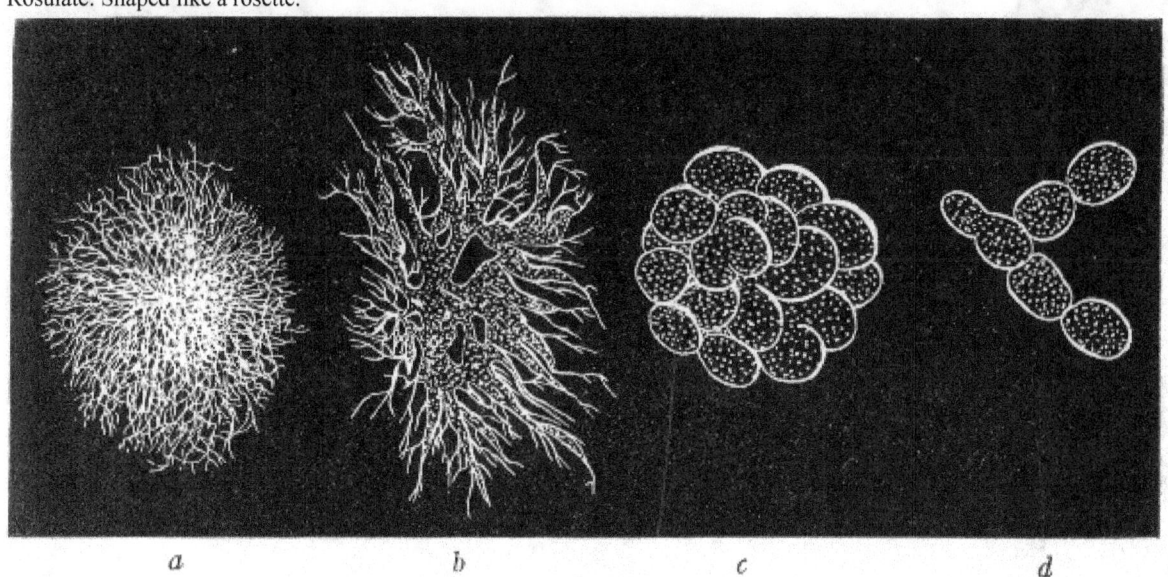

Fig. 142.—Types of colonies: a, Filamentous; b, rhizoid; c, conglomerate; d, toruloid.

(C) **Surface Elevation.**—
1. *General Character of Surface as a Whole*:
Flat: Thin, leafy, spreading over the surface (Fig. 143, *a*).
Effused: Spread over the surface as a thin, veily layer, more delicate than the preceding.
Raised: Growth thick, with abrupt terraced edges (Fig. 143, *b*).

Convex: Surface the segment of a circle, but very flatly convex (Fig. 143, c).
Pulvinate: Surface the segment of a circle, but decidedly convex (Fig. 143, d).
Capitate: Surface hemispherical (Fig. 143, e).
Umbilicate: Having a central pit or depression (Fig. 143, f).
Conical: Cone with rounded apex (Fig. 143, g).
Umbonate: Having a central convex nipple-like elevation (Fig. 143, h).

2. *Detailed Characters of Surface*:

Smooth: Surface even, without any of the following distinctive characters.
Alveolate: Marked by depressions separated by thin walls so as to resemble a honeycomb (Fig. 144).
Punctate: Dotted with punctures like pin-pricks.
Bullate: Like a blistered surface, rising in convex prominences, rather coarse.
Vesicular: More or less covered with minute vesicles due to gas formation; more minute than bullate.

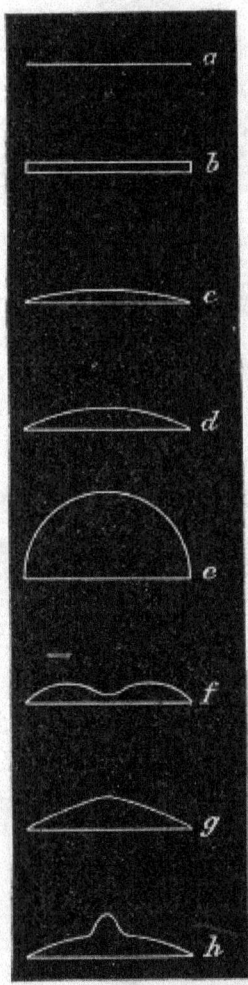

Fig. 143.—Surface elevation of colonies: a, Flat; b, raised; c, convex; d, pulvinate; e, capitate; f, umbilicate; g, conical; h, umbonate.

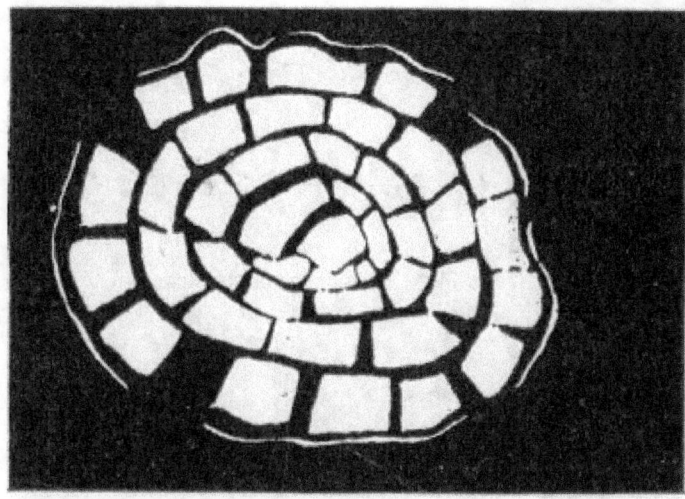

Fig. 144.—Types of colonies—alveolate.

Verrucose: Wart-like, bearing wart-like prominences.
Squamose: Scaly, covered with scales.
Echinate: Beset with pointed prominences.
Papillate: Beset with nipple or mamma-like processes.
Rugose: Short irregular folds, due to shrinkage of surface growth.
Corrugated: In long folds, due to shrinkage.
Contoured: An irregular but smoothly undulating surface, resembling the surface of a relief map.
Rimose: Abounding in chinks, clefts, or cracks.

(*D*) **Internal Structure of Colony** (*Microscopical*).—
Refraction Weak: Outline and surface of relief not strongly defined.
Refraction Strong: Outline and surface of relief strongly defined; dense, not filamentous colonies.

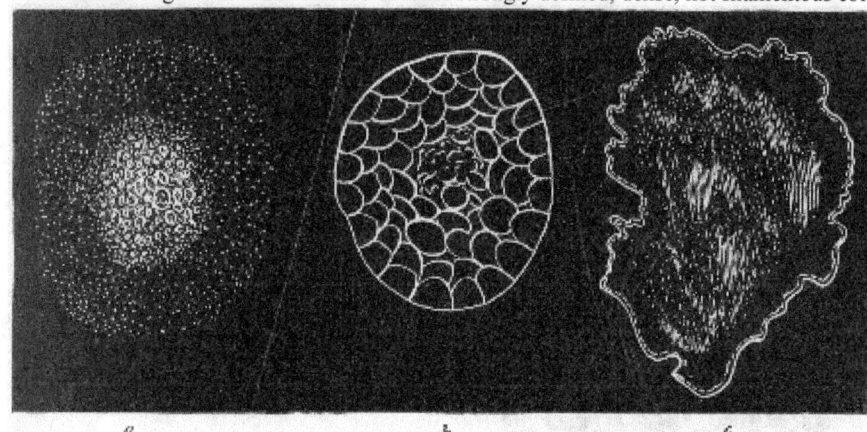

Fig. 145.—Types of colonies: a, Grumose; b, moruloid; c, clouded.

1. *General*:
Amorphous: Without any definite structure, such as is specified below.
Hyaline: Clear and colourless.
Homogeneous: Structure uniform throughout all parts of the colony.
Homochromous: Colour uniform throughout.

2. *Granulations or Blotchings*:
Finely granular.
Coarsely granular.
Grumose: Coarser than the preceding, with a clotted appearance, and particles in clustered grains (Fig. 145, *a*).
Moruloid: Having the character of a mulberry, segmented, by which the colony is divided in more or less regular segments (Fig. 145, *b*).
Clouded: Having a pale ground, with ill-defined patches of a deeper tint (Fig. 145, *c*).

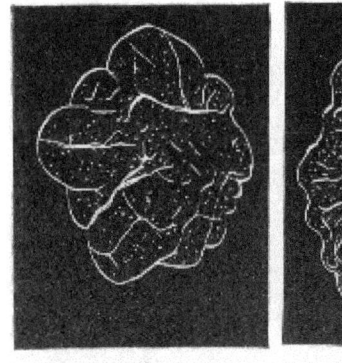 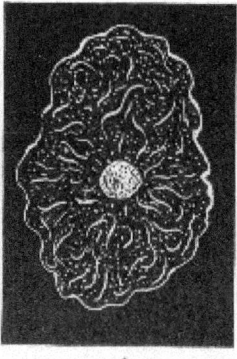 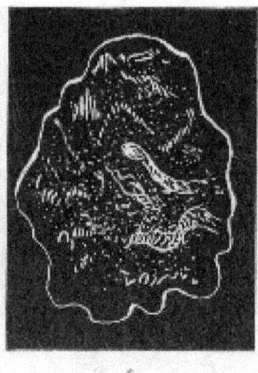

Fig. 146.—Types of colonies: a, Reticulate; b, gyrose; c, marmorated.

3. *Colony Marking or Striping*:
Reticulate: In the form of a network, like the veins of a leaf (Fig. 146, a).
Areolate: Divided into rather irregular, or angular, spaces by more or less definite boundaries.
Gyrose: Marked by wavy lines, indefinitely placed (Fig. 146, b).
Marmorated: Showing faint, irregular stripes, or traversed by vein-like markings, as in marble (Fig. 146, c).
Rivulose: Marked by lines like the rivers of a map.
Rimose: Showing chinks, cracks, or clefts.

Fig. 147.—Types of colonies—curled.

4. *Filamentous Colonies:*
Filamentous: As already defined.
Floccose: Composed of filaments, densely placed.
Curled: Filaments in parallel strands, like locks or ringlets (Fig. 147).

(*E*) **Edges of Colonies.**—
Entire: Without toothing or division (Fig. 148, a).
Undulate: Wavy (Fig. 148, b).
Repand: Like the border of an open umbrella (Fig. 148, c).
Erose: As if gnawed, irregularly toothed (Fig. 148, d).

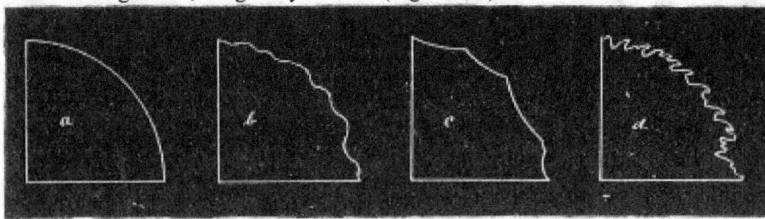

Fig. 148.—Edges of colonies: a, Entire; b, undulate; c, repand; d, erose.

Lobate.
Lobulate: Minutely lobate (Fig. 149, e).
Auriculate: With ear-like lobes (Fig. 149, f).
Lacerate: Irregularly cleft, as if torn (Fig. 149, g).
Fimbriate: Fringed (Fig. 149, h).
Ciliate: Hair-like extensions, radiately placed (Fig. 149, j).
Tufted.
Filamentous: As already defined.
Curled: As already defined.

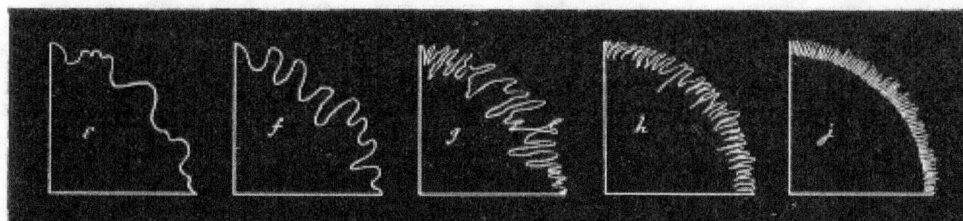

Fig. 149.—Edges of colonies: e, Lobar-lobulate; f, auriculate; g, lacerate; h, fimbriate; i, ciliate.

(*F*) **Optical Characters** (after Shuttleworth).—

1. *General Characters*:

Transparent: Transmitting light.
Vitreous: Transparent and colourless.
Oleaginous: Transparent and yellow; olive to linseed-oil coloured.
Resinous: Transparent and brown, varnish or resin-coloured.
Translucent: Faintly transparent.
Porcelaneous: Translucent and white.
Opalescent: Translucent; greyish-white by reflected light.
Nacreous: Translucent, greyish-white, with pearly lustre.
Sebaceous: Translucent, yellowish or greyish-white.
Butyrous: Translucent and yellow.
Ceraceous: Translucent and wax-coloured.
Opaque.
Cretaceous: Opaque and white, chalky.
Dull: Without lustre.
Glistening: Shining.
Fluorescent.
Iridescent.

2. *Chromogenicity*:

Colour of pigment.
Pigment restricted to colonies.
Pigment restricted to medium surrounding colonies.
Pigment present in colonies and in medium.

Streak or Smear Cultures.—

Gelatine and Agar.—Note general points as indicated under plate cultivations.

Inspissated Blood-serum.—Note the presence or absence of liquefaction of the medium. (The presence of condensation water at the bottom of the tube must not be confounded with liquefaction of the medium.)

All Oblique Tube Cultures.—

1. Colonies Discrete: Size, shape, etc., as for plate cultivations (*vide* page 261).
2. Colonies Confluent: Surface elevation and character of edge, as for plate cultivations (*vide* page 263).

Chromogenicity: As for plate cultures.

Gelatine Stab Cultures.—

(*A*) *Surface Growth.*—As for individual colonies in plate cultures (*vide* page 261).

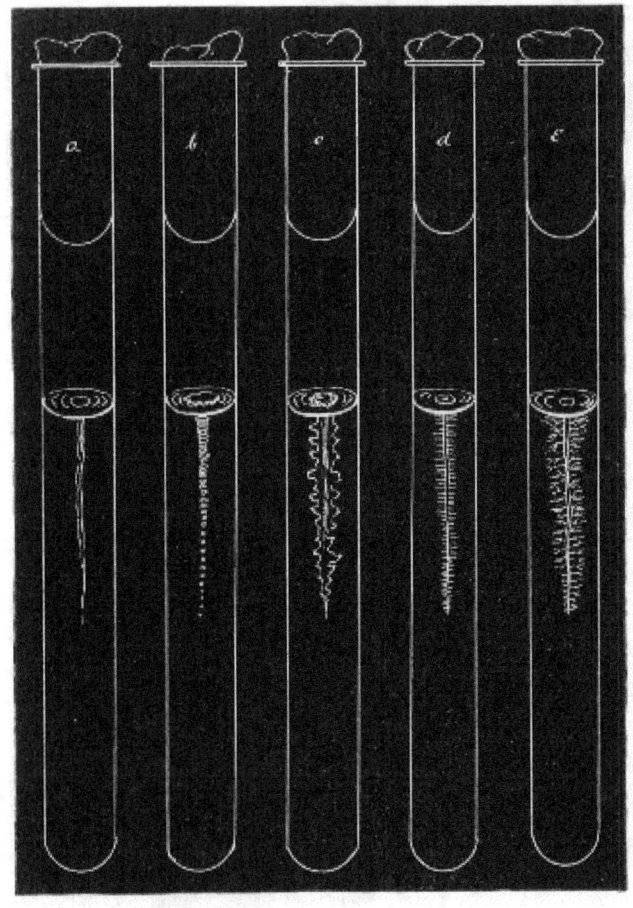

Fig. 150.—Stab cultivations—types of growth: a, Filiform; b, beaded; c, echinate; d, villous; e, arborescent.

(B) *Line of Puncture.*—
Filiform: Uniform growth, without special characters (Fig. 150, *a*).
Nodose: Consisting of closely aggregated colonies.
Beaded: Consisting of loosely placed or disjointed colonies (Fig. 150, *b*).
Papillate: Beset with papillate extensions.
Echinate: Beset with acicular extensions (Fig. 150, *c*).
Villous: Beset with short, undivided, hair-like extensions (Fig. 150, *d*).
Plumose: A delicate feathery growth.

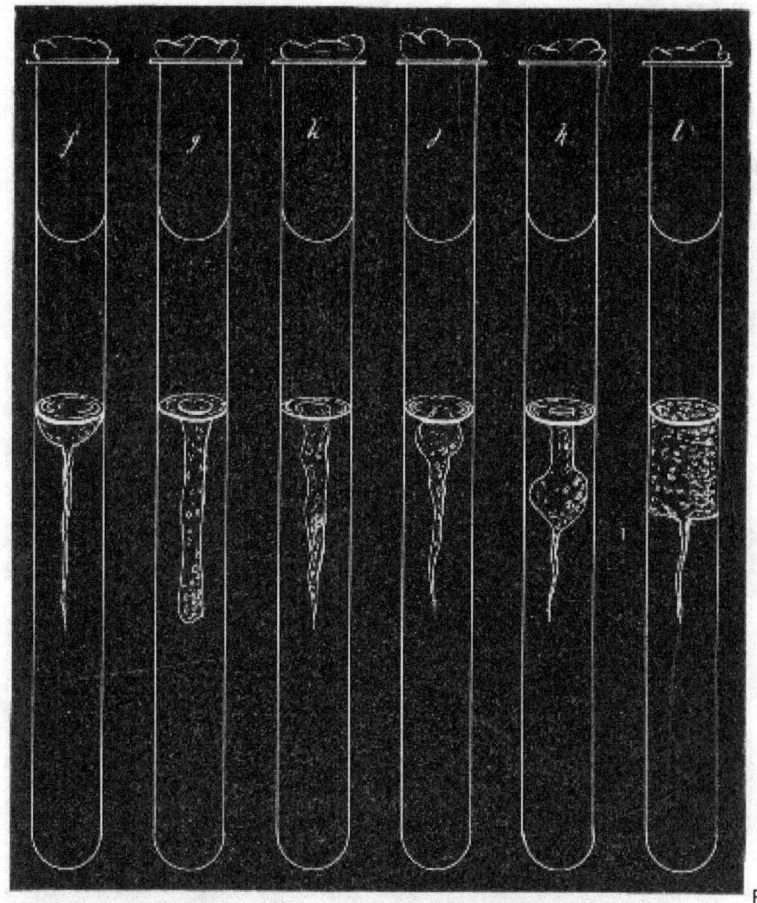

Fig. 151.—Stab cultivations—types of growth: f, Crateriform; g, saccate; h, infundibuliform; j, napiform; k, fusiform; l, stratiform.

Arborescent: Branched or tree-like, beset with branched hair-like extensions (Fig. 150, *e*).

(*C*) *Area of Liquefaction* (if present).—

Crateriform: A saucer-shaped liquefaction of the gelatine (Fig. 151, *f*).

Saccate: Shape of an elongated sack, tubular cylindrical (Fig. 151, *g*).

Infundibuliform: Shape of a funnel, conical (Fig. 151, *h*).

Napiform: Shape of a turnip (Fig. 151, *j*).

Fusiform: Outline of a parsnip, narrow at either end, broadest below the surface (Fig. 151, *k*).

Stratiform: Liquefaction extending to the walls of the tube and downward horizontally (Fig. 151, *l*).

(*D*) *Character of the Liquefied Gelatine.*—

1. Pellicle on surface.
2. Uniformly turbid.
3. Granular.
4. Mainly clear, but containing flocculi.
5. Deposit at apex of liquefied portion.

(*E*) *Production of Gas Bubbles.*

Shake Cultures.—

1. Presence or absence of liquefaction.
2. Production of gas bubbles.
3. Bulk of growth at the surface—aerobic.
4. Bulk of growth in depths—anaerobic.

Fluid Media.

1. Surface of the Liquid.—

Presence or absence of froth due to gas bubbles.

Presence or absence of pellicle formation.

Character of pellicle.

2. Body of the Liquid.—

Uniformly turbid.

Flocculi in suspension.

Granules in suspension.

Clear, with precipitate at bottom of tube.

Colouration of fluid, presence or absence of.
3. Precipitate.—
Character.
Amount.
Colour.
Carbohydrate Media.—
Growth.
Reaction.
Gas formation.
Coagulation or not of serum albumen (when serum water media are employed).
Litmus Milk Cultivations.—

1. Reaction: {Unaltered.
{Acid.
{Alkaline.

2. Odour.

3. Formation of gas.

4. Consistency: {Unaltered.
{Peptonised (character of solution).
{Coagulated.

5. Clot: Character {hard: solid.
{soft: floculent.
{ragged and broken up by gas bubbles.

(*a*) Coagulum undissolved.
(*b*) Coagulum finally peptonised, completely: incompletely.
Resulting solution, clear: turbid.

6. Whey: {Abundant.
{Scanty.
{Clear.
{Turbid.
{Coagulated by boiling, or not.

BY MICROSCOPICAL METHODS.
As a council of perfection preparations must be made from pure cultivations 4, 6, 8, 12, 18, and 24 hours; and subsequently at intervals of, say, twenty-four hours, during the entire period they are under observation, and examined—

(**A**) **Living.—1.** In **hanging drop**, to determine *motility* or *non-motility*.

In this connection it must be remembered that under certain conditions as to environment (*e. g.*, when examined in an unsuitable medium, atmosphere, temperature, etc.) motile bacilli may fail to exhibit activity. No organism, therefore, should be recorded as non-motile from one observation only; a series of observations at different ages and under varying conditions should form the basis of an opinion as to the absence of true locomotion.

Size.—In the case of non-motile or sluggishly motile organisms, endeavour to measure several individuals in each hanging drop by means of the eyepiece micrometer or the eikonometer (*vide* page 63), and average the results.

If the organism is one which forms spores, observe—

(*a*) *Spore Formation.*—Prepare hanging-drop cultivations (*vide* page 78) from vegetative forms of the organism, adding a trace of magenta solution (0.5 per cent.) or other intra vitam stain (see page 77) to the drop, on the point of the platinum needle, to facilitate the observation of the phenomenon by rendering the bacilli more distinct.

Place the preparation on the stage of the microscope; if necessary, using a warm stage.

Arrange illumination, etc., and select a solitary bacillus for observation, by the help of the 1/6-inch lens.

Substitute the 1/12-inch oil-immersion lens for the sixth, and observe the formation of the spore; if possible, measure any alteration in size which may occur by means of the Ramsden micrometer.

(*b*) *Spore Germination.*—Prepare hanging-drop cultivations from old cultivations in which no living vegetative forms are present, and observe the process of germination in a similar manner.

The comfort of the microscopist is largely enhanced in those cases where the period of observation is at all lengthy, by use of some form of eye screen before the unemployed eye, such as is figured on page 58 (Fig. 49).

If it is impossible to carry out the method suggested above, proceed as follows:

(*a*) *Spore Formation.*—Plant the organism in broth and incubate under optimum conditions.

At regular intervals, say every thirty minutes, remove a loopful of the cultivation and prepare a cover-slip film preparation.

Fix, while still wet, in the corrosive sublimate fixing solution.

Stain with aniline gentian violet, and partially decolourise with 2 per cent. acetic acid.

Mount and number consecutively; then examine.

(*b*) *Spore Germination.*—Expose a thick emulsion of the spores to a temperature of 80° C. for ten minutes in the differential steriliser (*vide* page 257).

Transfer the emulsion to a tube of sterile nutrient broth and incubate.

Remove specimens from the tube culture at intervals of, say, five minutes.

Fix, stain, etc., wet, as under (*a*), and examine.

(B) Fixed.—2. In stained preparations.

(*a*) To determine points in *morphology*:

Shape (*vide* classification, page 131).

Size:

(*a*) Prepare cover-slip film preparations at the various ages, and fix by exposure to a temperature of 115° C. for twenty minutes in hot-air oven.

(*b*) Stain the preparations by Gram's method (if applicable) or with dilute carbol-fuchsin, and mount in the usual way.

(*c*) Measure (*vide* page 66) some twenty-five individuals in each film by means of the Ramsden's or the stage micrometer and average the result.

Pleomorphism; If noted, record—

The predominant character of the variant forms. On what medium or media they are observed. At what period of development.

(*b*) To demonstrate details of *structure*:

Flagella: If noted, record—

Method of staining (*vide* page 101). Position and arrangement (*vide* page 136). Number.

Spores: If noted, record—

Method of staining. Shape. Size. Position within the parent cell. Condition, as to shape, of the parent cell (*vide* page 139). Optimum medium and temperature. Age of cultivation. Conditions of environment as to temperature, atmosphere. Method of germination (*vide* page 140).

Involution Forms: If noted, record—

Method of staining. Character (*e. g.*, if living or dead). Shape. On what medium they are observed. Age of medium. Environment.

Metachromatic Granules: If noted, record—

Method of staining. Character of granules. Number of granules. Colour of granules.

3. Staining Reactions.—

1. *Gram's Method.*—Positive or negative.
2. *Neisser's Method.*—If granules are noted, record—
1. Position. 2. Number.
3. *Ziehl-Neelsen's Method.*—Acid-fast or decolourised.
4. *Simple Aniline Dyes.*—(Noting those giving the best results, with details of staining processes.)

Methylene-blue }

Fuchsin } and their modifications.

Gentian violet }

Thionine blue }

BY BIOCHEMICAL METHODS.

Test cultivations of the organism for the presence of—

Soluble enzymes—proteolytic, diastatic, invertase.

Organic acids—(*a*) quantitatively—*i. e.*, estimate the total acid production; (*b*) qualitatively for formic, acetic, propionic, butyric, lactic.

Ammonia.

Neutral volatile substances—ethyl alcohol, aldehyde, acetone.

Aromatic products—indol, phenol.

Soluble pigments.

Test the power of reducing (*a*) colouring matters, (*b*) nitrates to nitrites.

Investigate the gas production—H_2S, CO_2, H_2. Estimate the ratio between the last two gases.

Prepare all cultivations for these methods of examination under *optimum* conditions, previously determined for each of the organisms it is intended to investigate, as to

(*a*) Reaction of medium; (*b*) Incubation temperature; (*c*) Atmospheric environment;

and keep careful records of these points, and also of the age of the cultivation used in the final examination.

Examine the cultivations for the various products of bacterial metabolism after forty-eight hours' growth, and **never omit to examine "control" (uninoculated) tube or flask of medium from the same batch, kept for a similar period under identical conditions**.

If the results are negative, test further cultivations at three days, five days, and ten days.

1. Enzyme Production.—

(*A*) *Proteolytic Enzymes.*—(Convert proteins into proteose, peptone and further products of hydrolysis; *e. g.*, B. pyocyaneus.)

Media Required:
Blood-serum and milk-serum which have been carefully filtered through a porcelain candle.
Reagents Required:
Ammonium sulphate. Thirty per cent. caustic soda solution. Copper sulphate, 0.5 per cent. aqueous solution. One per cent. acetic acid solution. Millon's reagent. Glyoxylic acid solution. Concentrated sulphuric acid.
Method.—
1. Prepare cultivations in bulk (50 c.c.) in a flask and incubate.
2. Make the liquid faintly acid with acetic acid, then boil. (This precipitates the unaltered proteins.)
3. Filter.
4. Take 10 c.c. of the filtrate in a test-tube and add 1 c.c. of the caustic soda, then add the copper sulphate drop by drop.
Pink colour which becomes violet with more copper sulphate = proteose and peptone.
5. Saturate the rest of the filtrate with ammonium sulphate.
Precipitate = proteose.
6. Filter and divide the filtrate into three parts *a*, *b* and *c*.
a. Repeat the copper sulphate test, using excess of caustic soda to displace the ammonia from the ammonium sulphate.
Pink colour = peptone.
b. Boil with Millon's reagent.
Red colour = tyrosine.
c. Add glyoxylic acid solution and run in concentrated sulphuric acid.
Violet ring at upper level of acid = tryptophane.
Both the tyrosine and tryptophane may be either in the free state or in combination as polypeptid or peptone.
(*B*) *Diastase*.—(Converts starch into sugar; *e. g.*, B. subtilis.)
Medium Required:
Inosite-free bouillon.
Reagents Required:
Starch. Thymol. Fehling's solution.
Method.—
1. Prepare tube cultivation and incubate.
2. Prepare a thin starch paste and add 2 per cent. thymol to it.
3. Mix equal parts of the cultivation to be tested and the starch paste, and place in the incubator at 37°C. for six to eight hours.
4. Filter.
Test the filtrate for sugar.
Boil some of the Fehling's solution in a test-tube.
Add the filtrate drop by drop until, if necessary, a quantity has been added equal in amount to the Fehling's solution employed, keeping the mixture at the boiling-point during the process.
Yellow or orange precipitate = sugar.
(*C*) *Invertase*.—(Convert saccharose into a mixture of dextrose and lævulose *e. g.*, B. fluorescens liquefaciens.)
Medium Required: Inosite-free bouillon.
Reagents Required: Cane sugar, 2 per cent. aqueous solution. Carbolic acid.
Method.—
1. Prepare tube cultivations and incubate.
2. Add 2 per cent. of carbolic acid to the sugar solution.
3. Mix equal quantities of the carbolised sugar solution and the cultivation in a test-tube; allow the mixture to stand for several hours.
4. Filter.
Test the filtrate for reducing sugar as in the preceding section.
(*D*) *Rennin and "Lab" Enzymes*.—(Coagulate milk independently of the action of acids; *e. g.*, B. prodigiosus.)
Media Required: Inosite-free bouillon. Litmus milk.
Method.—
1. Prepare tube cultivations and incubate.
2. After incubation heat the cultivation to 55° C. for half an hour, to sterilise.
3. By means of a sterile pipette run 5 c.c. of the cultivation into each of three tubes of litmus milk.
4. Place in the cold incubator at 22° C. and examine each day for ten days.
Absence of coagulation at the end of that period will indicate absence of rennin ferment formation.
Fermentation Reactions.
As tested upon carbohydrate substances and organic salts.
Media Required:
Peptone water containing various percentages (generally 2 per cent.) of each of the substances referred to under "sugar" media (page 177), also tubes of peptone water containing 1 per cent. respectively of each of the following:
Organic salts: Sodium citrate, formate, lactate, malate, tartrate.
Method.—
1. Prepare tube cultivations in each of the above media.
2. Observe from day to day up to the expiration of ten days if necessary.
3. Note growth, reaction, gas production.
2. Acid Production.

(*a*) *Quantitative.—*

Medium Required: Sugar (glucose) bouillon of known "optimum" reaction.

Apparatus and Reagents Required: As for estimating reaction of media (*vide* page 150).

Method.—

1. Prepare cultivation in bulk (100 c.c.) in a flask; also "control" flask of medium from same batch.

2. After suitable incubation, heat both flasks in the steamer at 100° C. for thirty minutes to sterilise.

3. Determine the *titre* of the medium in "inoculated" and "control" flasks as described in the preparation of nutrient media (*vide* page 151).

4. The difference between the titre of the medium in the two flasks gives the total acid production of the bacterium under observation in terms of normal NaOH.

Note.—If the growth is very heavy it may be a difficult matter to determine the end-point. The cultivation should then be filtered through a Berkefeld filter candle previous to step 2, and the filtrate employed in the titration.

(*b*) *Qualitative* (of all the organic acids present).—

Medium Required: Sugar (glucose or lactose) bouillon as in quantitative examination.

Reagents Required: Hydrochloric acid, concentrated. Hydrochloric acid, 25 per cent. Sulphuric acid, concentrated (pure). Phosphoric acid, concentrated solution. Ammonia. Ammonium sulphate. Baryta water. Sodium carbonate, saturated aqueous solution. Absolute alcohol. Ether. Calcium chloride. Calcium chloride solution. Zinc carbonate. Copper sulphate saturated aqueous solution. Alcoholic thiophene solution (0.15 c.c. in 100 c.c.). Animal charcoal. Five per cent. sodium nitroprusside solution. Potassium bichromate. Schiff's reagent. Arsenious oxide. Ferric chloride, 4 per cent. aqueous solution. Silver nitrate, 1 per cent. aqueous solution. Lugol's iodine. Ten per cent. caustic soda solution. Hard paraffin wax (melting-point about 52° C.).

Method.—

1. Prepare cultivation in bulk (500 c.c.) in a litre flask and add sterilised precipitated chalk, 10 grammes. Incubate at the optimum temperature.

2. After incubation throw a piece of paraffin wax (about a centimetre cube) into the cultivation and connect up the flask with a condenser.

The paraffin, which liquefies and forms a thin layer on the surface of the fluid, is necessary to prevent the cultivation frothing up and running unaltered through the condenser during the subsequent process of distillation.

3. Distill over 200 to 300 c.c.

Use a rose-top burner to minimise the danger of cracking the flask; and to the same end, well agitate the contents of the flask to prevent the chalk settling.

The distillate "**A**" will contain alcohol, etc. (*vide* page 285); the residue "**a**" will contain the volatile and fixed acids.

4. Disconnect the flask and filter. The residue "**a**" then = filtrate **B** and residue **b**.

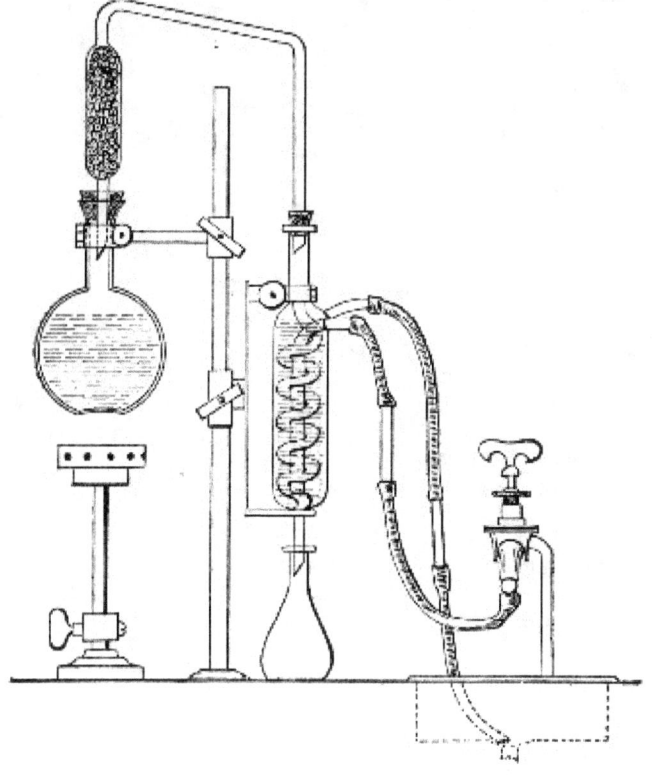

Fig. 152.—Arrangement of distillation apparatus for acids, etc.

5. Residue **b**. Wash the residue from the filter paper, dissolve by heating with dilute hydrochloric acid, and add calcium chloride solution and ammonia until alkaline.

White precipitate insoluble in acetic acid = oxalic acid.

6. Make up filtrate **B** to 500 c.c. with distilled water and divide into two parts.

7. Acidify 250 c.c. with 20 c.c. concentrated phosphoric acid (this liberates the volatile acids) and distil to small bulk.

The distillate "B" may contain formic, acetic, propionic, butyric and benzoic acids.

```
                    DISTILLATE "B."
                    (Volatile Acids.)

       1. Add baryta water till alkaline,
          and evaporate to dryness.

       2. Add 50 c.c. absolute alcohol and allow
          to stand, with frequent stirring, for
          two to three hours.

       3. Filter and wash with alcohol.
```

           FILTRATE   RESIDUE

may contain barium propionate,   may contain barium acetate,
barium butyrate.   barium formate, barium benzoate.

1. Evaporate to dryness.   1. Evaporate off alcohol and
                                        dissolve up the residue on
2. Dissolve residue in 150   the filter in hot water and
c.c. water.   neutralise.

3. Acidify with phosphoric   2. Divide the solution into
acid and distil.   four portions:

4. Saturate distillate with   (a) Add ferric chloride solution.
calcium chloride and distill
over a few c.c.   **Brown** colour = *acetic* or
                                         formic acids.
5. Test distillate for butyric
acid: **Buff ppt.** = *benzoic* acid
                                         (see ether soluble acids).
Add 3 c.c. alcohol and 4 drops
concentrated sulphuric acid.   (b) Add silver nitrate
                                     solution; then add one drop
   Smell of pineapple = butyric ammonia water, and boil.
   acid.
                                         Black precipitate of metallic
Propionic acid in small   silver = *formic* acid.
quantities cannot be
distinguished from butyric   (c) Evaporate to dryness; mix
acid by tests within the   with equal quantity of
scope of the bacteriological   arsenious oxide and heat
laboratory.   on platinum foil.

                                         Unpleasant **smell of cacodyl**
                                         = *acetic* acid.

                                     (d) Add a few drops of
                                     mercuric chloride solution
                                     in test-tube, and heat to

$70°$ C.

Precipitate of mercurous chloride which is slowly reduced to mercury = *formic* acid.

8. If the distillation of "B" is continued as long as acid comes over (distilled water being occasionally added to the distilling flask) the distillate can be measured and 50 c.c. used for titration. This will give the amount of volatile acid formation.

9. The second part of the filtrate "B" (see page 282) should be examined for lactic, oxalic, succinic, benzoic, salicylic, gallic and tannic acids, as follows:

Ether Soluble Acids.—

1. Evaporate to a thin syrup, acidify strongly with phosphoric acid.
2. Extract with five times its volume of ether by agitation in a separatory funnel.
3. Evaporate the ethereal extract to a thin syrup.
4. Add 100 c.c. water and mix thoroughly.
5. To a small portion of this solution add slight excess of sodium carbonate, evaporate to dryness on the water-bath, dissolve in 5-10 c.c. pure sulphuric acid, add 2 drops saturated copper sulphate solution, place in a test-tube and heat in a boiling water-bath for 2 minutes, cool, add 2 or 3 drops of the alcoholic thiophene and warm gently.

Cherry red colour = lactic acid.

If a brown colour is produced on the addition of sulphuric acid, another sample should be taken and boiled with animal charcoal before evaporating.

6. If lactic acid is definitely present, prepare zinc lactate by boiling part of the solution of the ether extract with excess of zinc carbonate, filtering and evaporating to crystallise. The crystals so obtained have a characteristic form, and if dried at $110°$ C, should contain 26.87 per cent. of zinc.

7. Test a portion of the rest of the solution of the ether extract for oxalic acid (page 282, step 5). Carefully neutralise the remainder and add ferric chloride solution.

Red brown gelatinous precipitate = succinic acid.

Buff precipitate = benzoic acid, and other acids related to benzoic acid.

Violet colour = salicylic acid.

Inky black colour or precipitate = gallic acid or tannic acid.

For further identification the melting-points of the crystalline acids, and the percentage of silver in their silver salts should be determined.

3. Ammonia Production.—

Medium Required: Nutrient bouillon.

Reagent Required: Nessler reagent.

Method.—

1. Prepare cultivation in bulk (100 c.c.) in a 250 c.c. flask and incubate together with a control flask.

Test the cultivation and the control for ammonia in the following manner:

2. To each flask add 2 grammes of calcined magnesia, then connect up with condensers and distil.
3. Collect 50 c.c. distillate, from each, in a Nessler glass.
4. Add 1 c.c. Nessler reagent to each glass by means of a clean pipette.

Yellow colour = ammonia.

The depth of colour is proportionate to the amount present.

4. Alcohol, etc., Production.—Divide the distillate "A" obtained in the course of a previous experiment (*vide* page 282, step 3) into four portions and test for the production of alcohol, acetaldehyde, acetone.

1. Add Lugol's iodine, then a little NaOH solution, and stir with a glass rod till the colour of the iodine disappears.

Pale-yellow crystalline precipitate of iodoform, with its characteristic smell, appearing in the cold, indicates acetaldehyde, or acetone; appearing only on warming indicates alcohol.

The precipitate may be absent even when the odour is pronounced.

2. Add Schiff's reagent.

Violet or red colour = aldehyde.

3. To 10 c.c. of solution add 2.5 c.c., 25 per cent. sulphuric acid, and a crystal or two of potassium bichromate and distil. Reduction of the bichromate to a green colour and a distillate, which smells of acetaldehyde and reacts with Schiff's reagent, shows the presence of alcohol in the original liquid.
4. Add a few drops of sodium nitroprusside solution, make alkaline with ammonia, then saturate with ammonium sulphate crystals. Acetone gives little colour on the addition of ammonia, but after the addition of ammonium sulphate a deep permanganate colour, which takes ten minutes to reach its full intensity. Aldehyde gives a carmine red unaltered by ammonium sulphate.

5. Indol Production.—

Media Required:

Inosite-free bouillon (*vide* page 183). Or peptone water (*vide* page 177).

Reagents Required:

Potassium persulphate, saturated aqueous solution. Paradimethylamino-benzaldehyde solution. This is prepared by mixing:

Paradimethylamino-benzaldehyde 4 grammes Absolute alcohol 380 c.c. Hydrochloric acid, concentrated 80 c.c.

Method.—

Prepare several test-tube cultivations of the organism to be tested, and incubate.

Test for indol by means of the Rosindol reaction in the following manner. (If the culture has been incubated at 37°C., it must be allowed to cool to the room temperature before applying the test.)

1. Remove 2 c.c. of the cultivation by means of a sterile pipette and transfer to a clean tube, then,
2. Add 2 c.c. paradimethylamino-benzaldehyde solution.
3. Add 2 c.c. potassium persulphate solution.

The presence of indol is indicated by the appearance of a delicate rose-pink colour throughout the mixture which deepens slightly on standing.

Indol is tested for in many laboratories by the ordinary nitrosoindol reaction which, however, is not so delicate a method as that above described. The test is carried out as follows:

1. Remove the cotton-wool plug from the tube, and run in 1 c.c. pure concentrated sulphuric acid down the side of the tube by means of a sterile pipette. Place the tube upright in a rack, and allow it to stand, if necessary, for ten minutes.

A rose-pink or red colour at the junction of the two liquids = indol (*plus a nitrite*).

2. If the colour of the medium remains unaltered, add 2 c.c. of a 0.01 per cent. aqueous solution sodium nitrite, and again allow the culture to stand for ten minutes.

Red colouration = indol.

Note.—In place of performing the test in two stages as given above, 2 c.c. concentrated *commercial* sulphuric, hydrochloric, or nitric acid (all of which hold a trace of nitrite in solution), may be run into the cultivation. The development of a red colour within twenty minutes will indicate the presence of indol.

5a. Phenol Production.—
Medium Required:
Nutrient bouillon.
Reagents Required:
Hydrochloric acid, concentrated. Millon's reagent. Ferric chloride, 1 per cent. aqueous solution.
Method.—
1. Prepare cultivation in a Bohemian flask containing at least 50 c.c. of medium, and incubate.
Test for phenol in the following manner:
2. Add 5 c.c., 25 per cent. sulphuric acid to the cultivation and connect up the flask with a condenser.
3. Distil over 15 to 20 c.c. Divide the distillate into three portions *a*, *b* and *c*.
4. Add to (*a*) 0.5 c.c. Millon's reagent and boil.
Red colour = phenol.
5. Add to (*b*) about 0.5 c.c. ferric chloride solution. Violet colour = phenol.
(If the distillate be acid the reaction will be negative.)
6. Add to (*c*) bromine water. Crystalline white ppt. of tribromo-phenol = phenol.

Note.—If both indol and phenol appear to be present in cultivations of the same organism, it is well to separate them before testing. This may be done in the following manner:

1. Prepare inosite-free bouillon cultivation, say 200 or 300 c.c., in a flask as before.
2. Render definitely acid by the addition of acetic acid and connect up the flask with a condenser.
3. Distil over 50 to 70 c.c.
Distillate will contain both indol and phenol.
4. Render the distillate strongly alkaline with caustic potash and redistil.
Distillate will contain indol; residue will contain phenol.
5. Test the distillate for indol (*vide ante*).
6. Saturate the residue, when cold, with carbon dioxide and redistil.
7. Test this distillate for phenol (*vide ante*).

6. Pigment Production.—
1. Prepare tube cultivations upon the various media and incubate under varying conditions as to temperature (at 37° C. and at 20°C.), atmosphere (aerobic and anaerobic), and light (exposure to and protection from).
Note the conditions most favorable to pigment formation.
2. Note the solubility of the pigment in various solvents, such as water (hot and cold), alcohol, ether, chloroform, benzol, carbon bisulphide.
3. Note the effect of acids and alkalies respectively upon the pigmented cultivation, or upon solutions of the pigment.
4. Note spectroscopic reactions.

7. Reducing Agent Formation.—
(*a*) *Colour Destruction.—*
1. Prepare tube cultivations in nutrient bouillon tinted with litmus, rosolic acid, neutral red, and incubate.
2. Examine the cultures each day and note whether any colour change occurs.
(*b*) *Nitrates to Nitrites.—*
Medium Required:
Nitrate bouillon (*vide* page 185). Or nitrate peptone solution (*vide* page 186).
Reagents Required:
Sulphuric acid (25 per cent.). Metaphenylene diamine, 5 per cent. aqueous solution.
Method.—

1. Prepare tube cultivations and incubate together with control tubes (*i. e.*, uninoculated tubes of the same medium, placed under identical conditions as to environment).

This precaution is necessary as the medium is liable to take up nitrites from the atmosphere, and an opinion as to the absence of nitrites in the cultivation is often based upon an equal colouration of the medium in the control tube.

Test both the culture tube and the control tube for the presence of nitrites.

2. Add a few drops of sulphuric acid to the medium in each of the tubes.
3. Then run in 2 or 3 c.c. metaphenylene diamine into each tube. Brownish-red colour = nitrites.

The depth of colour is proportionate to the amount present.

8. Gas Production.—

(*A*) *Carbon Dioxide and Hydrogen.—*

Apparatus Required:

Fermentation tubes (*vide* page 161) containing sugar bouillon (glucose, lactose, etc.). The medium should be prepared from inosite-free bouillon (*vide* page 183).

Reagent Required:

n/2 caustic soda.

Method.—

1. Inoculate the surface of the medium in the bulb of a fermentation tube and incubate.
2. Mark the level of the fluid in the closed branch of the fermentation tube, at intervals of twenty-four hours, and when the evolution of gas has ceased, measure the length of the column of gas with the millimetre scale.

Express this column of gas as a percentage of the entire length of the closed branch.

3. To analyse the gas and to determine roughly the relative proportions of CO_2 and H_2, proceed as follows:

Fill the bulb of the fermentation tube with caustic soda solution.

Close the mouth of the bulb with a rubber stopper.

Alternately invert and revert the tube six or eight times, to bring the soda solution into intimate contact with the gas.

Return the residual gas to the end of the closed branch, and measure.

The loss in volume of gas = carbon dioxide.

The residual gas = hydrogen.

Transfer gas to the bulb of the tube, and explode it by applying a lighted taper.

(*B*) *Sulphuretted Hydrogen.—*

Media Required:

Iron peptone solution (*vide* page 185). Lead peptone solution.

1. Inoculate tubes of media, and incubate together with control tubes.
2. Examine from day to day, at intervals of twenty-four hours.

The liberation of the H_2S will cause the yellowish-white precipitate to darken to a brownish-black, or jet black, the depth of the colour being proportionate to the amount of sulphuretted hydrogen present.

Quantitative: For exact quantitative analyses of the gases produced by bacteria from certain media of definite composition, the methods devised by Pakes must be employed, as follows:

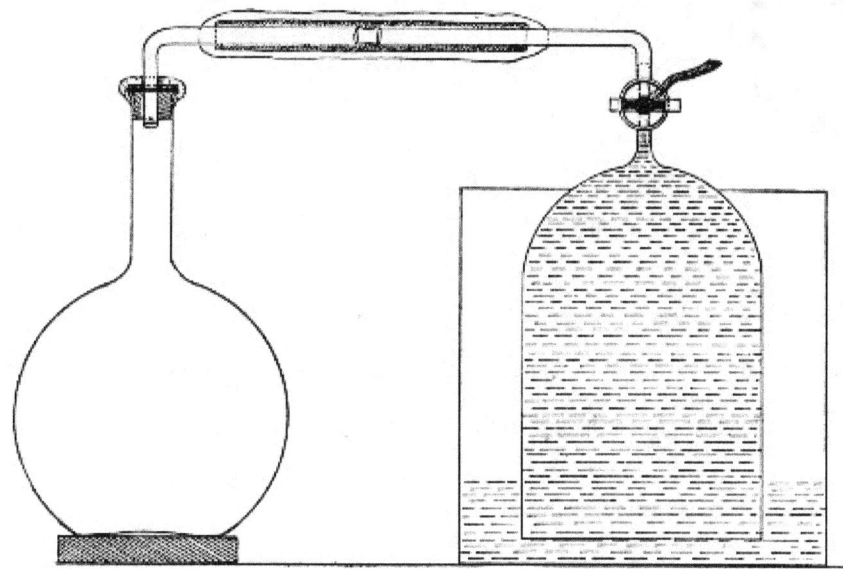

Fig. 153.—Gas-collecting apparatus.

Apparatus Required:

Bohemian flask (300 to 1500 c.c. capacity) containing from 100 to 400 c.c. of the medium. The mouth of the flask is fitted with a perforated rubber stopper, carrying an L-shaped piece of glass tubing (the short arm passing just through the stopper). To the long arm of the tube is attached a piece of pressure tubing some 8 cm. in length, plugged at its free end with a piece of cotton-wool. Measure accurately the total capacity of the flask and exit tube, also the amount of medium contained. Note the difference.

Gas receiver. This is a bell jar of stout glass, 14 cm. high and 9 cm. in diameter. At its apex a glass tube is fused in. This rises vertically 5 cm., and is then bent at right angles, the horizontal arm being 10 cm. in length. A three-way tap is let horizontally into the vertical tube just above its junction with the bell jar.

An iron cylinder just large enough to contain the bell jar.

About 15 kilos of metallic mercury.

Melted paraffin.

An Orsat-Lunge working with mercury instead of water, provided with two gas tubes of extra length (capacity 120 and 60 c.c. respectively and graduated throughout, both being water-jacketed) or other gas analysis apparatus, capable of dealing with CO_2, O_2, H_2, and N_2.

Method.—

1. Inoculate the medium in the flask in the usual manner, by means of a platinum needle, taking care that the neck of the flask and the rubber stopper are thoroughly flamed before and after the operation.

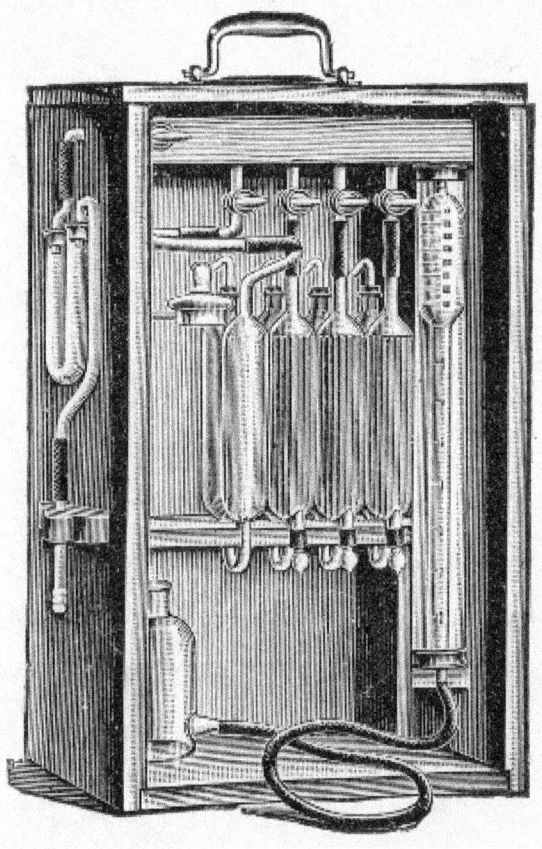

Fig. 154.—Orsat-Lunge gas analysis apparatus.

2. Fill the iron cylinder with mercury.

3. Place the bell jar mouth downward in the mercury—first seeing that there is free communication between the interior of the jar and the external air—and suck up the mercury into the tap; then shut off the tap.

4. Plug the open end of the three-way tap with melted wax.

5. Connect up the horizontal arm of the culture flask with that of the gas receiver by means of the pressure tubing (after removing the cotton-wool plug from the rubber tube), as shown in Fig. 153.

6. Give the three-way tap half turn to open communication between flask and receiver, and seal *all* joints by coating with a film of melted wax. When the tap is turned, the mercury in the receiver will naturally fall.

7. Place the entire apparatus in the incubator. (Two hours later, by which time the temperature of the apparatus is that of the incubator, mark the height of the mercury on the receiver.)

8. Examine the apparatus from day to day and mark the level of the mercury in the receiver at intervals of twenty-four hours.

9. When the evolution of gas has ceased, remove the apparatus from the incubator; clear out the wax from the nozzle of the three-way tap (first adjusting the tap so that no escape of gas shall take place) and connect it with the Orsat.

10. Remove, say, 100 c.c. of gas from the receiver, reverse the tap and force it into the culture flask. Remove 100 c.c. of mixed gases from the culture flask and replace in the receiver.

Repeat these processes three or four times to ensure thorough admixture of the contents of flask and receiver.

11. Now withdraw a sample of the mixed gases into the Orsat and analyse.

In calculating the results be careful to allow for the volume of air contained in the flask at the commencement of the experiment.

For the collection of gases formed under anaerobic conditions a slightly different procedure is adopted:

1. Fix a culture flask (500 c.c. capacity) with a perforated rubber stopper carrying an **L**-shaped piece of manometer tubing, each arm 5 cm. in length.

2. Prepare a second **L**-shaped piece of tubing, the short arm 5 cm. and the long arm 20 cm., and connect its short arm to the horizontal arm of the tube in the culture flask by means of a length of pressure tubing, provided with a screw clamp.

3. Fill the culture flask completely with boiling medium and pass the long piece of tubing through the plug of an Erlenmeyer flask (150 c.c. capacity) which contains 100 c.c. of the same medium.

4. Sterilise these coupled flasks by the discontinuous method, in the usual manner.

Immediately the last sterilisation is completed, screw up the clamp on the pressure tubing which connects them, and allow them to cool. As the fluid cools and contracts it leaves a vacuum in the neck of the flask below the rubber stopper.

5. To inoculate the culture flask, withdraw the long arm of the bent tube from the Erlenmeyer flask and pass it to the bottom of a test-tube containing a young cultivation (in a fluid medium similar to that contained in the culture flask) of the organism it is desired to investigate.

6. Slightly release the clamp on the pressure tubing to allow 4 or 5 c.c. of the culture to enter the flask.

7. Clamp the rubber tube tightly; remove the bent glass tube from the culture tube and plunge it into a flask containing recently boiled and quickly cooled distilled water.

8. Release the clamp again and wash in the remains of the cultivation until the culture flask and tubing are completely filled with water.

9. Clamp the rubber tubing tightly and take away the long-armed glass tubing.

10. Prepare the gas receiver as in the previous method (in this case, however, the mercury should be warmed slightly) and fill the horizontal arm of the receiver with hot water.

11. Connect up the culture flask with the horizontal arm of the gas receiver.

12. Remove the screw clamp from the rubber tubing, adjust the three-way tap, seal all joints with melted wax, and incubate.

13. Complete the investigation as described for the previous method.

BY PHYSICAL METHODS.

Examine cultivations of the organism with reference to its growth and development under the following headings:

Atmosphere:
(a) In the presence of oxygen.
(b) In the absence of oxygen.
(c) In the presence of gases other than oxygen.

Temperature:
(a) Range.
(b) Optimum.
(c) Thermal death-point:

Moist: Vegetative forms.

Spores.

Dry: Vegetative forms.

Spores.

Reaction of medium.

Resistance to lethal agents:
(a) Desiccation.
(b) Light: Diffuse.
Direct.
Primary colours.
(c) Heat.
(d) Chemical antiseptics and disinfectants.

Vitality in artificial cultures.

I. Atmosphere.—The question as to whether the organism under observation is (a) an obligate aerobe, (b) a facultative anaerobe, or (c) an obligate anaerobe is roughly decided by the appearance of cultivations in the fermentation tubes. Obvious growth in the closed branch as well as in the bulb or in the inverted gas tube as well as in the bulk of the medium will indicate that it is a facultative anaerobe; whilst growth only occurring in the bulb or in the closed branch shows that it is an obligate aerobe or anaerobe respectively. This method, however, is not sufficiently accurate for the present purpose, and the examination of an organism with respect to its behaviour in the absence of oxygen is carried out as follows:

Apparatus Required:

Buchner's tubes. Bulloch's apparatus. Exhaust pump. Pyrogallic acid. Dekanormal caustic soda.

Media Required:

Glucose formate agar. Glucose formate gelatine. Glucose formate bouillon.

Method.—

1. Prepare four sets of cultivations:

(A) Sloped glucose formate agar, and incubate aerobically at 37° C.

Sloped glucose formate gelatine, and incubate aerobically at 20° C.

(B) Sloped glucose agar to incubate anaerobically at 37° C.

Sloped glucose formate gelatine to incubate anaerobically at 20° C.

(C) Sloped glucose formate agar to incubate anaerobically at 37° C.

Glucose formate bouillon to incubate anaerobically at 37° C.

(D) Sloped glucose formate gelatine to incubate anaerobically at 20° C.

Glucose formate bouillon to incubate anaerobically at 20° C.

2. Seal the cultures forming set B in Buchner's tubes (*vide* page 239).

3. Seal the cultures forming set C in Bulloch's apparatus; exhaust the air by means of a vacuum pump, and provide for the absorption of any residual oxygen by the introduction of pyrogallic acid and caustic soda in solution (*vide* page 245). Treat set D in the same way.

4. Observe the cultivations macroscopically and microscopically at intervals of twenty-four hours until the completion, if necessary, of seven days' incubation.

5. Control these results.

Gases Other than Oxygen.—

Apparatus Required:

Bulloch's apparatus. Sterile gas filter (*vide* page 40). Gasometer containing the gas it is desired to test (SO_2, N_2O, NO, CO_2, etc.) or gas generator for its production.

Method.—

1. Prepare at least seven tube cultivations upon solid media and deposit them in Bulloch's apparatus.

2. Connect up the inlet tube of the Bulloch's jar with the sterile gas filter, and this again with the delivery tube of the gasometer or gas generator.

3. Open both stop-cocks of the Bulloch's apparatus and pass the gas through until it has completely replaced the air in the bell jar as shown by the result of analyses of samples collected from the exit tube.

4. Incubate under optimum conditions as to temperature.

5. Examine the cultivations at intervals of twenty-four hours, until the completion of seven days.

6. Remove one tube from the interior of the apparatus each day. If no growth is visible, incubate the tube under optimum conditions as to temperature *and* atmosphere, and in this way determine the length of exposure to the action of the gas necessary to kill the organisms under observation.

7. Control these results.

II. Temperature.—

(A) *Range.—*

1. Prepare a series of ten tube cultivations, in fluid media, of optimum reaction.

2. Arrange a series of incubators at fixed temperatures, varying 5° C. and including temperatures between 5° C. and 50° C.

(In the absence of a sufficient number of incubators utilise the water-bath employed in testing the thermal death-point of vegetative forms.)

3. Incubate one tube cultivation of the organism aerobically or anaerobically, as may be necessary, in each incubator, and examine at half-hour intervals for from five to eighteen hours.

4. Note that temperature at which growth is first observed macroscopically (Optimum temperature).

5. Continue the incubation until the completion of seven days. Note the extremes of temperature at which growth takes place (Range of temperature).

6. Control these results—if considered necessary arranging the series of incubators to include each degree centigrade for five degrees beyond each of the extremes previously noted.

(B) *Optimum.—*

1. Prepare a second series of ten tube cultivations under similar conditions as to reaction of medium.

2. Incubate in a series of incubators in which the temperature is regulated at intervals of 1° C. for five degrees on either side of optimum temperature observed in the previous experiment (A, step 4).

3. Observe again at half-hour intervals and note that temperature at which growth is first visible to the naked eye = Optimum temperature.

(C) *Thermal Death-point (t. d. p.)—*

Moist—Vegetative Forms:

The *t. d. p.* here is that **temperature** which with certainty kills a watery suspension of the organisms in question after an exposure of **10 minutes**.

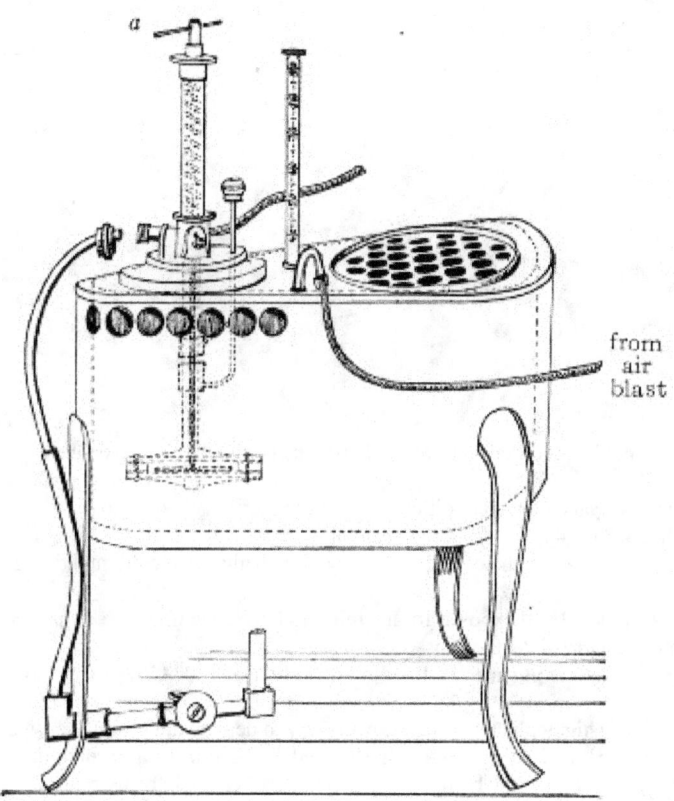

Fig. 155.—Hearson's water-bath.

Apparatus Required:

Water-bath. For the purpose of observing the thermal death-point a special water-bath is necessary. The temperature of this piece of apparatus is controlled by means of a capsule regulator that can be adjusted for intervals of half a degree centigrade through a range of 30°, from 50° C. to 80° C. by means of a spring, actuated by the handle *a*, which increases the pressure in the interior of the capsule. A hole is provided for the reception of the nozzle of a blast pump, so that a current of air may be blown through the water while the bath is in use, and thus ensure a uniform temperature of its contents. Through a second hole is suspended a certified centigrade thermometer, the bulb of which although completely immersed in the water is raised at least 2 cm. above the floor of the bath.

Sterile glass capsules.
Flask containing 250 c.c. sterile normal saline solution.
Case of sterile pipettes, 10 c.c. (in tenths of a cubic centimetre).
Special platinum loop.
Test-tubes, 18 by 1.5 cm., of thin German glass.
Case of sterile petri dishes.
Tubes of agar or gelatine.

Method.—

1. Prepare tube cultivations on solid media of optimum reaction; incubate forty-eight hours under optimum conditions as to temperature and atmosphere.
2. Examine preparations from the cultivation microscopically to determine the absence of spores.
3. Pipette 5 c.c. salt solution into each of twelve capsules.
4. Suspend three loopfuls of the surface growth (using a special platinum loop, *vide* page 316) in the normal saline solution by emulcifying evenly against the moist walls of each capsule.
5. Transfer emulsion from each capsule to sterile 250 c.c. flask, and mix.
6. Pipette 5 c.c. emulsion into each of twelve sterile test-tubes numbered consecutively.
7. Adjust the first tube in the water-bath, regulated at 40° C, by means of two rubber rings around the tube, one above and the other below the perforated top of the bath, so that the upper level of the fluid in the tube is about 4 cm. below the surface of the water in the bath, and the bottom of the tube is a similar distance above the bottom of the bath.
8. Arrange a control test-tube containing 5 c.c. sterile saline solution under similar conditions. Plug the tube with cotton-wool and pass a thermometer through the plug so that its bulb is immersed in the water.
9. Close the unoccupied perforations in the lid of the water-bath by means of glass balls.
10. Watch the thermometer in the test-tube until it records a temperature of 40° C. Note the time. Ten minutes later remove the tube containing the suspension, and cool rapidly by immersing its lower end in a stream of running water.
11. Pour three gelatine (or agar) plates containing respectively 0.2, 0.3, and 0.5 c.c. of the suspension, and incubate.
12. Pipette the remaining 4 c.c. of the suspension into a culture flask containing 250 c.c. of nutrient bouillon, and incubate.

13. Observe these cultivations from day to day. "No growth" must not be recorded as final until after the completion of seven days' incubation.

14. Extend these observations to the remaining tubes of the series, but varying the conditions so that each tube is exposed to a temperature 2° C. higher than the immediately preceding one—*i. e.*, 42° C., 44° C., 46° C., and so on.

15. Note that temperature, after exposure to which no growth takes place up to the end of seven days' incubation, = the thermal death-point.

16. If greater accuracy is desired, a second series of tubes may be prepared and exposed for ten minutes to fixed temperatures varying only 0.5° C., through a range of 5° C. on either side of the previously observed death-point.

Moist—Spores: The thermal death-point in the case of spores is that **time exposure** to a **fixed temperature of 100° C.** necessary to effect the death of all the spores present in a suspension.

Note.—If it is desired to retain the **time constant 10 minutes** and investigate the temperature necessary to destroy the spores, varying amounts of calcium chloride must be added to the water in the bath, when the boiling-point will be raised above 100° C. according to the percentage of calcium in solution. In such case use the bath figured on page 227; the bath figured on page 299 can only be used if the capsule is first removed.

It is determined in the following manner

Apparatus Required:
Steam-can fitted with a delivery tube and a large bore safety-valve tube.
Water-bath at 100° C.
Erlenmeyer flask, 500 c.c. capacity, containing 140 c.c. sterile normal saline solution and fitted with rubber stopper perforated with four holes.

The rubber stopper is fitted as follows:
(a) Thermometer to 120° C., its bulb immersed in the normal saline.
(b) Straight entry tube, reaching to the bottom of the flask, the upper end plugged with cotton-wool.
(c) Bent syphon tube, with pipette nozzle attached by means of rubber tubing and fitted with pinch-cock.
The nozzle is protected from accidental contamination by passing it through the cotton-wool plug of a small test-tube.
(d) A sickle-shaped piece of glass tubing passing just through the stopper, plugged with cotton-wool, to act as a vent for the steam.
Sterile plates.
Sterile pipettes.
Sterile test-tubes graduated to contain 5 c.c.

Media Required:
Gelatine or agar.
Culture flasks containing 200 c.c. nutrient bouillon.

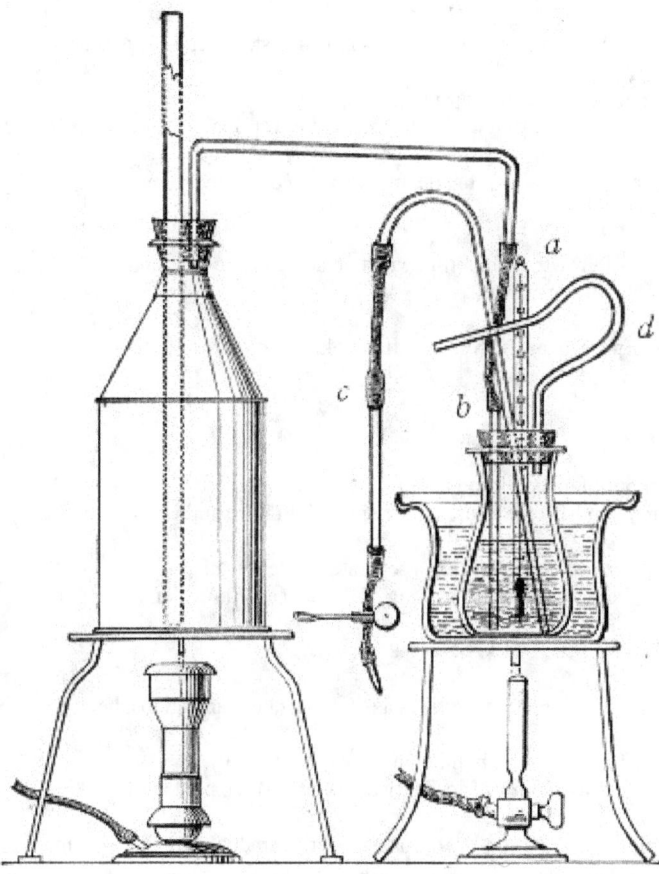

Fig. 156.—Apparatus arranged for the determination of the death-point of spores.

Method.—

1. Prepare twelve tube cultivations upon the surface (or two cultures in large flat culture bottles—*vide* page 5) of nutrient agar and incubate under the optimum conditions (previously determined), for the formation of spores.

Examine preparations from the cultures microscopically to determine the presence of spores.

2. Pipette 5 c.c. sterile normal saline into each culture tube or 30 c.c. into each bottle and by means of a sterile platinum spatula emulsify the entire surface growth with the solution.

3. Add the 60 c.c. emulsion to 140 c.c. normal saline contained in the fitted Erlenmeyer flask.

4. Place the flask in the water-bath of boiling water.

5. Connect up the straight tube, after removing the cotton-wool plug, with the delivery tube of the steam can; remove the plug from the vent tube.

6. When the thermometer reaches 100° C., open the spring clip on the *syphon*, discard the first cubic centimeter of suspension that syphons over (*i. e.*, the contents of the syphon tube); collect the next 5 c.c. of the suspension in the sterile graduated test-tube and pour plates and prepare flask cultures therefrom as in the previous experiments.

7. Repeat this process at intervals of twenty-five minutes' steaming.

8. Observe the inoculated plates and flasks up to the completion, if necessary, of seven days' incubation.

9. Control these experiments, but in this instance syphon off portions of the suspension at intervals of one-half to one minute during the five or ten minutes preceding the previously determined death-point.

Thermal Death-point.—

Dry—Vegetative Forms: The thermal death-point in this case is that **temperature** which with certainty kills a thin film of the organism in question after a time exposure of **ten minutes**.

Apparatus Required:

Hot-air oven, provided with thermo-regulator.
Sterile cover-slips.
Flask containing 250 c.c. sterile normal saline solution.
Case of sterile pipettes, 10 c.c. (in tenths of a cubic centimetre).
Case of sterile capsules.
Crucible tongs.
Method.—

1. Prepare an emulsion with three loopfuls from an optimum cultivation in 5 c.c. normal saline in a sterile capsule and examine microscopically to determine the absence of spore forms.

2. Make twelve cover-slip films on sterile cover-slips; place each in a sterile capsule to dry.

3. Expose each capsule in turn in the hot-air oven for ten minutes to a different fixed temperature, varying 5° C. between 60° C. and 120° C.

4. Remove each capsule from the oven with crucible tongs immediately after the ten minutes are completed; remove the cover-glass from its interior with a sterile pair of forceps.

5. Deposit the film in a flask containing 200 c.c. nutrient bouillon.

6. Prepare subcultivations from such flasks as show evidence of growth, to determine that no accidental contamination has taken place but that the organism originally spread on the film is responsible for the growth.

7. Control the result of these experiments.

Dry—Spores: The thermal death-point in this case is that **temperature** which with certainty kills the spores of the organism in question when present in a thin film after a time exposure of **10 minutes**.

Apparatus Required:

As for vegetative forms.

Method.—

1. Prepare a sloped agar tube cultivation and incubate under optimum conditions as to spore formations.

2. Pipette 5 c.c. sterile normal saline into the culture tube and emulsify the entire surface growth in it. Examine microscopically to determine the presence of spores in large numbers.

3. Spread thin even films on twelve sterile cover-slips and place each cover-slip in a separate sterile capsule.

4. Expose each capsule in turn for ten minutes to a different fixed temperature, varying 5°C, between 100° C. and 160°C.

5. Complete the examination as for vegetative forms.

III. Reaction of Medium.

(*A*) *Range.*—

1. Prepare a bouillon culture of the organism and incubate, under optimum conditions as to temperature and atmosphere, for twenty-four hours.

2. Pipette 0.1 c.c. of the cultivation into a sterile capsule; add 9.9 c.c. sterile bouillon and mix thoroughly.

3. Prepare a series of tubes of nutrient bouillon of varying reactions, from +25 to -30 (*vide* page 155), viz.: +25, +20, +15, +10, +5, neutral, -5, -10, -15, -20, -25, -30.

4. Inoculate each of the bouillon tubes with 0.1 c.c. of the diluted cultivation by means of a sterile graduated pipette and incubate under optimum conditions.

5. Observe the cultures at half-hourly intervals from the third to the twelfth hours. Note the reaction of the tube or tubes in which growth is first visible macroscopically (probably optimum reaction).

6. Continue the incubation until the completion, if necessary, of seven days. Note the extremes of acidity and alkalinity in which macroscopical growth has developed (Range of reaction).

7. Control the result of these observations.

(*B*) *Optimum Reaction.*—The optimum reaction has already been roughly determined whilst observing the range. It can be fixed within narrower limits by inoculating in a similar manner a series of tubes of bouillon which represent smaller variations in reaction than those previously employed (say, 1 instead of 5) for five points on either side of the previously observed optimum. For example, the optimum reaction observed in the set of experiments to determine the range was +10. Now plant tubes having reactions of +15, +14, +13, +12, +11, +10, +9, +8, +7, + 6, +5, and observe as before.

IV. Resistance to Lethal Agents.—

(*A*) *Desiccation.*—

Apparatus Required:

Mueller's desiccator. This consists of a bell glass fitted with an exhaust tube and stop-cock (*d*), which can be secured to a plate-glass base (*c*) by means of wax or grease. It contains a cylindrical vessel of porous clay (*a*) into the top of which pure sulphuric acid is poured whilst the material to be dried is placed within its walls on a glass shelf (*b*). The air is exhausted from the interior and the acid rapidly converts the clay vessel into a large absorbing surface (Fig. 157).

Exhaust pump.

Pure concentrated sulphuric acid.

Sterile cover-slips.

Sterile forceps.

Culture flask containing 200 c.c. nutrient bouillon.

Sterile ventilated Petri dish. This is prepared by bending three short pieces of aluminium wire into V shape and hanging these on the edge of the lower dish and resting the lid upon them (Fig. 158).

Method.—

1. Prepare a surface cultivation on nutrient agar in a culture bottle and incubate under optimum conditions for forty-eight hours.

2. Examine preparations from the cultivation, microscopically, to determine the absence of spores.

3. Pipette 5 c.c. sterile normal saline solution into the flask and suspend the entire growth in it.

4. Spread the suspension in thin, even films on sterile cover-slips and deposit inside sterile "plates" to dry.

5. As soon as dry, transfer the cover-slip films to the ventilated Petri dish by means of sterile forceps.

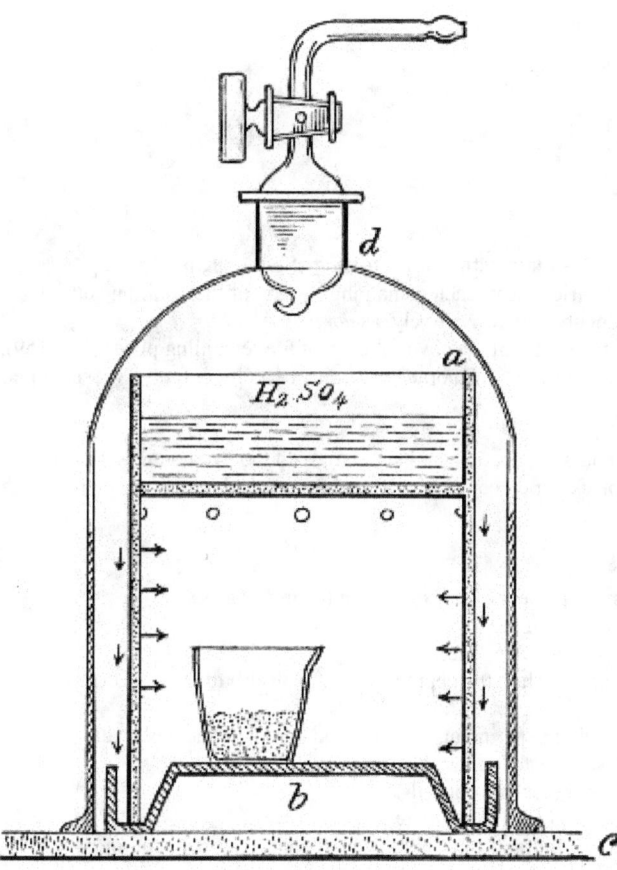

Fig. 157.—Mueller's desiccator.

6. Place the Petri dish inside the Mueller's desiccator; fill the upper chamber with pure sulphuric acid, cover with the bell jar, and exhaust the air from its interior. Ten minutes later connect up the desiccator to a sulphuric acid wash-bottle interposing an air filter so that only dry sterile air enters.

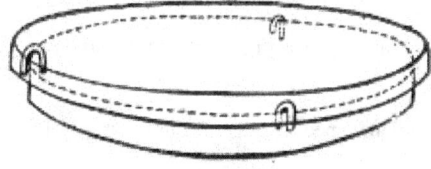

Fig. 158.—Petri dish for drying cultivations.

7. At intervals of five hours open the apparatus, remove one of the cover-slip films from the Petri dish, and transfer it to the interior of a culture flask, with every precaution against contamination. Reseal the desiccator and again exhaust, and subsequently admit dry sterile air as before.

8. Incubate the culture flask under optimum conditions until the completion of seven days, if necessary; and determine the time exposure at which death occurs.

9. Pour plates from those culture flasks which grow, to determine the absence of contamination.

10. Repeat these observations at hourly intervals for the five hours preceding and succeeding the death time, as determined in the first set of experiments.

(B) *Light.*—

(a) Diffuse Daylight:

1. Prepare a tube cultivation in nutrient bouillon, and incubate under optimum conditions, for forty-eight hours.

Fig. 159.—Plate with star for testing effect of light.

2. Pour twenty plate cultivations, ten of nutrient gelatine and ten of nutrient agar, each containing 0.1 c.c. of the bouillon culture.

3. Place one agar plate and one gelatine plate into the hot and cold incubators, respectively, as *controls*.

4. Fasten a piece of black paper, cut the shape of a cross or star, on the centre of the cover of each of the remaining plates (Fig. 159).

5. Expose these plates to the action of diffuse daylight (not direct sunlight) in the laboratory for one, two, three, four, five, six, eight, ten, twelve hours.

6. After exposure to light, incubate under optimum conditions.

7. Examine the plate cultivations after twenty-four and forty-eight hours' incubation, and compare with the two controls. Record results. If growth is absent from that portion of the plate unprotected by the black paper, continue the incubation and daily observation until the end of seven days.

8. Control the results.

(b) Direct Sunlight:

1. Prepare plate cultivations precisely as in the former experiments and place the two controls in the incubators.

2. Arrange the remaining plates upon a platform in the direct rays of the sun.

3. On the top of each plate stand a small glass dish 14 cm. in diameter and 5 cm. deep.

4. Fill a solution of potash alum (2 per cent. in distilled water) into each dish to the depth of 2 cm. to absorb the heat of the sun's rays and so eliminate possible effects of temperature on the cultivations.

5. After exposures for periods similar to those employed in the preceding experiment, incubate and complete the observation as above.

(c) Primary Colours: Each colour—violet, blue, green and red—must be tested separately.

1. Prepare plate cultivations, as in the previous "light" experiments, and incubate controls.

2. Fasten a strip of black paper, 3 cm. wide, across one diameter of the cover of each plate.

3. Coat the remainder of the surface of the cover with a film of pure photographic collodion which contains 2 per cent. of either of the following aniline dyes, as may be necessary:

Chrysoidin (for red). Malachite green (for green). Eosin, bluish (for blue). Methyl violet (for violet).

4. Expose the plates, thus prepared, to bright daylight (but not direct sunlight) for varying periods, and complete the observations as in the preceding experiments. The bactericidal action of light appears to depend upon the more refrangible rays of the violet end of the spectrum and is noted whether the red yellow rays are transmitted or not.

5. Control the results.

Note.—The ultra-violet rays obtained from a quartz mercury vapour lamp destroy bacterial life with great rapidity under laboratory conditions.

(*C*) *Heat.*—(*Vide* Thermal Death-point, page 298.)

(*D*) *Antiseptics and Disinfectants.*—The resistance exhibited by any given bacterium toward any specified disinfectant or germicide should be investigated with reference to the following points:

(A) **Inhibition coefficient**—*i. e.*, that *percentage of the disinfectant* present in the nutrient medium which is sufficient to prevent the growth and multiplication of the bacterium.

(B) **Inferior lethal coefficient**—*i. e.*, the *time exposure* necessary to kill *vegetative forms* of the bacterium suspended in water at 20° to 25° C, in which the disinfectant is present in *medium* concentration (concentration insufficient to cause plasmolysis). And if the bacterium is one which forms spores,

(C) **Superior lethal coefficient**—*i. e.*, the *time exposure* necessary to kill the *spores* of the bacterium under conditions similar to those obtaining in B.

The example here detailed only specifically refers to certain of the disinfectants:

viz:—Bichloride of mercury; Formaldehyde; Carbolic acid;

investigated with regard to B. anthracis, but the technique is practically similar for all other chemical disinfectants.

Inhibition Coefficient.—

Apparatus Required:

Case of sterile pipettes, 10 c.c. (in tenths).

Case of sterile pipettes, 1 c.c. (in tenths).

Sterile tubes or capsules for dilutions.

Tubes of nutrient bouillon each containing a measured 10 c.c. of medium.

Twenty-four-hour-old agar culture of a recently isolated **B.** Anthracis.

Germicides:
1. Five per cent. aqueous solution of carbolic acid.
2. One per cent. aqueous solution of perchloride of mercury.
3. One-tenth per cent. aqueous solution of formaldehyde.

Method.—
1. Number six bouillon tubes consecutively 1 to 6. Inoculate each from the stock cultivation of B. anthracis and at once add varying quantities of the carbolic acid solution, viz.:

To tube 1 add 2.0 c.c. (= 1:100) To tube 2 add 1.0 c.c. (= 1:200) To tube 3 add 0.6 c.c. (= 1:300) To tube 4 add 0.5 c.c. (= 1:400) To tube 5 add 0.4 c.c. (= 1:500) To tube 6 add 0.2 c.c. (= 1:1,000)

2. Prepare a similar series of tube cultivations numbered consecutively 7 to 12 and add varying quantities of the mercuric perchloride solution, viz.:

To tube 7 add 0.1 (= 1:1,000) To tube 8 add 0.05 (= 1:2,000) To tube 9 add 0.03 (= 1:3,000) To tube 10 add 0.025 (= 1:4,000) To tube 11 add 0.02 (= 1:5,000) To tube 12 add 0.01 (= 1:10,000)

3. Prepare a similar series of tube cultivations numbered consecutively 13 to 18 and add varying quantities of the formaldehyde solution, viz.:

To tube No. 13 add 1.0 c.c. (= 1:1,000) To tube No. 14 add 0.4 c.c. (= 1:2,500) To tube No. 15 add 0.2 c.c. (= 1:5,000) To tube No. 16 add 0.1 c.c. (= 1:10,000) To tube No. 17 add 0.075 c.c. (= 1:15,000) To tube No. 18 add 0.05 c.c. (= 1:20,000)

4. Incubate all three sets of cultivations under optimum conditions as to temperature and atmosphere.
5. Examine each of the culture tubes from day to day, until the completion of seven days, and note those tubes, if any, in which growth takes place.
6. From such tubes as show growth prepare subcultivations upon suitable media, and ascertain that the organism causing the growth is the one originally employed in the test and not an accidental contamination.

Inferior Lethal Coefficient.—
Apparatus Required:
Highly concentrated solutions of the disinfectants.
Sterile test-tubes in which to make dilutions from the concentrated solutions of the disinfectants.
Hanging-drop slides.
Cover-slips.
Erlenmeyer flask containing 100 c.c. sterile distilled water.
Case of sterile pipettes, 10 c.c. (in tenths of a cubic centimetre).
Case of sterile pipettes, 1 c.c. (in tenths of a cubic centimetre).

Method.—
1. Prepare a surface cultivation of the "test" organism B. anthracis upon nutrient agar in a culture bottle and incubate under optimum conditions for twenty-four hours; then examine the cultivation microscopically to determine the absence of spores.
2. Prepare solutions of different percentages of each disinfectant.
3. Make a series of hanging-drop preparations from the agar culture, using a loopful of disinfectant solution of the different percentages to prepare the emulsion on each cover-slip.
4. Examine microscopically and note the strongest solution which does not cause plasmolysis and the weakest solution which does plasmolyse the organism.
5. Make control preparations of these two solutions and determine the percentage to be tested.
6. Pipette 10 c.c. sterile water into the culture bottle and suspend the entire surface growth in it.
7. Transfer the suspension to the Erlenmeyer flask and mix it with the 90 c.c. of sterile water remaining in the flask.
8. Pipette 10 c.c. of the diluted suspension into each of ten sterile test-tubes.
9. Label one of the tubes "Control" and place it in the incubator at 18° C.
10. Add to each of the remaining tubes a sufficient quantity of a concentrated solution of the disinfectant to produce the percentage previously determined upon (*vide* step 5).
11. Incubate the tubes at 18° C. to 20° C.
12. At hourly intervals remove the control tube and one of the tubes with added disinfectant from the incubator.
13. Make a subcultivation from both the control and the test suspension, upon the surface of nutrient agar; incubate under optimum conditions.
14. Observe these culture tubes from day to day until the completion of seven days, and determine the shortest exposure necessary to cause the death of vegetative forms.

Superior Lethal Coefficient.—
1. Prepare surface cultivations of the "test" organisms upon nutrient agar in a culture bottle, and incubate under optimum conditions, for three days, for the formation of their spores.
2. Transfer the emulsion to a sterile test-tube and heat in the differential steriliser for ten minutes at 80° C. to destroy all vegetative forms.
3. Employing that percentage solution of the disinfectant determined in the previous experiment, and complete the investigations as detailed therein, steps 7 to 14, increasing the interval between planting the subcultivations to two, three, or five hours if considered advisable.

Note.—Where it is necessary to leave the organisms in contact with a strong solution of the disinfectant for lengthy periods, some means must be adopted to remove every trace of the disinfectant from the bacteria before transferring them to fresh culture media; otherwise, although not actually killed, the presence of the disinfectant may prevent their development, and so give rise to an erroneous conclusion. Consequently it is essential in all germicidal experiments to determine first of all the inhibition coefficient of the germicide employed.

Under the circumstances referred to above it is usually sufficient to prepare the subcultures in such a volume of fluid nutrient medium as would suffice to reduce the concentration of the germicide to about one hundredth of the inhibition percentage, assuming that the entire bulk of inoculum was made up of that strength of germicide employed in the test. In some cases it is a simple matter to neutralise the germicide and render it inert by washing the organisms in some non-germicidal solution (such for example as ammonium sulphide when using mercurial salts as the germicide). When, however, it is desired to remove the last traces of germicide proceed as follows:

1. Transfer the suspension of bacteria to sterile centrifugal tubes; add the required amount of disinfectant, and allow it to remain in contact with the bacteria for the necessary period.

2. Centrifugalise thoroughly, pipette off the supernatant fluid; fill the tube with sterile water and distribute the deposit evenly throughout the fluid.

3. Centrifugalise again, pipette off the supernatant fluid; fill the tube with sterile water; distribute the deposit evenly throughout the fluid, and transfer the suspension to a litre flask.

4. Make up to a litre by the addition of sterile water; filter the suspension through a sterile porcelain candle.

5. Emulsify the bacterial residue with 5 c.c. sterile bouillon.

6. Prepare the necessary subcultivations from this emulsion.

PATHOGENESIS.

Living Bacteria.—

(a) Psychrophilic Bacteria: When the organism will only grow at or below 18° to 20° C.,

1. Prepare cultivations in nutrient broth and incubate under optimum conditions.

2. After seven days' incubation inject that amount of the culture corresponding to 1 per cent. of the body-weight of a healthy frog, into the reptile's dorsal lymph sac.

3. Observe until death takes place, or, in the event of a negative result, until the completion of twenty-eight days (*vide* Chapter XVIII).

4. If, and when, death occurs, make a careful post-mortem examination (*vide* Chapter XIX).

(b) Mesophilic Bacteria: When the organism grows at 35° to 37° C.,

1. Prepare cultivations in nutrient broth and incubate under optimum conditions for forty-eight hours.

2. Select two white mice, as nearly as possible of the same age, size, and weight.

3. Inoculate the first mouse, subcutaneously at the root of the tail, with an amount of cultivation equivalent to 1 per cent. of its body-weight.

4. Inoculate the second mouse intraperitoneally with a similar dose.

5. Observe carefully until death occurs, or until the lapse of twenty-eight days.

6. If the inoculated animals succumb, make complete post-mortem examination.

If death follows shortly after the injection of cultivations of bacteria, the inoculation experiments should be repeated two or three times. Then, if the organism under observation invariably exhibits pathogenic effects, steps should be taken to ascertain, if possible, the minimal lethal dose (*vide infra*) of the growth upon solid media for the frog or white mouse respectively. Other experimental animals—e. g., the white rat, guinea-pig, and rabbit—should next be tested in a similar manner.

7. If the inoculated mice are unaffected, test the action of the organism in question upon white rats, guinea-pigs, rabbits, etc.

Minimal Lethal Dose (m. l. d.); If the purpose of the inoculation is to determine the minimal lethal dose, a slightly different procedure must be followed. For this and other exact experiments a special platinum loop is manufactured, some 2.5 mm. by 0.75 mm., with parallel sides, and calibrated by careful weighing, to determine approximately the amount of moist bacterial growth, the loop will hold when filled.

1. The cultivation must be prepared on a solid medium of the optimum reaction, incubated at the optimum temperature, and injected at the period of greatest activity and vigour, of the particular organism it is desired to test.

2. Arrange four sterile capsules in a row and label them I, II, III, and IV. Into the first deliver 10 c.c. sterile bouillon by means of a sterile graduated pipette; and into each of the remaining three, 9.9 c.c.

3. Remove one loopful of the bacterial growth from the surface of the medium in the culture tube, observing the usual precautions against contamination, and emulsify it evenly with the bouillon in the first capsule. Each cubic centimetre of the emulsion will now contain one-tenth of the organisms contained in the original loopful (written shortly 0.1 loop).

4. Remove 0.1 c.c. of the emulsion in the first capsule by means of a sterile graduated pipette and transfer it to the second capsule and mix thoroughly. Drop the infected pipette into a jar of lysol solution. This makes up the bulk of the fluid in the second capsule to 10 c.c., and therefore every cubic centimetre of bouillon in capsule II contains 0.001 loop.

5. Similarly, 0.1 c.c. of the mixture is transferred from capsule II to capsule III (1 c.c. of bouillon in capsule III contains 0.00001 loop), and then from capsule III to capsule IV (1 c.c. of bouillon in capsule IV contains 0.0000001 loop).

The dilutions thus prepared may be summarised in a table;

Capsule I = 1 loopful + 10 c.c. water ∴ 1 c.c.=0.1 loop. Capsule II = 0.1 c.c. capsule I + 9.9 c.c. water ∴ 1 c.c.=0.001 loop. Capsule III = 0.1 c.c. capsule II + 9.9 c.c. water ∴ 1 c.c.=0.00001 loop. Capsule IV = 0.1 c.c. capsule III + 9.9 c.c. water ∴ 1 c.c. = 0.0000001 loop.

6. With sterile graduated pipettes remove the necessary quantity of bouillon corresponding to the various divisors of ten of the loop from the respective capsules, and transfer each "dose" to a separate sterile capsule and label; and to such doses as are small in bulk, add the necessary quantity of sterile bouillon to make up to 1 c.c.

7. Multiples of the loop are prepared by emulsifying 1, 2, 5, or 10 loops each with 1 c.c. sterile bouillon in separate sterile capsules.

8. Inoculate a series of animals with these measured doses, filling the syringe first from that capsule containing the smallest dose, then from the capsule containing the next smallest, and so on. If care is taken, it will not be found necessary to sterilise the syringe during the series of inoculations.

9. Plant tubes of gelatine or agar, liquefied by heat, from each of the higher dilutions, say from 0.0000001 loop to 0.01 loop; pour plates and incubate. When growth is visible enumerate the number of organisms present in each, average up and calculate the number of bacteria present in one loopful of the inoculum.

10. The smallest dose which causes the infection and death of the inoculated animal is noted as the minimal lethal dose.

Toxins.—

Prepare flask cultivations of the organism under observation in glucose formate broth, and incubate for fourteen days under optimum conditions.

(a) Intracellular or Insoluble Toxins:

1. Heat the fluid culture in a water-bath at 60° C. for thirty minutes. (The resulting sterile, turbid fluid is often spoken of as "killed" culture,)

2. Inoculate a tube of sterile bouillon with a similar quantity, and incubate under optimum conditions. This "control" then serves to demonstrate the freedom of the toxin from living bacteria.

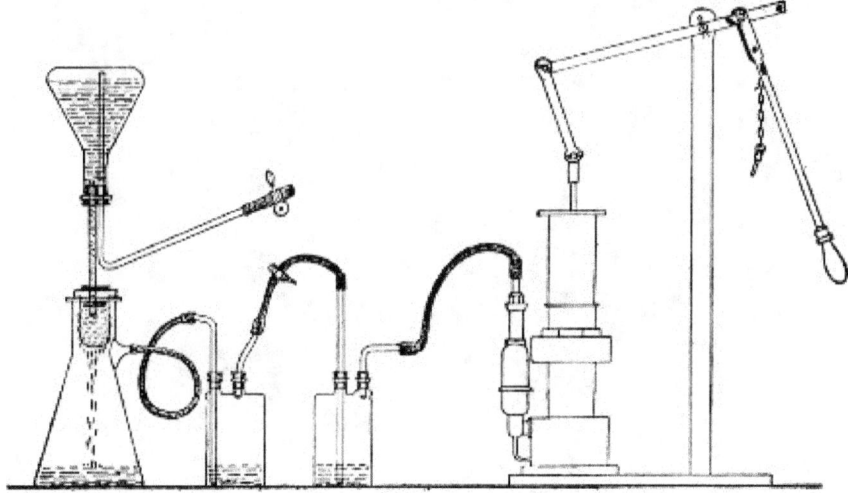

Fig. 160.—Apparatus arrange for toxin filtration.

3. Inject intraveneously that amount of the cultivation corresponding to 1 per cent. of the body-weight of the selected animal, usually one of the small rodents.

4. Observe during life or until the completion of twenty-eight days, and in the event of death occurring during that period, make a complete post-mortem examination.

5. Repeat the experiment at least once. In the event of a positive result estimate the minimal lethal dose of "killed" culture for each of the species of animals experimented upon.

(b) Extracellular or Soluble Toxins:

1. Filter the cultivation through a porcelain filter candle (Berkefeld) into a sterile filter flask, arranging the apparatus as in the accompanying figure (Fig. 160).

2. Inoculate mice, rats, guinea-pigs, and rabbits subcutaneously with that quantity of toxin corresponding to 1 per cent. of the body-weight of each respectively, and observe, if necessary, until the completion of one month.

3. Inoculate a "control" tube of bouillon with a similar quantity and incubate, to determine the freedom of the filtered toxin from living bacteria.

4. In the event of a fatal termination make complete and careful post-mortem examinations.

5. Repeat the experiments and, if the results are positive, ascertain the minimal lethal dose of toxin for each of the susceptible animals.

The estimation of the *m. l. d.* of a toxin is carried out on lines similar to those laid down for living bacteria (*vide* page 316) merely substituting 1 c.c. of toxin as the unit in place of the unit "loopful" of living culture.

It frequently happens, during the course of casual investigations that a bouillon-tube culture is available for a toxin test whilst a flask cultivation is not. In such cases, Martin's small filter candle and tube (Fig. 161) specially designed for the filtration of small quantities of fluid, is invaluable. This consists of a narrow filter flask just large enough to accommodate an ordinary 18 × 2 cm. test-tube. The mouth of the tubular Chamberland candle 15 × 1.5 cm. is closed by a perforated rubber cork into which fits the end of the stem of a thistle headed funnel, whilst immediately below the butt of the funnel is situated a rubber cork to close the mouth of the filter flask. When the apparatus is fixed in position and connected to an exhaust pump, the cultivation is poured into the head of the funnel and owing to the relatively large filtering surface the germ free filtrate is rapidly drawn through into the test-tube receiver.

Raising the Virulence of an Organism.—If it is desired to raise or "exalt" the virulence of a feebly pathogenic organism, special methods of inoculation are necessary, carefully adjusted to the exigencies of each individual case. Among the most important are the following:

1. *Passage of Virus.*—The inoculation of pure cultivations of the organism into highly susceptible animals, and passing it as rapidly as possible from animal to animal, always selecting that method of inoculation-*e. g.*, intraperitoneal—which places the organism under the most favorable conditions for its growth and multiplication.

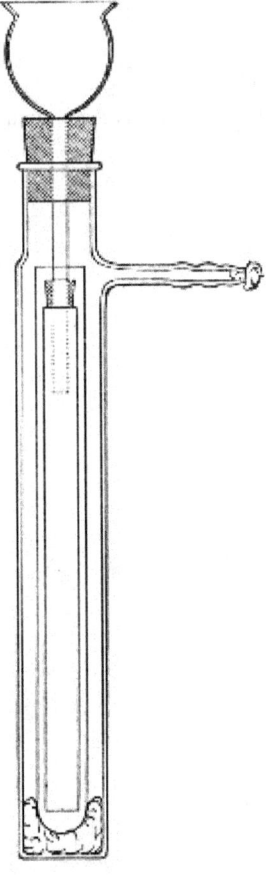

Fig. 161—Martin's filtering apparatus for small quantities of fluid.

2. *Virus Plus Virulent Organisms.*—The inoculation of pure cultivations of the organism together with pure cultivations of some other microbe which in itself is sufficiently virulent to ensure the death of the experimental animal, either into the same situation or into some other part of the body. By this association the organism of low virulence will frequently acquire a higher degree of virulence, which may be still further raised by means of "passages" (*vide supra*).

3. *Virus Plus Toxins.*—The inoculation of pure cultivations of the organism into some selected situation, together with the subcutaneous, intraperitoneal, or intravenous injection of a toxin—*e. g.*, one of those elaborated by the proteus group—either simultaneously with, before, or immediately after, the injection of the feeble virus. By this means the natural resistance of the animal is lowered, and the organism inoculated is enabled to multiply and produce its pathogenic effect, its virulence being subsequently exalted by means of "passages."

Attenuating the Virulence of an Organism.—Attenuating or lowering the virulence of a pathogenic microbe is usually attained with much less difficulty than the exaltation of its virulence, and is generally effected by varying the environment of the cultivations, as for example:

1. Cultivating in such media as are unsuitable by reason of their (*a*) composition or (*b*) reaction.
2. Cultivating in suitable media, but at an unsuitable temperature.
3. Cultivating in suitable media, but in an unsuitable atmosphere.
4. Cultivation in suitable media, but under unfavorable conditions as to light, motion, etc.

Attenuation of the virus can also be secured by

5. Passage through naturally resistant animals.
6. Exposure to desiccation.
7. Exposure to gaseous disinfectants.
8. By a combination of two or more of the above methods.

IMMUNISATION.

The further study of the pathogenetic powers of any particular bacterium involves the active immunisation of one or more previously normal animals. This end may be attained by various means; but it must be remembered that immunisation is not carried out by any hard and fast rule or by one method alone, but usually by a combination of methods adapted to the exigencies of each particular case. The ordinary methods include:

A. Active Immunisation.
I. By inoculation with dead bacteria (*i. e.*, bacteria killed by heat; the action of ultra-violet rays, of chemical germicides, or by autolysis).
II. By the inoculation of attenuated strains of bacteria.
III. By the inoculation of living virulent bacteria (exalted in virulence if necessary).
B. Combined Active and Passive Immunisation:
IV. By the inoculation of toxin-antitoxin mixtures.

ACTIVE IMMUNISATION.

The immunisation of the rabbit against the Diplococcus pneumoniæ may be instanced as an example of the general methods of immunisation of laboratory animals.

1. Take a full grown rabbit weighing not less than 1200 to 1500 grammes (large rabbits of 2000 grammes and over are the most suitable for immunising experiments). Observe weight and temperature carefully during the few days occupied in the following steps.

2. Inoculate a small rabbit intraperitoneally with one or two loopfuls of a twenty-four-hour-old blood agar cultivation of a *virulent* strain of Diplococcus pneumoniæ.

Death should follow within twenty-four hours, and in any case will not be delayed beyond forty-eight hours.

3. Under aseptic precautions, at the post-mortem, transfer a loopful of heart blood to an Erlenmeyer flask containing 50 c.c. sterile nutrient broth. Incubate at 37° C. for twenty-four hours.

4. Prepare also several blood agar cultures from the heart blood of the rabbit, label them all O.C. (original culture). After twenty-four hours incubation at 37° C. place an india-rubber cap over the plugged mouth of the tube of all but one of these cultures and paint the cap with Canada balsam or shellac varnish, dry, and replace in the hot incubator.

This will prevent evaporation, and cultures thus sealed will remain unaltered in virulence for a considerable time.

5. Make a fresh subcultivation on blood agar from the uncapped O.C. cultivation and after twenty-four hours incubation at 37° C. determine the minimal lethal dose of this strain upon a series of mice (see page 316).

6. Suspend the flask containing the twenty-four-hour-old broth culture (step 3) in the water-bath at 60° C. for one hour. Cool the flask rapidly under a stream of cold water.

7. Determine the sterility of this (?) killed cultivation by transferring one cubic centimetre to each of several tubes of nutrient broth, and incubate at 37° C. for twenty-four hours. If growth of Diplococcus pneumoniæ occurs, again heat culture in water-bath at 60° C. for one hour and again test for sterility.

8. Inject the selected rabbit intravenously (see page 363) with 2 c.c. of the killed cultivation, and inject a further 10 c.c. into the peritoneal cavity.

During the next few days the animal will lose some weight and perhaps show a certain amount of pyrexia.

9. When the temperature and weight have again returned to normal—generally about seven days after the inoculation—again inject killed cultivation, this time giving a dose of 5 c.c. intravenously and 20 c.c. intraperitoneally. A temperature and weight reaction similar to, but less marked than that following the first injection will probably be observed, but after about a week's interval the animal will be ready for the next injection.

10. When ready to give the third injection prepare a fresh blood agar subculture from another O.C. tube and after twenty-four hours incubation prepare a minimal lethal dose (as determined in 5) and inject it subcutaneously into the rabbit's abdominal wall.

A slight local reaction will probably be observed as well as the weight and temperature reactions.

11. A week to ten days later inject a similar minimal lethal dose into the peritoneal cavity.

12. Observe the weight and temperature of the rabbit very carefully, and regulating the dates of inoculation by the animal's general condition, continue to inject living cultivations of the pneumococcus into the peritoneal cavity, gradually increasing the dose by multiples of ten.

13. At intervals of two months samples of blood may be collected from the posterior auricular vein and the serum tested for specific antibodies.

14. Under favourable conditions it will be found after some six months steady work that the rabbit may be injected intraperitoneally with an entire blood agar cultivation without any ill effects being apparent; and this characteristic—resistance to the lethal effects of large doses of the virus—is the sole criterion of *immunity*. Further, the serum separated from blood withdrawn from the animal about a week after an injection, if used in doses of .01 c.c., will protect a mouse against the lethal effects of at least ten minimal lethal doses of living pneumococci.

In the foregoing illustration it has been assumed that complete acquired active immunity has been conferred upon the experimental rabbit in consequence of the formation of antibody, specific to the diplococcus pneumoniac, sufficient in amount to ensure the destruction of enormous doses of the living cocci—the *antigen* (that is the substance injected in response to which *antibody* has been elaborated) in this particular case being the bacterial protoplasm of the pneumococcus with its endo-toxins.

But provided death does not immediately follow the injection of the antigen, specific antibody is always formed in greater or lesser amount; and in experimental work a sufficient amount of any required antibody can often be obtained without carrying the process of immunisation to its logical termination.

For instance, if the immunisation of a rabbit toward Bacillus typhosus is commenced on the lines already set out it will often be found, after a few injections of "killed" cultivation that the blood serum of the animal (even when diluted with several hundred times its volume of normal saline) contains specific agglutinin for B. typhosus—and if the sole object of the experiment has been the preparation of agglutinin the inoculations may well be stopped at this point, although the animal is not yet immune in the strict meaning of the word.

Again, antibodies may be formed in response to antigens other than infective particles—thus the injection into suitable animals of foreign proteins such as egg albumin, heterologous blood sera or red blood discs from a different species of animal, will result in the formation of specific antibodies possessing definite affinities for their respective antigens.

The most important antibody of this latter type is Hæmolysin, a substance that makes its appearance in the blood serum of an animal previously injected with washed blood cells from an animal of a different species. The serum from such an animal possesses the power of disintegrating red blood discs of the variety employed as antigen and causing the discharge of their contained hæmoglobin, and is specific in its action to the extent of failing to exert any injurious effect upon the red blood cells of any other species of animal.

The action of this serum is due to the presence of two distinct bodies, complement and hæmolysin.

Complement (or alexine) is a thermo-labile readily oxidised body present in variable but unalterable amount in the normal serum of every animal. It is a substance which exerts a lytic effect upon all foreign matter introduced into the blood or tissues; but by itself is a comparatively inert body, and is only capable of exerting its maximum lytic effect in the presence of and in combination with a specific antibody, or immune body.

Complement is obtained (unmixed with antibody) by collecting fresh blood serum from any healthy normal (that is uninoculated) animal. Guinea-pigs' serum is that most frequently employed for experimental work.

Hæmolysin (immune body, copula, sensitising body, amboceptor) is a *thermostable* antibody formed in response to the injection of red cells which although in itself inert is capable of linking up complement present in the normal serum to the red cells of the variety used as antigen—a combination resulting in hæmolysis.

Hæmolysin is obtained by collecting fresh blood serum from a suitably inoculated animal and exposing it to a temperature of 56° C. (to destroy the thermo-labile complement) for 15 to 30 minutes before use. It is then referred to as *inactivated*, and is *reactivated* by the addition of fresh normal serum—that is serum containing complement.

Hæmolysin is of importance academically owing to the fact that many of the problems of immunity have been elucidated by its aid; but its present practical importance lies in the application of the *hæmolytic system* (that is hæmolysin, corresponding erythrocyte solution and complement) to certain laboratory methods having for their object either the identification of the infective entity or the diagnosis of the existence of infection.

For use in these laboratory methods of diagnosis it is most convenient to prepare hæmolytic serum specific for human blood—whether the laboratory is isolated or attached to a large hospital. Ox blood, sheep blood or goat blood if readily obtainable, may however be used instead, and although the following method is directed to the preparation of human hæmolysin the same procedure serves in all cases.

THE PREPARATION OF HÆMOLYTIC SERUM.

Apparatus Required:

Small centrifuge, preferably electrically driven, with two receptacles for tubes, and enclosed in a safety shield (Fig. 162). Sterile centrifuge tubes (10 c.c. capacity), Fig. 163. Sterile pipettes (10 c.c. graduated) in case. Sterile glass capsules (in case). Sterile test-tubes. Sterile all glass syringe (5 c.c. or 10 c.c. capacity) and needle.

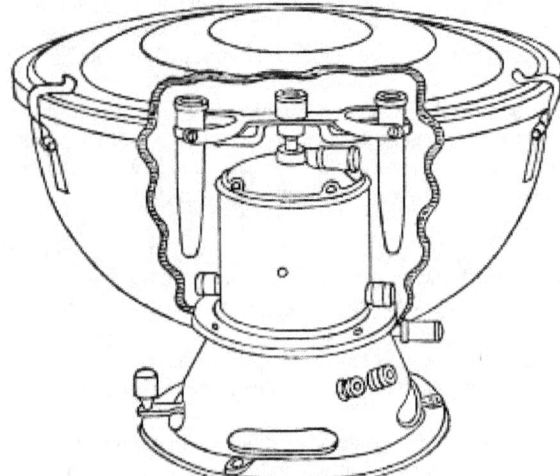

Fig. 162.—Small electrical centrifuge.

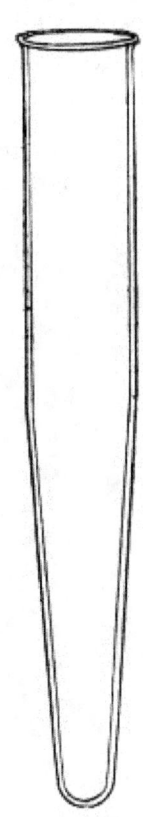 Fig. 163.—Centrifuge tube.

Reagents Required:
Normal saline solution. 10 per cent. sodium citrate solution in normal saline. Human blood (*vide infra*).
Method.—
1. Select a healthy full-grown rabbit of not less than 2500 grammes weight in accordance with the directions already given (page 322) and prepare it for intraperitoneal inoculation.
2. Measure out 2 c.c. citrated human blood (collected at a surgical operation or a venesection, or withdrawn by venipuncture from the median basilic or median cephalic vein of a normal adult) into a centrifuge tube and centrifugalise thoroughly.
3. Wash with three changes of normal saline (*vide* also page 388).
4. Transfer the washed cells to a sterile capsule by means of a sterile pipette. Add 5 c.c. of normal saline and mix thoroughly.
5. Take up the mixture of cells and saline in the all-glass syringe and inject into the peritoneal cavity of the rabbit.
6. Seven days later inject intraperitoneally the washed cells from 5 c.c. human blood mixed with 5 c.c. normal saline.
7. Seven days later inject the washed cells from 10 c.c. human blood mixed with 5 c.c. normal saline.
8. After a further interval of seven days repeat the injection of washed cells from 10 c.c. human blood mixed with 5 c.c. normal saline.
Note.—Better results are obtained if the second and subsequent injections are made intravenously, even when smaller quantities of washed red cells are employed. If, however, the intravenous route is selected exceeding great care must be exercised to avoid the introduction of air into the vein—an accident which is followed, within a few minutes, by the death of the rabbit from pulmonary embolism.
9. Allow five days to elapse, then collect a preliminary sample of blood, say about 2 c.c., from the rabbit's ear. Allow it to clot, separate off the serum and transfer to a sterile test-tube. Place the test-tube in a water-bath at 56° C. for fifteen minutes (to inactivate) and test the serum quantitatively for hæmolytic properties in the following manner:

THE TITRATION OF HÆMOLYTIC SERUM.

Apparatus Required:
Electrical centrifuge. Sterile centrifuge tubes. Water-bath regulated at 56°C. Sterilised pipettes 10 c.c. graduated in tenths. Sterilised pipettes 1 c.c. graduated in tenths. Sterile test-tubes, 16 × 2 cm. Small sterile test-tubes, 9 × 1 cm. Small test-tube rack, or roll of plasticine. Capillary teat pipettes. Stout rubber band or length of small rubber tubing.

Reagents Required and Method of Preparation:
1. Normal saline solution.
2. Hæmolytic serum inactivated by preliminary heating to 56° C. for 15 minutes (*vide supra*) in test-tube labelled **H. S.**
3. Complement. Fresh guinea-pig serum in test-tube labelled **C.**
Kill a normal guinea-pig with chloroform vapour.
Open the thorax with all aseptic precautions, and collect as much blood as possible from the heart with a sterile Pasteur pipette.
Transfer it to a sterile centrifuge tube and place the tube in the incubator at 37° C. Two hours later separate the clot from the sides of the tube, and centrifugalise thoroughly.
Pipette off the clear serum to a clean sterilised test-tube.

4. Erythrocyte solution, in test-tube labelled E.

Collect and wash human red blood cells (see page 388, 1-8). Measure the volume of red cells available and prepare a 2 per cent. suspension in normal saline solution.

Method.—

1. Take two test-tubes and number them 1 and 2, and pipette into each 9 c.c. of normal saline solution.
2. Add 1 c.c. of hæmolytic rabbit serum to tube No. 1 and mix thoroughly: take up 1 c.c. of the mixture and add it to tube No. 2; mix thoroughly.
3. Set up ten small test-tubes in test-tube rack or in roll of plasticine, and number 1 to 10.
4. Pipette into tube No. 1 0.5 c.c. = 0.5 c.c. hæmolytic serum}
Pipette into tube No. 2 0.1 c.c. = 0.1 c.c. hæmolytic serum}   From tube H. S.

Pipette into tube No. 3 0.5 c.c. = 0.05 c.c. hæmolytic serum}
Pipette into tube No. 4 0.3 c.c. = 0.03 c.c. hæmolytic serum}
Pipette into tube No. 5 0.2 c.c. = 0.02 c.c. hæmolytic serum}   From tube 1.
pipette into tube No. 6 0.1 c.c. = 0.01 c.c. hæmolytic serum}

Pipette into tube No. 7 0.5 c.c. = 0.005 c.c. hæmolytic serum}
Pipette into tube No. 8 0.3 c.c. = 0.003 c.c. hæmolytic serum}
Pipette into tube No. 9 0.2 c.c. = 0.002 c.c. hæmolytic serum}   From tube 2.
Pipette into tube No. 10 0.1 c.c. = 0.001 c.c. hæmolytic serum}

5. To each tube add 1 c.c. of erythrocyte solution.
6. When necessary (that is to say in tubes 2, 4, 5, 6, 8, 9 and 10) add normal saline solution to the mixture in the test-tubes till the column of fluid in each reaches to the same level.
7. Shake each tube in turn, so as to thoroughly mix its contents. Plug the mouth of each tube with cotton wool, and place entire set in the incubator at 37°C. for one hour.
8. Remove the tubes from the incubator and into each tube pipette 0.1 c.c. complement (guinea-pig's serum) and replace tubes in incubator at 37° C. for further period of one hour.
9. Remove the tubes from the incubator, and if complete hæmolysis has not taken place in every tube, stand on one side, preferably in the ice chest, for an hour.
10. Then examine the tubes.

Complete hæmolysis is indicated by a clear red solution, with no deposit of red cells at the bottom of the test-tube.

Absence of hæmolysis is indicated by a clear or turbid colourless fluid, with a deposit of red cells at the bottom of the test-tubes.

The smallest amount of hæmolytic serum that has caused complete hæmolysis is known as the minimal hæmolytic dose (*M. H. D.*) and if hæmolysis has occurred in all the tubes down to No. 7—the m. h. d. of this particular serum is .005 c.c. = 200 minimal hæmolytic doses per cubic centimetre. Such a serum is strong enough for experimental work; indeed, for many purposes, complete hæmolysis down to tube 6 will indicate a serum sufficiently strong(= 100 m. h. d. per cubic centimetre). If, however, only the first one or two tubes are completely hæmolysed, this is an indication that the rabbit should receive further injections in order to raise the hæmolytic power to a sufficiently high level.

STORAGE OF HÆMOLYSIN.

If, and when the hæmolysin content of the rabbit's serum is found to be sufficient, destroy the animal by chloroform vapour.

Remove as much of its blood as possible from the heart under aseptic precautions into sterilized centrifuge tubes.

Transfer the tubes of blood to the incubator at 37° C. for two hours—then centrifugalize thoroughly.

Pipette off the clear serum, and fill in quantities of 1 c.c., into small glass ampoules or pipettes, and hermetically seal in the blowpipe flame, care being taken to avoid scorching the serum.

Place the ampoules when filled with serum and sealed, in a water-bath at 56° C. for 30 minutes. This destroys the complement, *i. e.*, inactivates the serum, and at the same time, provided the various operations have been carried out under aseptic precautions, ensures its sterility. A longer exposure reduces the hæmolytic power.

Place the ampoules in a closed metal box and store in the ice chest for future use.

FOOTNOTES:

The quantities here given are not absolutely correct. If exactitude is essential the student must calculate the amount required by the aid of the Percentage Formula, Appendix, page 496.

See Percentage Formula, Appendix, page 496.

XVII. EXPERIMENTAL INOCULATION OF ANIMALS.

The use of living animals for inoculation experiments may become a necessary procedure in the Bacteriological Laboratory for some one or more of the following reasons:

A. Determination of Pathogenetic Properties of Bacteria already Isolated in Pure Culture (see page 315).

The exact study of the conditions influencing the virulence (including its maintenance, exaltation and attenuation) of an organism, and precise observations upon the pathogenic effects produced by its entrance into, and multiplication within the body tissues can obviously only be carried out by means of experimental inoculation; whilst many points relating to vitality, longevity, etc., can be most readily elucidated by such experiments.

B. Isolation of Pathogenetic Bacteria.

Certain highly parasitic bacteria (which grow with difficulty upon the artificial media of the laboratory) can only be isolated with considerable difficulty from associated saprophytic bacteria when cultural methods alone are employed; but if the mixture of parasite and saprophytes is injected into an animal susceptible to the action of the former, the pathogenic organism can readily be isolated from the tissues of the infected animal. The pneumococcus for example occurs in the sputum of patients suffering from acute lobar pneumonia, but usually in association with various saprophytes derived from the mouth and pharynx. The optimum medium for the growth of the pneumococcus, blood agar, is also an excellent pabulum for the saprophytes of the mouth, and plate cultures are rapidly overgrown by them to the destruction of the more delicate pneumococcus. But inoculate some of the sputum under the skin of a mouse and three or four days later the pneumococcus will have entered the blood stream (leaving the saprophytes at the seat of inoculation) and killed the animal. Cultivations made at the post-mortem (see page 398) from the mouse's heart blood will yield a pure growth of the pneumococcus.

C. Identification of Pathogenetic Bacteria.

The resemblances, morphological and cultural, existing between certain pathogenetic bacteria are in some cases so great as to completely overwhelm the differences; again the same bacterium may under varying conditions assume appearances so different from those regarded as typical or normal as to throw doubt on its identity. In each case a simple inoculation experiment may decide the point at once. As a concrete example may be instanced an autopsy on an animal dead from an unknown infection. Cultivations from the heart blood gave a pure growth of a typical (capsulated) pneumococcus. Cultivations from the liver gave a pure growth of what appeared to be a typical (non-capsulated) Streptococcus pyogenes longus. The latter inoculated into a rabbit caused the death of the animal from pneumococcic septicæmia, and cultures from the rabbit's blood gave a pure growth of a typical (capsulated) pneumococcus.

D. Study of the Problems of Immunity.

It is only by a careful and elaborate study of the behaviour of the animal cell and the body fluids vis-à-vis with the infecting bacterium that it becomes possible to throw light upon the complex problem whereby the cell opposes successful resistance to the diffusion of the invading microbe, or succeeds in driving out the microbe subsequently to the occurrence of that diffusion.

At the moment, however, our attention is directed to the first of these broad headings, for it is by the application of the knowledge acquired in its pursuit that we are able to deal with problems arising under any of the remainder.

For whatever purpose the inoculation is performed, it is essential that the experiment should be planned to secure the maximum amount of information and the minimum of discomfort to the animal used. Every care therefore must be taken to ensure that the virus is introduced into the exact tissue or organ selected; and the operation itself must be carried out with skill and expedition, and under strictly aseptic conditions.

In the course of inoculation studies many instances of natural immunity, both racial and individual, will be met with; but it must be recollected that natural immunity is relative only and never absolute, and care be taken not to label an organism as *non-pathogenic* until many different methods of inoculation have been performed upon different species of animals, combined when necessary with various procedures calculated to overcome any apparent immunity, and have invariably given negative results.

In some countries experiments upon animals are only permitted under direct license from the Government, and then only within premises specially licensed for the purpose. In England this license is in the grant of the Home Secretary, and confers the permission to experiment upon animals under general anæsthesia, provided that after the experiment is completed the animal must be destroyed before regaining consciousness. If it is intended to carry out simple hypodermic inoculations and superficial venesections, Certificate A, granting this specific permission and dispensing with the necessity for general anæsthesia must be obtained *in addition to the license*; whilst if the inoculation entails more extensive operative procedures, and it is necessary to observe the subsequent course of the infection, should such occur, the license must be *coupled with Certificate B*—since this certificate removes the compulsion to destroy the animal whilst under the anæsthetic. Further special certificates and combinations of certificates are required if cats, dogs, horses, asses or cattle are to be the subjects of experiment. Under every certificate it is expressly stipulated that if the animal shows signs of pain it must be destroyed immediately.

The animals generally employed in the study of the pathogenic properties of the various micro-organisms are:

Cold Blooded.	*Warm Blooded.*	*Hot Blooded.*
Frog.	Mouse.	Fowl.
Toad.	Rat.	Pigeon.
Lizard.	Guinea pig.	
	Rabbit.	

Monkey.

Preparation.—Before inoculation, the experimental animals should be carefully examined, to avoid the risk of employing such as are already diseased: since it must be remembered that in a state of nature, as well as in captivity, the animals employed for laboratory inoculations are subject to infection by various animal and vegetable parasites, and in some instances such infection presents no symptoms which are obvious to the casual examination; the sex should be noted, the weight recorded, and the rectal temperature taken. The remaining items of importance are the time of the inoculation, the material that is inoculated, and the method of inoculation, and finally under what authority the experiment is performed. In the author's laboratory these data are entered upon a pink card which forms part of a card index system. The card further provides space for notes on the course of the resulting infection, and carries on the reverse the weight and temperature chart (Figs. 164 and 165).

	INOCULATION CARD.	Laboratory No.
Date	o'clock........ m.	
Animal	No. in Series	♂ ♀
Weight	Temp.	
Character and Dose of Inoculum.		
Method of Inoculation.		Under Certificate.
Notes on Course of Infection.		
Result.		Signed

Fig. 164.—Front of inoculation card.

Preliminary Inspection and Examination.—The preliminary examination should comprise observation of the animal at rest and in motion; the appearance of the fur, feathers or scales, inspection of the eyes, and of external orifices of the body; tactile examination of the body and limbs, and palpation of the groins and abdomen; and in many cases the microscopical examination of fresh and stained blood-films.

Some of the commoner forms of naturally acquired infection may be briefly mentioned, without however touching upon the various fleas, lice and ticks which at times infect the ordinary laboratory animals.

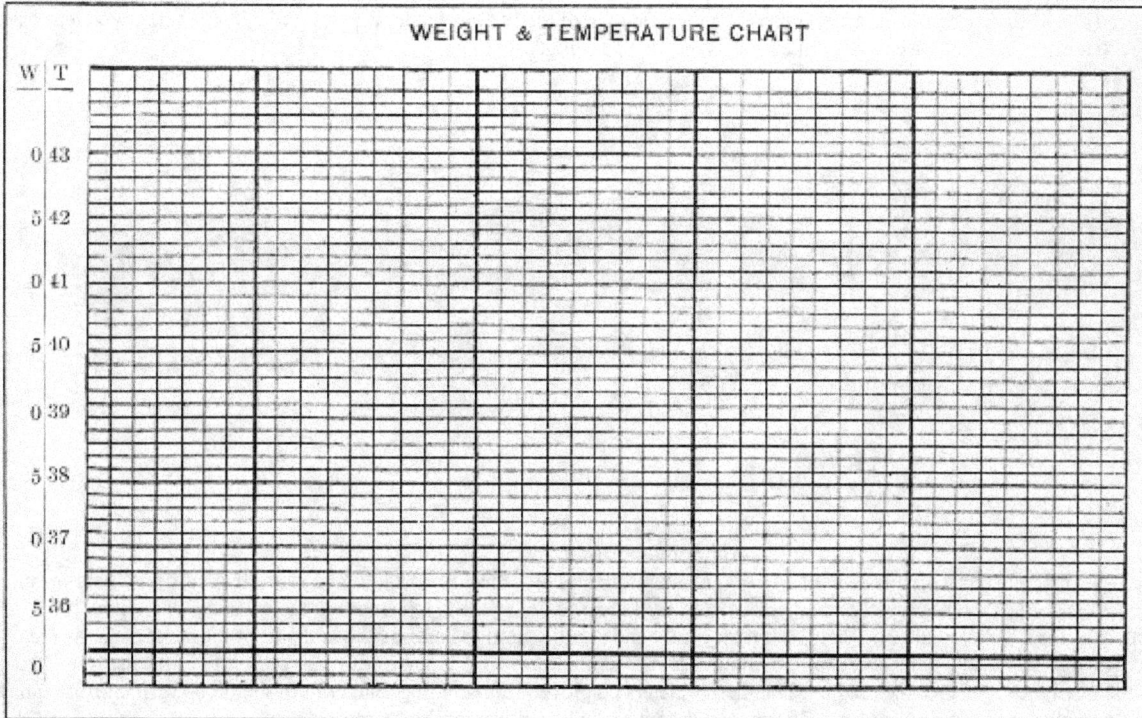

Fig. 165.— Back of inoculation card.

The Rabbit, particularly in captivity, is subject to attacks of Psoric Acari, and the infection is readily transmitted to rabbits in neighbouring cages and also to guinea pigs, but not to rats and mice. One species (*Sarcoptes minor* var. *cuniculi*) gives rise to the ordinary mange. The infection first shows itself as thick yellowish scales and crusts around the nose, mouth and eyes, spreads to the bases and outer surfaces of the ears (never to the inside of the concha), to the fore and hind legs and into the groins and around the genitals. The acari can be readily demonstrated microscopically in scrapings of the skin, treated with liquor potassæ. Another form of scabies (due to Psoroptes *communis cuniculi*) commences at the bottom of the concha, which is filled with whitish-yellow masses consisting of dried crusts, scales, fæces, and dead acari. The base of the ear is hard and swollen, and lifting the animal by the ears—as is usually done—gives rise to considerable pain; indeed this symptom may be the one which first attracts attention to an infection, which causes progressive wasting and terminates in death. A mixed infection—sarcoptic plus psorotic acariasis—is sometimes seen.

If it is decided to try and save animals suffering from infection by these parasites, they must be segregated, the scabs carefully cleaned from the infected areas and the denuded surfaces washed with 5 per cent. solution of Potassium persulphate (a few drops being allowed to run into the concha), or with a preparation containing equal parts of soft paraffin and vaseline with a few drops of lysol. This treatment should be repeated daily until the acarus is destroyed and the animal has regained its normal condition. The cages should be disinfected and all neighbouring animals carefully examined, and any which show signs of infection should be treated in a similar manner. Favus also attacks the rabbit, and the typical spots are first noted around the base of the ear.

Infection by *Coccidium oviforme* is very common, without however presenting any symptoms by which the infection may be recognised. Usually the condition is only noted post-mortem, when the liver is found to be studded with numerous cascating tubercles, which on examination prove to be cystic areas crowded with coccidia. Sometimes too the liver of a rabbit dead from some intentional or accidental bacterial infection is found at the post-mortem to be marked by fine yellowish streaks and small tubercles due to the embryos of *Tænia serrata*, while the cystic form (*Cysticercus pisiformis*) is often noted free in the peritoneal cavity, or invading the mesentery.

Abscess formation from infection with ordinary pyogenic bacteria occurs naturally in the rabbit, and frequently the animal house of a laboratory is decimated by an infective septicæmia due to *B. cuniculicida*.

The **Mouse** and **Rat** suffer from septicæmia, and from the cysticercus form of *Tænia murina*; the cystic form (*Cysticercus fasciolaris*) of *T. crassicollis* has its habitat in their livers. These small rodents are frequently infected with scabies, but if freely provided with clean straw will clean themselves by rubbing through it. The mouse is also attacked by favus, and the rat is often infected with *Trypanosoma Lewisi*.

The **Guinea pig**, like the rabbit, suffers from scabies and coccidiosis. In addition it is often naturally infected with *B. tuberculosis*, and it is a wise precaution to test animals as soon as they reach the laboratory by injecting Koch's Old Tuberculin—0.5 c.c. causing death in the tuberculous cavy within 48 hours.

The **Monkey** is naturally prone to tuberculosis, and should be injected with 1 c.c. Old Tuberculin on arrival in the laboratory. The tissues of the monkey also serve as the habitat for a Nematode worm parasitic in cattle (*Œsophagostoma inflatum*) resembling the Anchylostomum, and this parasite frequently bores through the intestinal wall, and provokes the formation of small cysts in the immediately adjacent mesentery. The presence of these cysts may give rise to considerable speculation at the post-mortem.

The **Pigeon** may be infected by *Hæmosporidia*, and its blood show the presence of halteridia. This bird may also be the subject of a bacterial infection known as pigeon diphtheria; while the fowl may be subject to scabies and ringworm, or suffer from fowl cholera or fowl septicæmia—infections due to members of the hæmorrhagic septicæmia group.

Weighing.—The larger animals are most conveniently weighed in a decimal scale provided with a metal cage for their reception instead of the ordinary pan (Fig. 166). Mice and rats are weighed in a modification of the letter balance, weighing to 250 grammes, which has a conical wire cage, (carefully counterpoised) substituted for its original pan (Fig. 167).

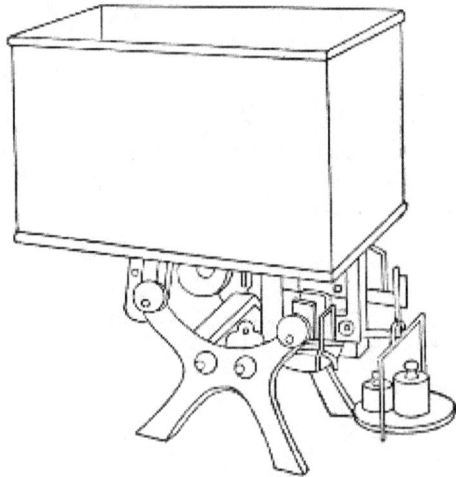

Fig. 166.—Rabbit scales.

Temperature.—To take the rectal temperature of any of the laboratory animals, the animal should be carefully and firmly held by an assistant. Introduce the bulb of an ordinary clinical thermometer, well greased with vaseline, just within the sphincter ani. Allow it to remain in this position for a few seconds, and then push it on gently and steadily until the entire bulb and part of the stem, as far as the constriction, have passed into the rectum. Three to five minutes later, the time varying of course with the sensibility of the thermometer used, withdraw the instrument and take the reading. The thermometers employed for recording temperature should be verified from time to time by comparison with a standard Kew certified Thermometer kept in the laboratory for that purpose.

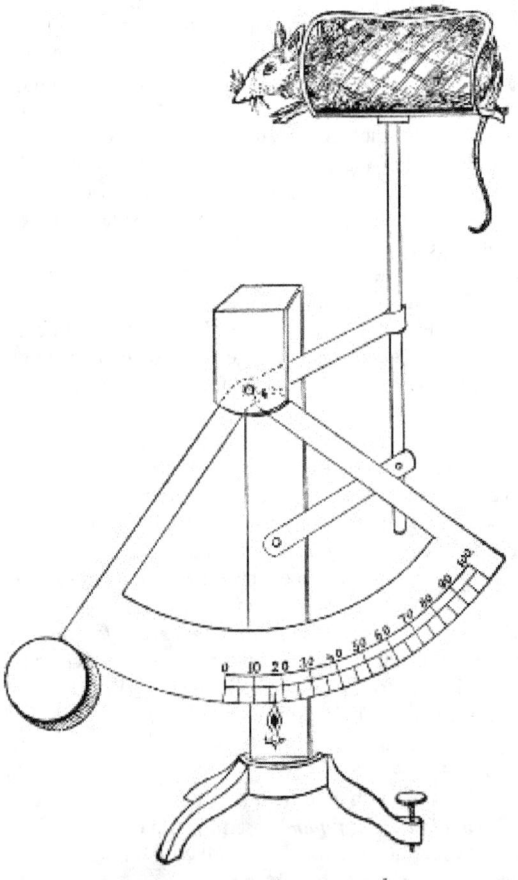

Fig. 167.—Mouse scales

Cages.—During the period which elapses between inoculation and death, or complete recovery, the experimental animals must be kept in suitable receptacles which can easily be kept clean and readily disinfected.

The *mouse* is usually stored in a glass jar (Fig. 168) 11 cm. high and 11 cm. in diameter, closed by a wire gauze cover which is weighted with lead or fastened to the mouth of the jar by a bayonet catch. A small oblong label, 5 cm. by 2.5 cm., sand-blasted on the side of the cylinder, is a very convenient device as notes made upon this with an ordinary lead pencil show up well and only require the use of a damp cloth to remove them (Fig. 168).

The *rat* is kept under observation in a glass jar similar, but larger, to that used for the mouse.

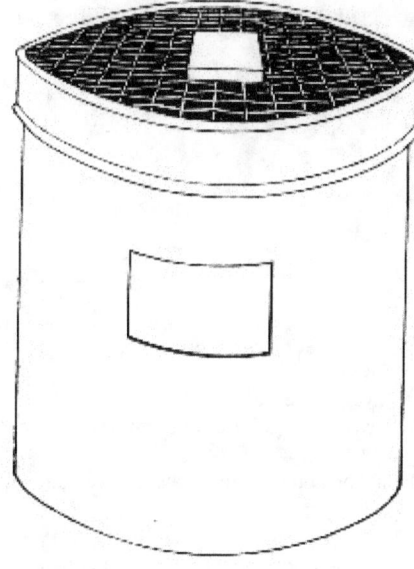

Fig. 168.—Mouse jar.

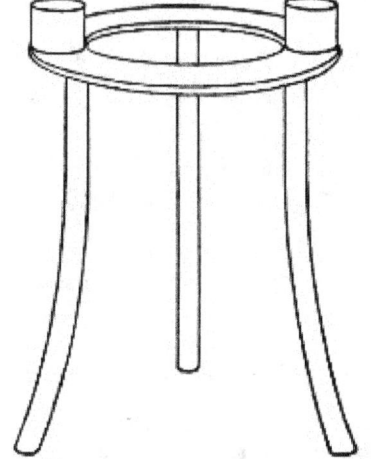

Fig. 169.—Tripod.

A layer of sawdust at the bottom of the jar absorbs any moisture and cotton-wool or paper shavings should be provided for bedding. The food should consist of bran and oats with an occasional feed of bread-and-milk sop.

The use of a metal tripod, on the platform of which are soldered two small cups for the reception of the food, inside the cage, prevents waste of food or its contamination with excreta (Fig. 169).

After use the jars and tripods are sterilised either by chemical reagents or by autoclaving.

The *rabbit* and the *guinea-pig* are confined in cages of suitable size, made entirely of metal (Fig. 170). The sides and top and bottom are of woven wire work; beneath the cage is a movable metal tray filled with sawdust, for the reception of the excreta. The cage as a whole is raised from the ground on short legs. The sides, etc., are generally hinged so that the cage packs up flat, for convenience of storing and also of sterilising.

The ordinary rat cage, a rectangular wire-work box, 30 cm. from front to back, 20 cm. wide, and 14 cm. high, makes an excellent cage for guinea-pigs if fitted with a shallow zinc tray, 35 cm. by 24 cm., for it to stand upon.

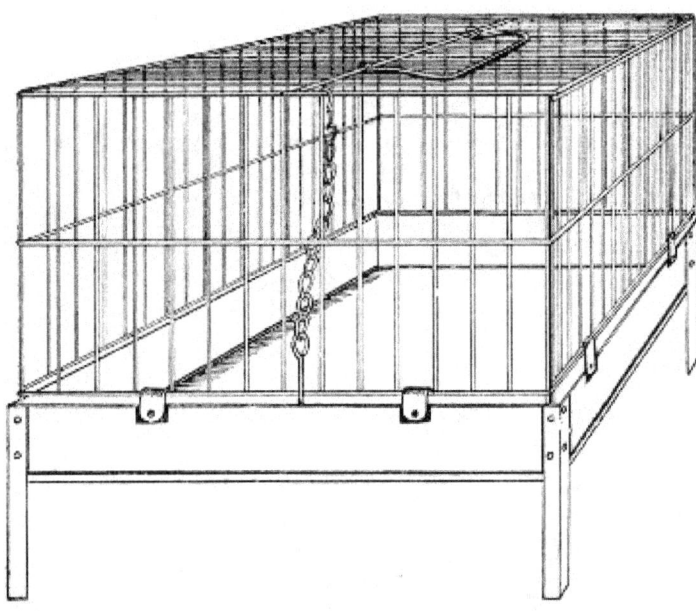

Fig. 170.—Metal rabbit rage.

A plentiful supply of straw should be provided for bedding and the food should consist of fresh vegetables, cabbage leaves, carrot and turnip tops and the like for the morning meal and broken animal biscuits for the evening meal. Occasionally a little water may be placed in the cage in an earthenware dish.

The tray which receives the dejecta should be cleaned out and supplied with fresh sawdust each day, and the soiled sawdust, remains of food, etc., should be cremated.

These cages are sterilised after use either by autoclaving or spraying with formalin.

As **animal inoculation** is purely a surgical operation, the necessary instruments will be similar to those employed by the surgeon, and, like them, must be sterile. In the performance of the inoculation strict attention must be paid to asepsis, and suitable precautions adopted to guard against accidental contamination of the material to be introduced into the animal. In addition, the hands of the operator should be carefully disinfected.

The list of apparatus used in animal inoculations given below comprises practically everything needed for any inoculation. Needless to remark, all the apparatus will never be required for any one inoculation.

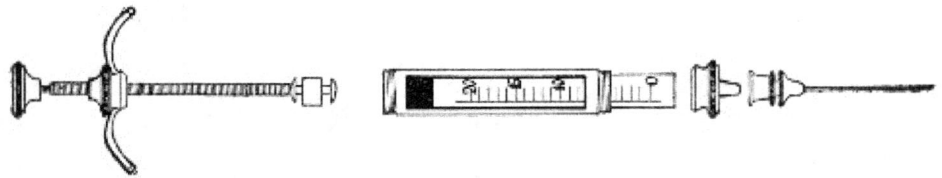

Fig. 171.—Hypodermic syringe with finger rests.

Apparatus Required for Animal Inoculation:

1. Water steriliser (*vide* page 33). It is also convenient to have a second water steriliser, similar but smaller (23 by 7 by 5 cm.), for the sterilisation of the syringes.

2. Injection syringe. The best form is one of the ordinary hypodermic pattern, 1 c.c. capacity graduated in twentieths of a cubic centimeter (0.05 c.c.), fitted with finger rests, but with the leather washers and the packing of the piston replaced by those made of asbestos (Fig. 171). The instrument must be easily taken to pieces, and spare parts should be kept on hand to replace accidental breakage or loss. Other useful syringes are those of 2 c.c., 5 c.c., 10 c.c., and 20 c.c. capacity. A good supply of needles must be kept on hand, both sharp-pointed and with blunt ends. To sterilise the syringe, fill it with water, loosen the packing of the piston and all the screw joints, place it in the steriliser and boil for at least five minutes. Disinfect the syringe *after use*, in a similar manner. The needles, which are exceedingly apt to rust after being boiled, should be stored in a pot of absolute alcohol when not in use.

3. Operating table.

4. Surgical instruments. Sterilise these before use by boiling, and disinfect them *after use* by the same means. Wipe perfectly dry immediately after the disinfection is completed.

Scissors, probe and sharp-pointed.
Dissecting forceps of various patterns.
Pressure forceps.
Retractors (small self retaining Fig. 172).
Aneurism needles, sharp and blunt.
Scalpels, } Keratomes, } with metal handles. Trephines, }

Michel's steel clips and special forceps for applying the same. These small steel clips enable the operator to easily and rapidly close skin incisions and are most satisfactory for animal operations.

Surgical needles.
Needle holder.
Soft rubber catheters, various sizes.
Gum elastic œsophageal bougies with connection to fit syringe.

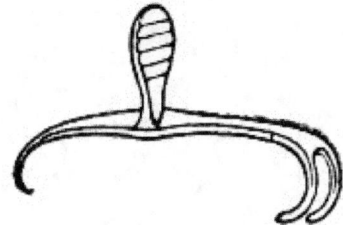

Fig. 172. Small self retaining retractors.

5. Anæsthetic.

(*a*) General: The safest general anæsthetic for animals is an A. C. E. mixture, freshly prepared, containing by volume alcohol 1 part, chloroform 2 parts, ether 6 parts, and should be administered on a "cone" formed by twisting up one corner of a towel and placing a wad of cotton-wool inside it, or from a saturated cotton-wool pad packed into the bottom of a small beaker.

(*b*) Local:

1. Cocaine hydrochloride, 2 per cent. in adrenalin 1 per mille solution. 2. Beta-eucaine, 2 per cent. in adrenalin, 1 per mille solution. 3. Ethyl chloride jet.

6. Sterile glass capsules of various sizes.

7. Cases of sterile pipettes { 10 c.c. (in tenths of a cubic centimetre).
{ 1 c.c. (in hundredths of a cubic centimetre).

8. Flasks (75 c.c.) containing sterilised normal saline solution (or sterile bouillon).

9. Sterilised cotton-wool. Cotton-wool (absorbent) is packed loosely in a copper cylinder similar to that used for storing capsules, and sterilised in the hot-air oven.

10. Sterilised gauze. Gauze is sterilised in the same way as cotton-wool.

11. Sterilised silk and catgut for sutures. These are sterilised, as required, by boiling for some ten minutes in the water steriliser.

12. Flexible collodion (or compound tincture of benzoin).

13. Grease pencil.

14. Tie-on celluloid labels, to affix to the cages.

15. Razor.

16. Small pot of warm water.

17. Liquid soap. Liquid soap is prepared as follows: Measure out 100 grammes of soft soap and add to 500 c.c. of 2 per cent. lysol solution in a large glass beaker; dissolve by heating in a water-bath at about 90° C. Bottle and label "Liquid Soap."

18. In place of the liquid soap and razor it is sometimes convenient to use a Depilatory powder.

Barium sulphide 1 part Rice starch 3 parts

Dust the powder thickly over the area to be denuded of hair, sprinkle with water and mix into a thin paste *in situ*; allow the paste to act for three minutes, then scrape off with a bone spatula—the hair comes away with the paste and leaves a perfectly bare patch. This process is preferably carried out, the day previous to the operation.

Material Utilised for Inoculation.—The material inoculated may be either—

1. Cultures of bacteria—grown in fluid media, or on solid media.

2. Metabolic products of bacterial activity—*e. g.*, toxins in solution.

3. Pathological products (fluid secretions and excretions, solid tissues).

The Preparation of the Inoculum.—

(*a*) *Cultivations in Fluid Media.*—

1. Flame the plug of the culture tube.

2. Remove the plug and flame the mouth of the tube.

3. Slightly raise the lid of a sterile capsule, insert the mouth of the culture tube into the aperture and pour some of the cultivation into the capsule.

4. Remove the mouth of the culture tube from the capsule, replace the lid of the latter, flame the mouth of the tube, and replug.

5. Remove the syringe from the steriliser, squirt out the water from its interior, and allow to cool.

6. Raise the lid of the capsule sufficiently to admit the needle of the syringe and draw the required amount of the cultivation into the barrel of the syringe.

(Or, remove a definite measured quantity of the cultivation directly from the tube or flask by means of a sterile graduated pipette, discharge the measured amount into a sterile capsule, and fill into the syringe; or take up the required quantity of the cultivation directly into the graduated syringe from the tube or flask.)

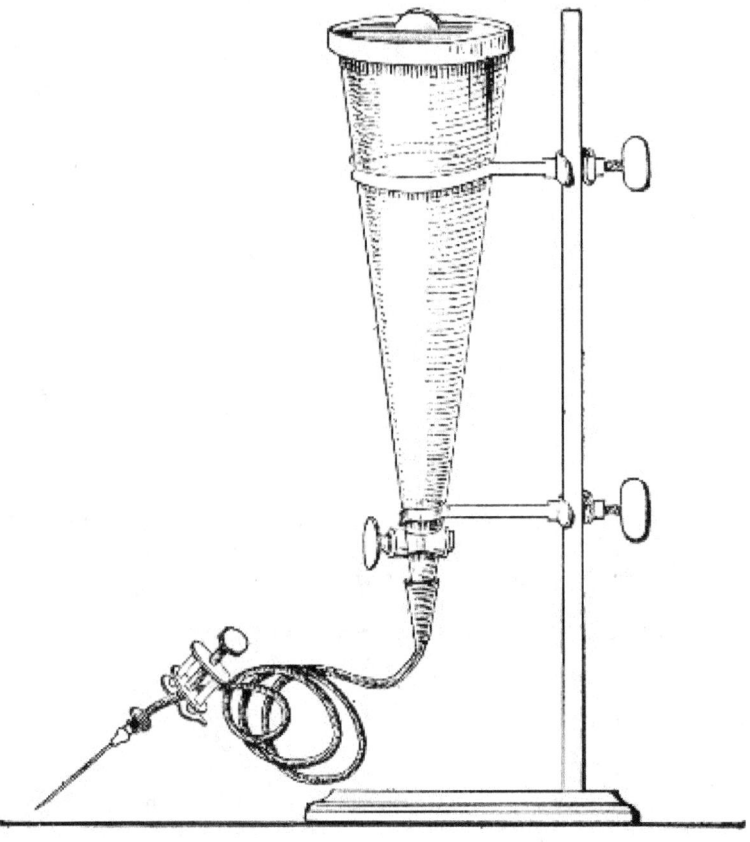

Fig. 173.—Conical separatory funnel, fitted for injection of fluid cultivations.

If it is necessary to introduce a large bulk of fluid into the animal, the cultivation should be transferred with aseptic precautions, to a sterile separatory funnel, preferably of the shape shown in figure 173, and graduated if necessary. This is supported on a retort stand and raised sufficiently above the level of the animal to be injected, so as to secure a good "fall." A piece of sterilised rubber tubing of suitable length, fitted with an injection needle and provided with a screw clamp, is now attached to the nozzle of the funnel and the operation completed according to the requirements of the particular case.

This method is quite satisfactory when the injection is made into the pleural or abdominal cavities or directly into a vein but if the injection has to be made into the subcutaneous tissue the "fall" may not be sufficient to force the fluid in. In this case it will be necessary to transfer the culture to a sterile wash-bottle and fasten a rubber hand bellows to the air inlet tube (interposing an air filter) and attach the tubing with the injection needle to the outlet tube (Fig. 174). By careful use sufficient force can be obtained to drive the injection in.

(b) *Cultivations on Solid Media (e. g., Sloped Agar)*.—

1. By means of a sterile graduated pipette introduce a suitable small quantity of sterile bouillon (or sterile normal saline solution) into the culture tube.

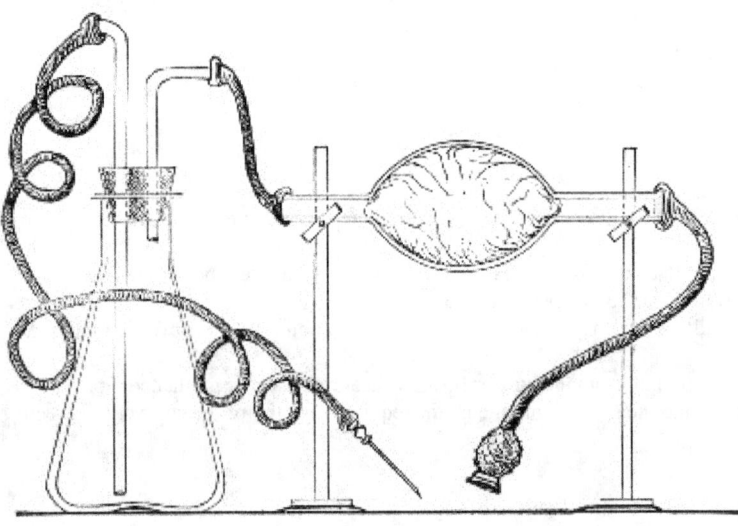

Fig. 174.—Arrangement of pressure injection

apparatus.

2. With a sterile platinum loop or spatula scrape the bacterial growth off the surface of the medium, and emulsify it with the bouillon. It then becomes to all intents and purposes a fluid inoculum.

3. Pour the emulsion into a sterile capsule and fill the syringe therefrom.

(*c*) *Toxins.*—Prepared by previously described methods (*vide* page 318), are manipulated in a similar manner to cultivations in fluid media.

(*d*) *Pathological Products.*—Fluid secretions, excretions, etc., such as serous exudation, pus, blood, etc., are treated as fluid cultivations; but if the material is very thick or viscous, a small quantity of sterile bouillon or normal saline solution may be used to dilute it, and thorough incorporation effected by the help of a sterile platinum rod.

Solid tissues, such as spleen, lymph glands, etc., may be divided into small pieces by sterile instruments and rubbed up in a sterilised agate mortar (using an agate pestle), with a small quantity of sterile bouillon, and the syringe filled from the resulting emulsion.

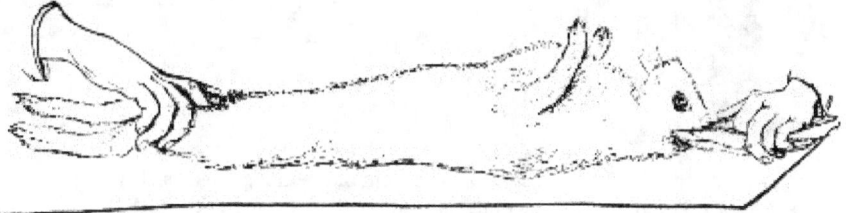

Fig. 175.—Holding rabbit for shaving.

If it is desired to inoculate tissue *en masse*, remove from the material a small cube of 1 or 2 mm. and introduce it into a wound made by sterile instruments in a suitable situation, and occlude the wound by means of Michel's steel clips and a sealed dressing.

Method of Securing Animals During Inoculation.—

For the majority of inoculations, especially when no anæsthetic is administered, it is customary to employ an assistant to hold the animal (see Fig. 175).

If working single handed Voge's holder for guinea-pigs, is a useful piece of apparatus the method of using which is readily seen from the accompanying figures (Figs. 176, 177).

The instrument itself consists of a hollow copper cylinder, one end of which is turned over a ring of stout copper wire, and from this open end a slot is cut extending about half way along one side of the cylinder. The opposite end is closed by a "pull-off" cap and is perforated around its edge by a row of ventilating holes, which correspond with holes cut in the rim of the cap. In the event of the animal resisting attempts to remove it from the holder backwards, this cap is taken off and the holder placed on the table and the guinea-pig allowed to walk out.

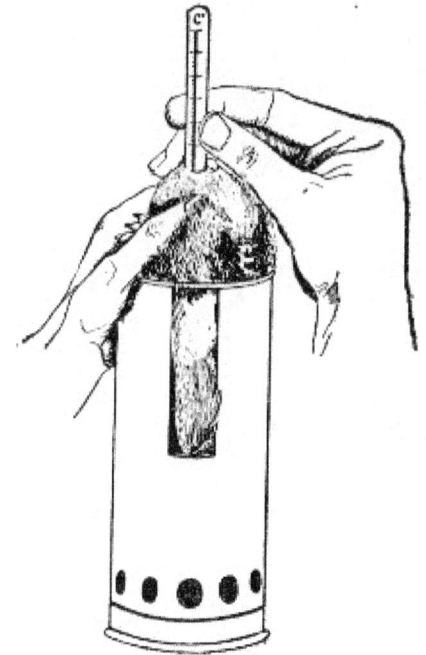

Fig. 176.—Taking guinea-pig's temperature.

To provide for different-sized animals, two sizes of this holder will be found useful:

1. Length, 16 cm.; breadth, 6 cm.; size of slot, 8 cm. by 2.5 cm.
2. Length, 20 cm.; breadth, 8 cm.; size of slot, 10 cm. by 2.5 cm.

A convenient holder for mice and even small rats is shown in figure 178, the tail being securely held by the spring clip. Needless to say, the holder should be entirely of metal, and the wire cage detachable and easily renewed.

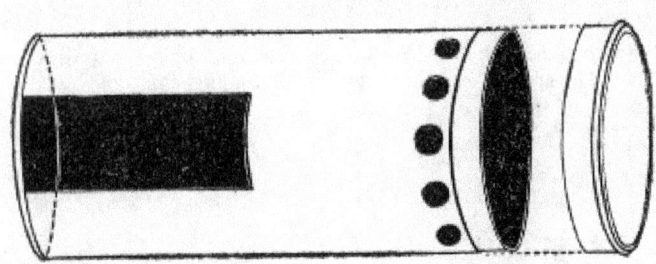

Fig. 177.—Voge's holder.

When the animal is anæsthetised, it is more convenient to secure it firmly to some simple form of operating table, such as Tatin's (Fig. 179), which will accommodate rabbits, guinea-pigs, and rats: or to the more elaborate table devised by the author (Fig. 180).

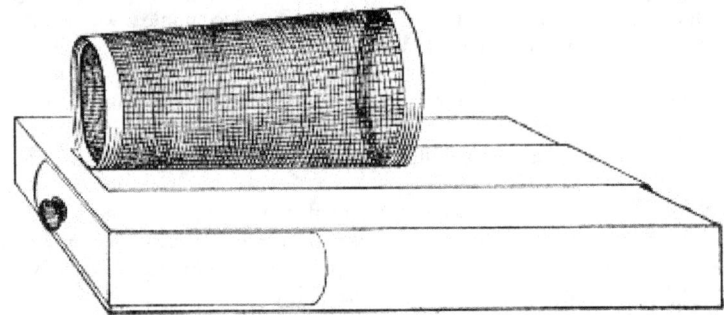

Fig. 178.—Mouse holder.

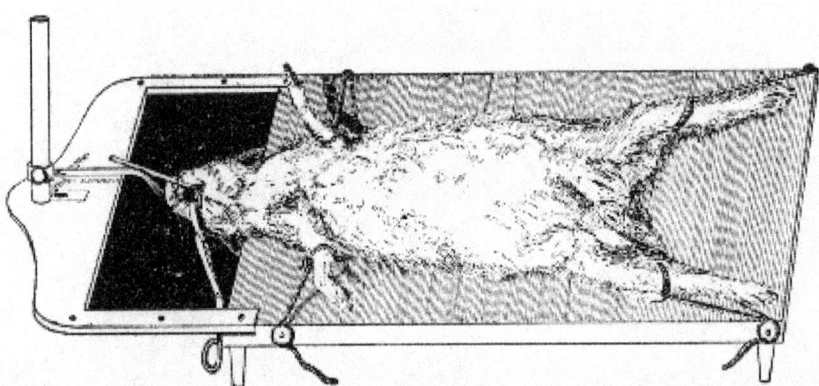

Fig. 179.—Tatin's operation table.

Operation Table.—This is a table of the "aseptic" type, composed of steel tubing, nickel-plated or enamelled. The table-top frame is sufficiently large to accommodate rabbits, dogs and monkeys; and is supported upon telescopic uprights, so that it is adjustable as to height; in its long axis it can be inclined (at either end) to 45° from the horizontal. Further it can be completely rotated about its long axis. The table-top itself is composed of a sheet of copper wire gauze loosely suspended from the long sides of the tubular frame. The slackness of the gauze bed permits of an india rubber hot water bottle, or an electrotherm being placed under the animal, and if during the course of an experiment it is necessary to reverse the animal, the table-top frame is completely rotated, the device adopted for suspending the gauze is detached and the gauze reversed also, so that it again supports the animal from below.

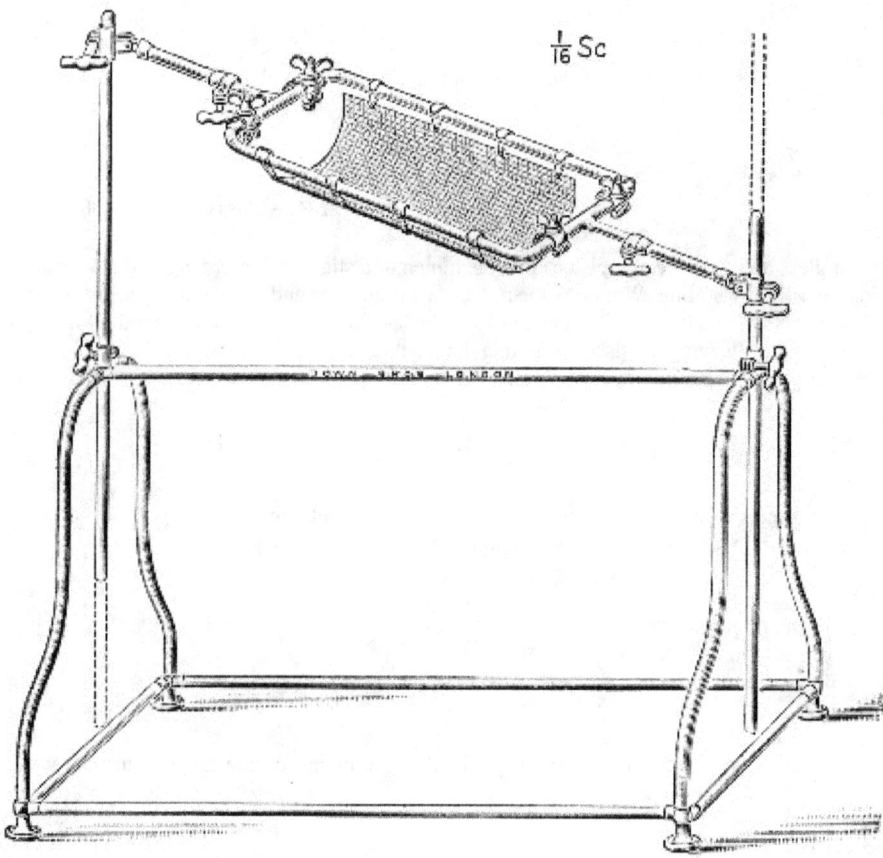

Fig. 180.—Author's operating table

METHODS OF INOCULATION.

The following methods of inoculation apply more particularly to the rabbit, but from them it will readily be seen what modifications in technique, if any, are necessary in the case of the other experimental animals.

1. Cutaneous Inoculation.—(*Anæsthetic, none.*)

1. Have the animal firmly held by an assistant (or secured to the operating table).
2. Apply the liquid soap to the fur, over the area selected for inoculation, with a wad of cotton-wool, and lather freely by the aid of warm water; shave carefully and thoroughly; or apply the depilatory powder.
3. Wash the denuded area of skin thoroughly with 2 per cent. lysol solution.
4. Wash off the lysol with ether and allow the latter to evaporate.
5. Make numerous short, parallel, superficial incisions with the point of a sterile scalpel.
6. When the oozing from the incisions has ceased, rub the inoculum into the scarifications by means of the flat of a scalpel blade, or a sterile platinum spatula.
7. Cover the inoculated area with a pad of sterile gauze secured *in situ* by strips of adhesive plaster or by sealing down the edges of the gauze with collodion.
8. Release the animal, place it in its cage, and affix a label upon which is written:
(a) Distinctive name or number of the animal. (b) Its weight. (c) Particulars as to source and dose of inoculum. (d) Date of inoculation.

2. Subcutaneous Inoculation.—

(a) *Fluid Inoculum.*—(*Anæsthetic, none.*)

Steps 1-4. As for cutaneous inoculation.

5. Pinch up a fold of skin between the forefinger and thumb of the left hand; take the charged hypodermic syringe in the right hand, enter the needle into a ridge of skin raised by the left finger and thumb, and push it steadily onward until about 2 cm. of the needle are lying in the subcutaneous tissue. Now release the grasp of the left hand and slowly inject the fluid contained in the syringe.
6. Withdraw the needle, and at the same moment close the puncture with a wad of cotton wool, to prevent the escape of any of the inoculum. The injected fluid, unless large in amount, will be absorbed within a very short time.
7. Label, etc.

(b) *Solid Inoculum.*—(*Anæsthetic, none; or Ethyl chloride spray.*)

Steps 1-4. As for cutaneous inoculation.

5. Raise a small fold of skin in a pair of forceps, and make a small incision through the skin with a pair of sharp-pointed scissors or with the point of a scalpel.
6. Insert a probe through the opening and push it steadily onward in the subcutaneous tissue, and by lateral movements separate the skin from the underlying muscles to form a funnel-shaped pocket with its apex toward the point of entrance.

7. By means of a pair of fine-pointed forceps introduce a small piece of the inoculum into this pocket and deposit it as far as possible from the point of entrance.

Fig. 181.—Glass tube syringe for subcutaneous "solid" inoculation.

Or, improvise a syringe by sliding a piece of glass rod (to serve as a piston) into the lumen of a slightly shorter length of glass tubing and secure in position by a band of rubber tubing. Sterilise by boiling. Withdraw the rod a few millimetres and deposit the piece of tissue within the orifice of the tube, by means of sterile forceps. Now pass the tube into the depths of the "pocket," push on the glass rod till it projects beyond the end of the tube, and withdraw the apparatus, leaving the tissue behind in the wound.

8. Close the wound in the skin with Michel's clips and a dressing of gauze sealed with collodion (or Tinct. benzoin).

9. Label, etc.

3. Intramuscular.—

(a) *Fluid Inoculum.—(Anæsthetic, none.)*

Steps 1-4. As for cutaneous inoculation.

5. Steady the skin over the selected muscle or muscles with the slightly separated left forefinger and thumb.

6. Thrust the needle of the injecting syringe boldly into the muscular tissue and inject the inoculum slowly.

7. Label, etc.

(b) *Solid Inoculum.—(Anæsthetic, A. C. E.)*

1. Secure the animal to the operation table and anæsthetise.

2. Shave and disinfect the skin at the seat of operation.

3. Surround the field of operation by strips of gauze wrung out in 2 per cent. lysol solution.

4. Incise skin, aponeurosis, and muscle in turn.

5. Deposit the inoculum in the depths of the incision.

6. Close the wound in the muscle with buried sutures and the cutaneous wound with either continuous or interrupted sutures or with Michel's steel clips.

7. Apply a sealed dressing of gauze and collodion.

8. Remove the animal from the operating table.

9. Label, etc.

4. Intraperitoneal.—

(a) *Fluid Inoculum.—(Anæsthetic, none.)*

Steps 1-4. As for cutaneous inoculation. Shave a fairly broad transverse area, stretching from flank to flank.

5. Place the left forefinger on one flank and the thumb on the opposite, and pinch up the entire thickness of the abdominal parietes in a triangular fold. Now, by slipping the peritoneal surfaces (which are in apposition) one over the other, ascertain that no coils of intestine are included in the fold.

6. Take the syringe in the right hand and with the needle transfix the fold near its base (Fig. 182).

7. Now release the fold, but hold the syringe steady; as the parietes flatten out, the point of the needle is left free in the peritoneal cavity (see Fig. 183).

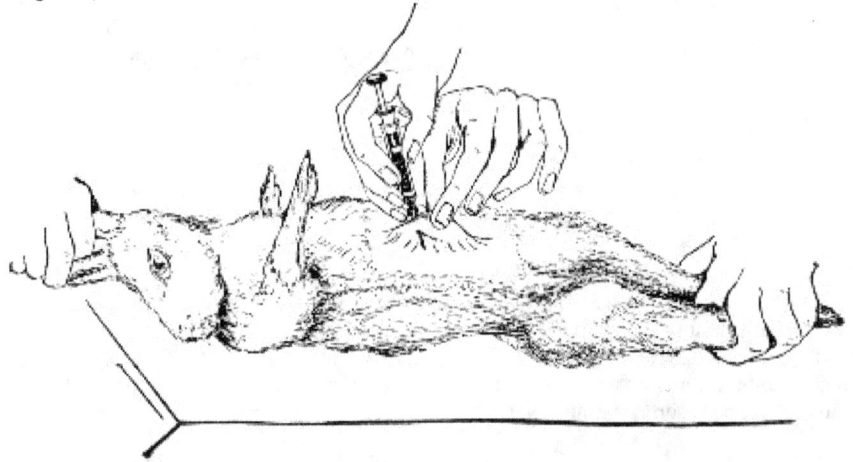

Fig. 182.—Intraperitoneal inoculation—fluid.

8. Inject the fluid from the syringe.

9. Label, etc.

Fig. 183.—Section of abdominal wall, etc., showing point of needle lying free in the peritoneal cavity above the coils of intestine.

Second Method:

Steps 1-4. As in the first method.

5. Anæsthetise a small selected area of skin by spraying it with ethyl chloride.

6. Heat platinum searing wire (0.5 mm. wire, twisted to the shape indicated in figure 184, mounted in an aluminium handle) to redness, and with it burn a hole through the anæsthetic area of skin and abdominal muscle down to, but not through, the visceral peritoneum.

7. Fix a blunt-ended needle on to the charged syringe, and by pressing the rounded end firmly against the peritoneum it can easily be pushed through into the peritoneal cavity.

8. Inject the fluid from the syringe.

9. Label, etc.

This method is especially useful when it is desired to collect samples of the peritoneal fluid from time to time during the period of observation, as fluid can be removed from the peritoneal cavity, at intervals, through this aperture in the abdominal parietes, by means of a sterile capillary pipette.

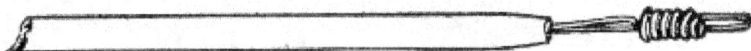

Fig. 184.—Platinum wire for burning hole through parietes.

(b) *Solid Inoculum* (or the implantation of capsules containing fluid cultivations).—(*Anæsthetic, A. C. E.*)

1. Anæsthetise the animal and secure it to the operating table.
2. Shave a large area of the abdominal parietes.
3. Make an incision through the skin in the middle line about 2 cm. in length, midway between the lower end of the sternum and the pubes.
4. Divide the aponeuroses between the recti upon a director.
5. Divide the peritoneum upon a director.
6. Introduce the inoculum into the peritoneal cavity.
7. Close the peritoneal cavity with Lembert's sutures.
8. Close the skin and aponeurosis incisions together with interrupted sutures or Michel's steel clips, and apply a sealed dressing.
9. Release the animal from the operating table.
10. Label, etc.

Suitable sacs may be readily prepared by either of the following methods:

A. **Collodion Sacs.**

1. Dip a small test-tube (5 by 0.5 cm.), bottom downward, into a beaker of collodion, and dry in the air; repeat this process three or four times.
2. Dip the tube, with its coating of collodion, alternately into a beaker of alcohol and one of water. This loosens the collodion and allows it to be peeled off in the shape of a small test-tube.
3. Take a 20 cm. length of glass tubing, of about the diameter of the test-tube used in forming the sac, and insert one end into the open mouth of the sac.
4. Suspend the glass tube with attached sac, inside a larger test-tube, by packing cotton-wool in the mouth of the test-tube around the glass tubing, and place in the incubator at 37° C. for twenty-four hours. When removed from the incubator, the sac will be firmly adherent to the extremity of the glass tubing.
5. Plug the open end of the glass tubing with cotton-wool, and sterilise the test-tube and its contents in the hot-air oven.

To use the sac, remove the plug from the glass tubing, partly fill the sac with cultivation to be inoculated, by means of a sterile capillary pipette, and replug the tubing. When the abdominal cavity has been opened, remove the tubing and attached sac from the protecting test-tube, close the sac by tying a sterilised silk thread tightly around it a little below the end of the glass tubing, and separate it from the tubing by cutting through the collodion above the ligature, and the sac is ready for insertion in the peritoneal cavity.

B. **Celloidin Sacs** (*Harris*).

Materials Required.

Quill glass tubing.

Gelatine capsules such as pharmacists prepare for the exhibition of bulky powders.

Various grades of celloidin, thick and thin, in wide-mouthed bottles.

1. Take a piece of quill glass tubing some 4 cm. long by 5 mm. diameter; heat one end in the bunsen flame.
2. Thrust the heated end of the tube just through one end of a gelatine capsule and allow it to cool (Fig. 185).
3. Remove any gelatine from the lumen of the tube with a heated platinum needle; paint the joint between capsule and tube with moderately thick celloidin and allow to dry.

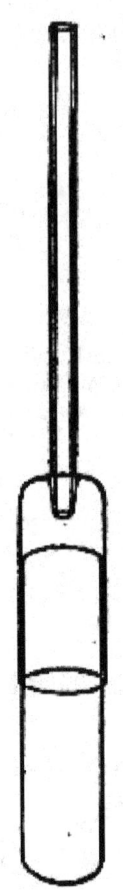

Fig. 185.—Making celloidin capsules.

4. Dip the capsule into a beaker containing thin celloidin, beyond the junction with the glass and after removal rotate it in front of the blowpipe air blast to dry it evenly. Repeat these manœuvres until a sufficiently thick coating is obtained.

5. Apply thick celloidin to the tube-capsule joint, the opposite end of the capsule, and the line of junction of the capsule with its cap; dry thoroughly.

6. With a teat pipette fill the capsule (through the attached tube) with hot water, and stand the capsule in a beaker of boiling water for a few minutes to melt the gelatine.

7. Remove the solution of gelatine from the interior of the celloidin case with a pipette.

8. Fill the sac with nutrient broth and place it, *glass tube downward*, in a tube containing sufficient sterile nutrient broth to cover the sac to the depth of 1 cm. Plug the tube and sterilise in the steamer in the usual manner.

9. To prepare the sac for use, empty it out of the broth tube into a sterile glass dish.

10. Grasp the tube near its junction with the sac in the jaws of sterile forceps, and with a teat pipette remove sufficient of the contained broth to leave a small space in the sac. Introduce the inoculum in the form of an emulsion by means of another pipette.

11. Still holding the tube in the forceps, draw it out and seal off near the sac in the blowpipe flame.

12. When cool wash the sac in sterile water, then transfer to a tube of nutrient broth and incubate over night to determine its impermeability to bacteria.

13. If the broth outside the sac remains sterile, insert the sac in the peritoneal cavity of the experimental animal.

5. Intracranial.—(*Anæsthetic, A. C. E.*)

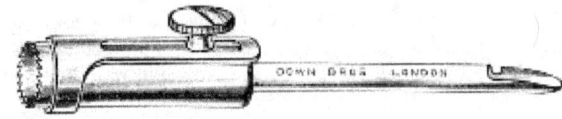

Fig. 186.—Guarded trephine.

Trephines and Surgical Engine.—The most useful instrument for intracranial operations upon animals is the small nasal trephine (Curtis) having a tooth cutting circle of 7 mm. The addition of an adjustable collar guard—secured by a screw—prevents accidental laceration of the dura mater or brain substance (Fig. 186). This size is suitable for monkeys, dogs, cats and large rabbits. Other smaller sizes which will be found useful for guinea pigs and other small animals cut circles of 6 and 4 mm.; for very small animals—young guinea pigs and rats—a small dental drill or screw will make a sufficiently large hole to admit the syringe needle. The trephine can be set in ordinary metal handles and rotated by hand, but a surgical engine of some kind is much preferable on the score of rapidity and safety to the animal. The Guy's electrical Dental engine (Fig. 187) which can be connected to a lamp socket or wall plug, and is operated by a foot switch, although inexpensive is eminently satisfactory.

Note.—A fine dental drill attached to the dental engine renders the manufacture of aluminium handles needles (see page 71) quite an easy matter.

(*a*) *Subdural.*
1. Anæsthetise the animal and secure it to the operating table, dorsum uppermost.
2. Shave a portion of the scalp immediately in front of the ears.

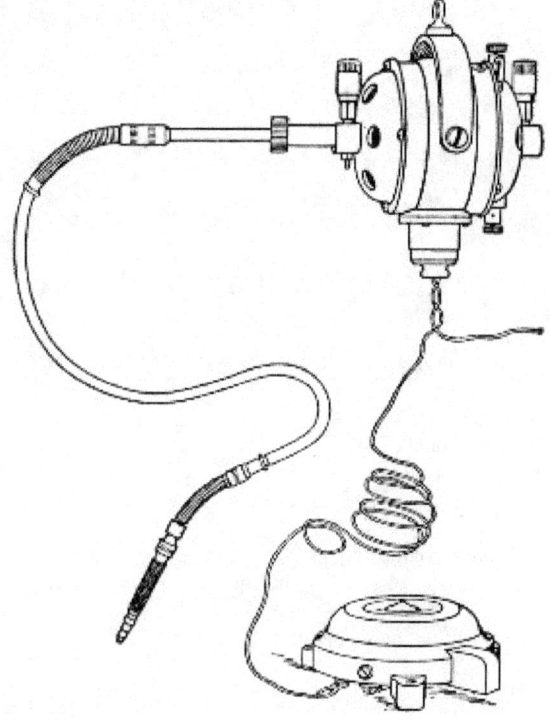

Fig. 187.—Guy's electrical dental engine.

3. Mark out with a sharp scalpel a crescentic flap of skin muscle, etc., convexity forward, commencing 0.5 cm. in front of the root of one ear and terminating at a similar spot in front of the other ear. Reflect the marked flap.
4. Make a corresponding incision through the periosteum and raise it with a blunt dissector.
5. With a small trephine (diameter 6 mm.) remove a circular piece of bone from the parietal segment. The centre of the trephine hole should be at the intersection of the median line and a line joining the posterior canthi (Fig. 188).
6. Introduce the inoculum by means of a hypodermic syringe, perforating the dura mater with the needle and depositing the material immediately below this membrane, at the same time taking care to avoid injuring the sinuses.
7. Turn back the flap of skin and secure it in position with Michel's steel clips.
8. Dress with sterile gauze and wool and seal the dressing with collodion.
9. Label, etc.

(*b*) *Intracerebral.*—This inoculation is performed precisely as for subdural save in step 6 the needle after perforating the dura mater is pushed onward into the substance of one or other cerebral hemispheres before the contents are ejected.

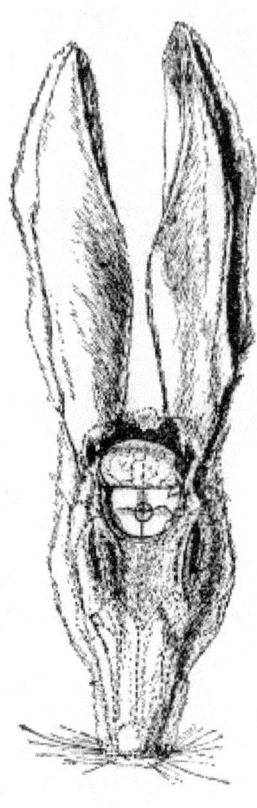

Fig. 188.—Intracranial inoculation of rabbit. The circle indicates the situation of the trephine hole.

6. Intraocular.—

(*a*) *Fluid Inoculum.*—(*Anæsthetic, cocaine.*)

1. Instil a few drops of a sterile solution of cocaine, and repeat the instillation in two minutes.
2. Five minutes later have the animal firmly held by an assistant as in intravenous injection (see Fig. 189), the head being steadied by the assistant's hands.
3. Select two needles to accurately fit the same syringe and sterilise.
4. Attach one needle to the syringe and take up the required dose of inoculum and remove the needle.
5. Steady the eye with fixation forceps; then pierce the cornea with the other syringe needle and allow the aqueous to escape through the needle.
6. Without removing the needle from the cornea attach the syringe and make the injection into the anterior chamber.
7. Irrigate the conjunctival sac with sterile saline solution.
8. Label, etc.

(*b*) *Solid Inoculum.*—(*Anæsthetic, A. C. E.*)

1. Anæsthetise the animal and secure it firmly to the operating table.
2. Irrigate the conjunctival sac thoroughly with sterile saline solution.
3. Make an incision through the upper quadrant of the cornea into the anterior chamber by means of a triangular keratome.
4. Separate the lips of the corneal wound with a flexible silver spatula; seize the solid inoculum in a pair of iris forceps, introduce it through the corneal wound, and deposit it on the anterior surface of the iris; withdraw the forceps.
5. Again irrigate the sac and the surface of the cornea.
6. Release the animal from the operating table.
7. Label, etc.

7. Intrapulmonary.—

Fluid Inoculum.—(*Anæsthetic, none.*)

1. Have the animal firmly held by an assistant. (In this case the foreleg of the selected side is drawn up by the assistant and held with the ear of that side.)
2. Shave carefully in the axillary line and disinfect the denuded skin.
3. Thrust the needle of the syringe boldly through the fifth or sixth intercostal space into the lung tissue.
4. Inject the contents of the syringe slowly.
5. Label, etc.

8. Intravenous.—

Fluid Inoculum.—(*Anæsthetic, none.*)

The site selected for the injection in the rabbit is the posterior auricular vein (see Fig. 192). Although this is smaller than the median vein, it is firmly bound down to the cartilage of the ear by dense connective tissue, and is therefore more readily accessible. (In the guinea-pig the jugular vein must be utilised, and in order to perform the inoculation satisfactorily a general anæsthetic must be administered to the animal. In the monkey or the dog, the internal saphenous vein is the most convenient and before puncturing should be distended or rendered prominent by compressing the vein above the selected site.)

Preparation of the Inoculum.—Care must be taken in preparing the inoculum, as the injection of even small fragments may cause fatal embolism. To obviate this risk the fluid should, if possible, be filtered through sterile filter paper before filling into the syringe.

Air bubbles, when injected into a vein, frequently cause immediate death. To prevent this, the syringe after being filled should be held in the vertical position, needle uppermost. A piece of sterile filter paper is then impaled on the needle and the piston of the syringe pressed upward until all the air is expelled from the barrel and needle. Should any drops of the inoculum be forced out, they will fall on the filter paper, which should be immediately burned.

1. Have the animal firmly held by an assistant. The selected ear is grasped at its root and stretched forward toward the operator.
2. Shave the posterior border of the dorsum of the ear.
3. Disinfect the skin over the vein, rubbing it vigorously with cotton-wool soaked in lysol. The friction will make the vein more conspicuous. Wash the lysol off with ether and allow the latter to evaporate.
4. Direct the assistant to compress the vein at the root of the ear. This will cause its peripheral portion to swell up and increase in calibre.
5. Hold the syringe as one would a pen and thrust the point of the needle through the skin and the wall of the vein till it enters the lumen of the vein (Fig. 189). Now press it onward in the direction of the blood stream—*i. e.*, toward the body of the animal.
6. Direct the assistant to cease compressing the root of the ear, and *slowly* inject the inoculum. (If the fluid is being forced into the subcutaneous tissue, a condition which is at once indicated by the swelling that occurs, the injection must be stopped and another attempt made at a spot closer to the root of the ear or at some point on the corresponding vein on the opposite ear.)
7. Withdraw the needle and press a pledget of cotton-wool over the puncture to ensure closure of the aperture in the vein wall.
8. Label, etc.

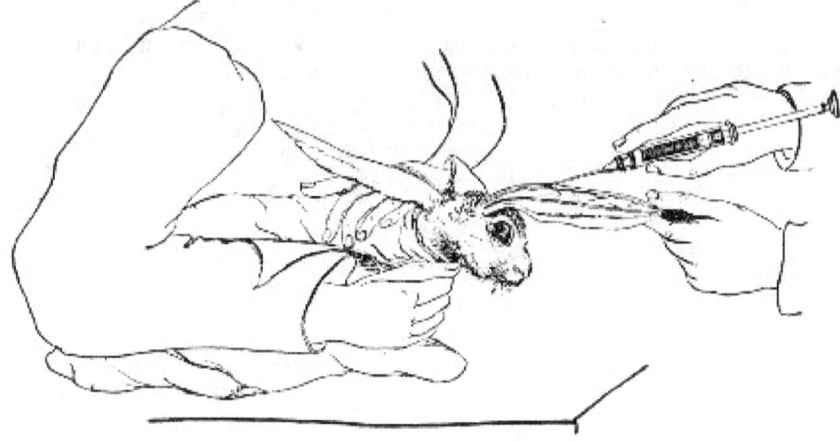

Fig. 189.—Intravenous inoculation.

9. **Inhalation.**—

(*a*) *Fluid Inoculum.*—(*Anæsthetic, none.*)
1. Place the animal in a closed metal box.
2. Through a hole in one side introduce the nozzle of some simple spraying apparatus, such as is used for nasal medicaments.
3. Fill the reservoir of the instrument (previously sterilised) with the fluid inoculum, and having attached the bellows, spray the inoculum into the interior of the box.
4. On the completion of the spraying, open the box, spray the animal thoroughly with a 10 per cent. solution of formaldehyde (to destroy any of the virus that may be adhering to fur or feathers).
5. Transfer the animal to its cage.
6. Label, etc.
7. Thoroughly disinfect the inhalation chamber.

(*b*) *Fluid or Powdered Inoculum.*—Anæsthetic, A. C. E.
1. Anæsthetise the animal and secure it firmly to the operating table.

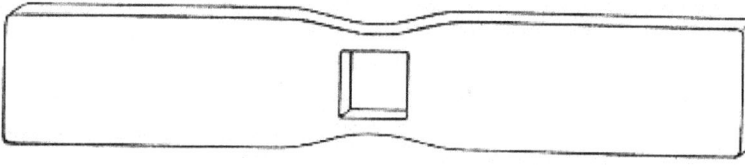

Fig. 190.—Gag for rabbits.

2. Prop open the mouth by means of some form of gag; seize the tongue with a pair of forceps and draw it forward.

The most convenient form of gag for the rabbit or cat is that shown in Fig. 190. It is simply a strip of hard wood shaped at the middle and provided with a square orifice through which a tracheal or œsophageal tube can be passed.

3. Pass a previously sterilised glass tube (17 cm. long, 0.5 cm. diameter, with its terminal 2 cm. slightly curved) down through the larynx into the trachea.
4. Connect the straight portion of a **Y**-shaped piece of tubing to the upper end of the sterilised tube and couple one branch of the **Y** to a separatory funnel containing the fluid inoculum, or insufflator containing the powdered inoculum, and the other to a hand bellows.
5. Allow the fluid inoculum to run into the lungs by gravity, or blow in the powdered inoculum by means of a rubber-ball bellows.

6. Remove the intratracheal tube; release the animal from the table.

7. Label, etc.

As an alternative method in the case of fairly large animals, such as rabbits, etc., a sterile piece of glass tubing of suitable diameter may be passed through the larynx down the trachea almost to its bifurcation. Fluid cultivations may then be literally poured into the lungs, or cultivations, dried and powdered, may be blown into the lung by the aid of a small hand bellows or even a teat pipette.

10. Intragastric Inoculation.—*Fluid or semi-fluid inoculum. (Anæsthetic none.)*

The method of performing the operation is varied slightly according to the size of the experimental animal.

A. Monkey, Rabbit, Guinea-pig.

1. Secure the animal to the operating table ventral surface uppermost.
2. Prop the mouth open with a gag; draw the tongue forward with forceps.
3. Sterilise a soft rubber catheter (No. 10 or 8 English scale, or No. 18 or 15 French) and lubricate it with sterile glycerine.
4. Pass it to the back of the pharynx, keeping the end in the middle line.
5. Gently assist the progress of the catheter down the œsophagus until it passes the cardiac orifice of the stomach. Do not use any force.
6. Take up the required dose of inoculum into a sterilised pipette. Insert the point of the pipette into the open end of the catheter and allow the fluid to run down into the stomach. Remove the pipette and drop it into a jar of lysol.
7. With another sterile pipette run one cubic centimetre of sterile saline solution through the catheter to wash out the last traces of the inoculum.
8. Withdraw the catheter.
9. Label, etc.

B. Rats and Mice (Mark's Method).

1. Secure the animal in the vertical position.

(a) *Rat.*—Take a pair of catch sinus forceps about 22 cm. in length and seize the animal by the loose skin of the head as far forward as possible—fix the forceps, and holding the instrument vertically upward, transfer to the left hand of an assistant who secures the animal's tail between the fingers grasping the handle of the forceps. (See Fig. 191.)

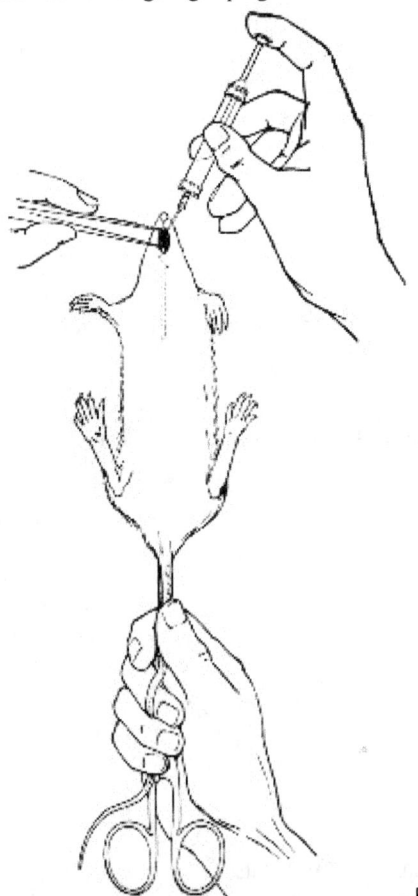

Fig. 191.—Intragastric inoculation of rat.

(b) *Mouse.*—An assistant grasps the loose skin between the ears as far forwards as possible between the forefinger and thumb of the left hand. He now grasps the tail with the right hand, draws the mouse straight and passes the tail between the fourth and little fingers of the left hand and secures it there.

2. The assistant takes a closed pair of thin-bladed forceps in his right hand, passes the ends into the animal's mouth, then allows the blades to separate. This opens the animal's jaw and serves as a gag.

3. Moisten the sterilised œsophageal tube with sterile water. (This tube is of silk rubber, 6.5 cm. in length, with the distal end rounded, the proximal end mounted in a syringe needle head, which fits the nozzles of the two sterile syringes to be used.)

4. Grasp the tube about its middle and pass it into the animal's mouth, downwards and a little to one side or the other until its length is lost in the digestive tract and mouth. Gentle guidance is alone necessary. Do not use any force.

5. Take up the required dose of inoculum into the syringe; insert the nozzle of the syringe into the needle-mount, and force the piston down.

6. Steadying the needle-mount with the left hand, detach the syringe.

7. Draw up some sterile water in the second (sterile) syringe, and inserting its nozzle into the needle-mount force a few drops of water through the tube to wash it out.

8. With one quick upward movement remove the tube from the animal's mouth.

9. Label, etc.

One other method of inoculation remains to be described, which does not require operative interference.

11. Feeding.—

1. *Fluid Inoculum.*—Small pieces of sterilised bread or sop (sterilised in the steamer at 100° C.) are soaked in the fluid inoculum and offered to the animals in a sterile Petri dish or capsule.

2. *Solid Inoculum.*—Small pieces of tissue are placed in sterile vessels and offered to the animals.

FOOTNOTES:

This table is made by Messrs. Down Bros., St. Thomas's Street, London, S. E.
This modification is made for the author by Messrs. Down Bros., St. Thomas's Street, London, S. E.
Manufactured by Messrs. Francis Lepper, 56, Great Marlborough Street, London, W.

XVIII. THE STUDY OF EXPERIMENTAL INFECTIONS DURING LIFE.

The possession of pathogenetic properties by an organism under study is indicated by the "infection" of the experimental animal—a term which is employed to summarise the condition resulting from the successful invasion of the tissues of the experimental animal by the micro-organisms inoculated and by their multiplication therein. Infection is considered to have taken place:

1. When the death of the animal is produced as a direct consequence of the inoculation.
2. When without necessarily producing death the inoculation causes local or general changes of a pathological character.
3. When either with or without death, or local or general changes occurring, certain substances make their appearance in the body fluids, which can be shown (*in vitro* or *in vivo*) to exert some profound and specific effect when brought into contact with subcultivations of the organism originally inoculated.

The important factors in the production of infection are:

A. Seed. Virulence of organism.
Dose of organism.

B. Soil. Resistance offered by the cells of the experimental animal.

The first two factors, although variable, are to a certain extent under the control of the experimenter. Thus by suitable means the virulence of an organism can be exalted or attenuated, whilst the size of the dose may be increased or diminished. The third factor also varies, not only amongst different species of animals, but also amongst different individuals of the same species. The essential causes of this variation are not so obvious, so that beyond selecting the animals intended for similar experiments with regard to such points as age, size or sex, but little can be done to standardise cell resistance.

Immediately an animal has been inoculated a period of clinical observation must be entered upon, which should only terminate with the death of the animal. The general observations should at first and if the infection is an acute one, be made daily—later, and if the animal appears to be unaffected or if the infection is chronic, both general and special observations should be carried out at weekly intervals. If the animal appears to be still unaffected, it should be killed with chloroform vapour at the end of two or three months and a complete post-mortem carried out.

A. The **general observations** should take cognisance of:

1. *General appearance.* The experimental animal should be inspected daily, not only with a view to detecting symptoms due to the experimental infection, but also to prevent any intercurrent infection, naturally acquired, from escaping notice (*vide* page 337).

2. *The weight* of the inoculated animal should be observed and recorded each day during the course of an experimental infection at precisely the same hour, preferably just before the morning feed.

3. *The temperature* should similarly be recorded daily, if not more frequently, during the whole period the animal is under observation, and carefully charted—individual variations will at once become apparent. It should be borne in mind that the temperature regarded as normal for man (37.5° C.) is not the normal average temperature of any of the lower animals save the rat and mouse. The accompanying table of normal averages for the animals usually employed in bacteriological research may be of use in preventing the erroneous assumption that pyrexia is present in an animal, which merely shows its own normal temperature.

NORMAL AVERAGES.

Animal.	Rectal Temp. °C.	Pulse.	Respirations.
		Rate per minute.	
Frog	8.9-17.2	80	12
Mouse	37.4	120	...
Rat	37.5	...	210
Guinea pig	38.6	150	80
Rabbit	38.7	135	55
Cat	38.7	130	24
Dog	38.6	95	15
Goat	40.0	75	16
Ox	38.8	45	..
Horse	37.9	38	11
Monkey (Rhesus)	38.4	100	19
Pigeon	40.9	136	30
Fowl	41.6	140	12

B. **Special observations** comprise some or all of the following, according to the method of inoculation and the character of the virus.

1. *The site of inoculation* should be minutely examined at least at weekly intervals, and the neighbouring lymphatic glands palpated.

2. Any *local reaction* at the site of inoculation and any other readily accessible lesion should be carefully investigated. Any suppurative process which may occur, whether in the subcutaneous tissues or in joints, should be explored and the pus carefully examined both microscopically and culturally.

Fluid secretions and excretions, such as pus or serous exudates when accessible are collected direct from the body in sterile capillary pipettes (*vide* Fig. 13a,) in the following manner:

1. Open the case containing the pipettes, grasp one by the plugged end, remove it from the case, and replace the lid of the latter.
2. Attach a rubber teat (*vide* page 10) to the plugged end of the pipette and use the teat as the handle of the pipette.
3. Pass the entire length of the pipette twice or thrice through the flame of the Bunsen burner.
4. Snap off the sealed end of the pipette with a pair of sterile forceps.
5. Compress the india-rubber teat, thrust the point of the pipette into the secretion; now relax the pressure on the teat and allow the pipette to fill.
6. Remove the point of the pipette from the secretion, allow the fluid to run a short distance up the capillary stem and seal the point of the pipette in the flame. (If using a pipette with a constriction below the plugged mouthpiece (Fig 13b), this portion of the pipette may also be sealed in the flame.)

When ready to examine the morbid material snap off the sealed end of the pipette with sterile forceps and eject the contents of the pipette into a sterile capsule. The material can now be utilized for cover-slip preparations, cultivations and inoculation experiment.

3. *The peripheral blood* should be examined from time to time for from this tissue is often obtained the fullest information as to the course and progress of an infection.

a. The **histological examination of the blood** should be directed chiefly to observations on the number and kind of white cells; and since but few bacteriologists are at the same time expert comparative hæmatologists, some notes on the normal characters of the blood of the commoner laboratory animals, contrasted with those of man, are inserted for reference. These have been very kindly compiled for me by my friend and one time colleague Dr. Cecil Price Jones.

COMPARATIVE HÆMOCYTOLOGY OF LABORATORY ANIMALS.

Animal	Totals		Percentages					
	Red cells	White cells	Hb, per cent.	Lymphocytes, per cent.	Large monos, per cent.	Polymorph, per cent.	Eosinoph, per cent.	Mast cells, per cent.
Frog	490,000	8,000	58	40	10.0	22.0	15	13
Mouse	8,700,000	8,000	78	60	21.5	17.0	1.4	0.1
Rat	9,000,000	9,000	85	54	7.0	37.5	1.3	0.2
Guinea-pig	5,700,000	10,000	99	55	9.0	32.8	3.0	0.2

Rabbit	6,000,000	7,000	70	50	2.0	46.0	0.6	1.4
Rhesus	4,500,000	13,000	77	43	5.0	50.0	1.3	0.7
Goat	14,600,000	15,000	58	35	6.3	56.7	1.25	0.75
Fowl	3,500,000	30,000	100	49	3.0	42.0	1.0	5.0
Pigeon	3,500,000	20,000	101	43	9.0	43.0	3.0	2.0
Man(adult) Normal limits.	5,000,000 (4.5-5) millions.	7,500 (7-9) thousands.	100 (95-101)	25 (20-30)	5.5 (4-8)	65 (55-68)	4.0 (3-5)	0.5 (0.5-2)

The above table represents in each case the average of a large number of counts.

Remarks.

Frog.—The *red cells* are large oval nucleated (20-25μ by 12-15μ) discs, the nucleus relatively small and irregularly elongated or oval, about 10μ in length. Many primitive and developing forms are usually observed—also free nuclei and many cells in various stages of degeneration. Hæmoglobin estimation is difficult owing to turbidity of the blood after dilution with water. The *polymorphonuclear* leucocytes are large cells, about 20μ; no definite granules can be observed. The *eosinophile* cells contain large deeply staining coccal-shaped granules.

Mouse.—The granules of the *polymorphonuclear* leucocytes are usually not stained, or only very faintly so. The nucleus of the *eosinophile cell* is ring-shaped or much divided, and the granules are coccal and stain oxyphile. The remarkable character of the blood is the high percentage of large *mononuclear* cells.

Rat.—The fine rod-shaped granules of the *polymorphonuclear* leucocytes are usually very faintly stained. The granules of *eosinophile* cells are well stained and coccal-shaped, the nucleus is often ring shaped. The *basophile* granular cells are few—but the granules are large, and stain deeply basophile.

Guinea-pig.—Polychromasia and punctate basophilia of *red cells* are very commonly observed—nucleated red cells are also frequent. The large *mononuclear* cells often contain vacuoles—"Kurlow cells"—possibly of a parasitic nature.

Rabbit.—It is not uncommon to find nucleated *red cells* in films from quite healthy animals. The granules of the *polymorphonuclear* leucocytes stain oxyphile. The coarse granules of the *eosinophile* cells appear to stain less deeply oxyphile, probably owing to the basophile staining of the cytoplasm.

Rhesus monkey.—The blood cells resemble those met with in human blood. The minute neutrophile granules of the *polymorphonuclear* leucocytes are often very scanty, and sometimes apparently absent. The *eosinophile* cells are not so densely packed with coarse oxpyhile granules as in the human eosinophile, and the nuclei of these cells are usually much divided, or polymorphous.

Goat.—The *red cells* are small, nonnucleated discs, only about 4.5μ diameter, not much more than half that of the human red cell. The *polymorphonuclear* leucocytes have only a few very minute coccal-shaped oxyphile granules, the nucleus is polymorphous. The *eosinophile* cells are large cells up to 20μ, the cytoplasm is basophile and contains coarse coccal-shaped oxyphile granules, and the nucleus is often much divided.

Fowl.—The *red cells* are oval nucleated discs about 12μ by 6μ, the nucleus being relatively small (about 4μ long), irregularly elongated or oval; round, more deeply stained cells with round or diffuse nuclei, also free nuclei and degenerated forms of red cells are often present. The granules of the cells corresponding to the *polymorphonuclear* leucocytes are rod-shaped, often beaded or with clubbed ends. The nucleus is not polymorphous, but usually divided into two, though it may be single. The cells probably corresponding to *eosinophile* leucocytes have fine coccal-shaped granules, faintly staining eosinophile or neutrophile. The basophile granules of the "mast" cells are coccal-shaped, of various size—often quite powdery.

Pigeon.—*Red cells* resemble those of the fowl, and similar varieties of appearance may be noted. The granules of those cells which correspond to *polymorphonuclear* leucocytes are rod-shaped, but smaller and finer than in the fowl, and do not show clubbed appearances. The nucleus is not polymorphous, and only occasionally divided. The coccal-shaped granules of the *eosinophile* cells are stained more deeply oxyphile than those of the corresponding cells of the fowl.

The preparation of dried films for this histological examination of the blood is carried out as follows:

1. Small samples of blood for the preparation of blood films are most conveniently obtained from the veins of the ear in most of the ordinary laboratory animals, viz., monkey, goat, dog, cat, rabbit, guinea-pig; in the pigeon and fowl the axillary vein should be punctured; in the rat and mouse either a vein in the ear or preferably by wounding the tip of the tail; in the frog, the web of the foot should be selected.

2. Puncture the selected vein with a sharp needle. A flat Hagedorn needle (size No. 8) with a cutting edge is the most useful for this purpose. If the vein cannot be distended by proximal compression, vigourous friction with a piece of dry lint may have the desired effect—or a test-tube full of water at about 40°C. may be placed close to the vein. Failing these methods, a drop or two of xylol may be dropped on the skin just over the vein, left on for a few seconds and then wiped off with a piece of dry lint.

3. One of the short ends of a 3 by 1 glass slip is brought into contact with the exuding drop of blood, so that it picks up a small drop.

4. The slide is then lowered transversely on to the surface of a second 3 by 1 slip, which rests on the bench near to one end at an angle of about 45°, and retained in this position for a few seconds, while the drop of blood spreads along the whole of the line of contact (see also Fig. 69).

5. Draw the first slide firmly and evenly along the entire length of the lower slide, leaving a thin regular film which will probably show the blood cells only one layer thick.

6. Allow the film to dry in the air.

7. Stain with one of the polychrome blood stains (see page 97).

8. Examine microscopically.

b. The **bacteriological examination of the blood** is directed solely to the demonstration of the presence in the circulating blood of the organisms previously injected into the animal. For this purpose several cubic centimetres of blood should be taken in an all-glass syringe from an accessible vein corresponding to one of those suggested as the site of intravenous inoculation—and under similar aseptic precautions.

1. Sterilise an all-glass syringe of suitable size, and when cool draw into the syringe some sterile sodium citrate solution and moisten the whole of the interior of the barrel; then eject all the citrate solution if less than 5 c.c. blood are to be withdrawn; if more than 5 c.c. are required retain about half a cubic centimetre of the fluid in the syringe. This prevents coagulation of the blood.

The sodium citrate solution is prepared by dissolving:

Sodium citrate	10 gramme.
Sodium chloride	0.75 grammes.
In distilled water	100 c.c.

Sterilise by boiling.

2. Prepare the animal as for intravenous inoculation (see page 363) and introduce the syringe needle into the lumen of the selected vein.
3. Slowly withdraw the piston of the syringe. When sufficient blood has been collected direct the assistant to release the proximal compression of the vein; and withdraw the needle.
4. Remove the needle from the nozzle of the syringe and deliver the citrated blood into a small Ehlenmeyer flask containing about 250 c.c. of nutrient broth.
5. Label, incubate and examine daily until growth occurs or until the expiration of ten days.

c. The **serological examination of the blood** is directed to the demonstration of the presence of certain specific antibodies in the sera of experimentally infected animals, and within certain limits to an estimation of their amounts.

The chief of these bodies are:

Antitoxin. Agglutinin. Precipitin. Opsonin. Immune body or Bacteriolysin.

None of these substances are capable of isolation in a state of purity apart from the blood serum, consequently special methods have been elaborated to permit of their recognition. In every instance the behaviour of serum from the experimental animal, which may be termed "specific" serum, is studied in comparison with that of serum from an uninoculated animal of the same species, and which is termed "normal" serum. In view of minor differences in constitution exhibited by the serum of various individuals of the same series, it is usual to employ a mixture of sera obtained from several different normal animals of the same species as the inoculated animal, under the term "pooled serum." The method of collecting blood (*e. g.*, from the rabbit) for serological tests is as follows:

Collection of Serum.

Apparatus required:

Razor. Liquid soap. Cotton-wool. Lysol 2 per cent. solution, in drop bottle. Ether in drop bottle. Flat Hagedorn needles. Blood pipettes (Fig. 16, page 12). Centrifugal machine. Centrifuge tubes. Glass cutting knife. Bunsen flame. Writing diamond or grease pencil.

Method.

1. Shave the dorsal surface of the ear over the course of the posterior auricular vein (see Fig. 192).
2. Sterilise the skin by washing with lysol.

The lysol should be applied with sterile cotton-wool and the ear vigourously rubbed, not only to remove superficial scales of epithelium, but also to render the ear hyperæmic and the vein prominent.

3. Remove the lysol with ether dropped from a drop bottle, and allow the ether to evaporate.
4. Puncture the vein with a sterile Hagedorn needle.
5. Take a small blood-collecting pipette (Fig. 161) and hold it at an angle to the ear, one end touching the issuing drop of blood, the other depressed.

The blood will now enter the pipette at first by capillarity; afterward gravity will also come into play and the pipette can be two-thirds filled without difficulty.

6. Hold the tube by the end containing the blood, the clean end pointing obliquely upward—warm this end at the bunsen flame to expel some of the contained air; then seal the clean point in the flame.

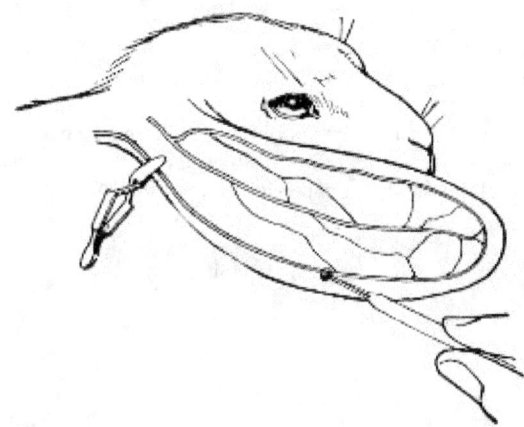

Fig. 192.—Collecting blood from rabbit.

7. Place the pipette down on a cool surface (*e. g.*, a glass slide). The rapid cooling of the air in the clean end of the pipette creates a negative pressure, and the blood is sucked back into the pipette, leaving the soiled end free from blood. Seal this end in the bunsen flame.

8. Mark the distinctive title of the specimen (*e. g.*, animal's number) upon the pipette with a writing diamond or grease pencil.

9. When the sealed ends are cold and the blood has clotted, place the pipette on the centrifuge, clean end downward; counterpoise and centrifugalise thoroughly. On removing the pipette from the centrifuge, the red cells will be collected in a firm mass at one end, and above them will appear the clear serum.

10. By marking the blood pipette above the level of the serum with the glass cutting knife and snapping the tube at that point, the blood-serum becomes readily accessible for testing purposes.

If larger quantities of blood are required, the animal, after puncturing the vein, should be inverted, an assistant holding it up by the legs. Blood to the volume of several cubic centimetres will now drop from the punctured vein, and should be caught in a tapering centrifuge tube, the tube transferred to the incubator at 37° C. for two hours, then placed in the centrifugal machine, counterpoised and centrifugalised thoroughly. The three most important of the antibodies referred to which can be demonstrated with a certain amount of facility are agglutinin, opsonin and bacteriolysin; and the methods of testing for these bodies will now be considered.

AGGLUTININ.

Agglutinin is the name given to a substance present in the blood-serum of an animal that has successfully resisted inoculation with a certain micro-organism. This substance possesses the power of collecting together in clumps and masses, or agglutinating watery suspensions of that particular microbe.

Dilution of the Specific Serum:

Apparatus required:

Sterile graduated capillary pipettes to contain 10 c. mm. (Fig. 17).

Sterile graduated capillary pipettes to contain 90 c. mm. (Fig. 17).

Small sterile test-tubes 5 × 0.5 cm.

Normal saline solution in flask or test-tube.

Pipette of specific serum.

Glass cutting knife, or three-square file.

Glass capsule, nearly full of dry silver sand, or roll of plasticine.

Grease pencil.

Method.—

1. Take three sterile test-tubes and number them 1, 2 and 3.

2. Pipette 0.9 c.c. sterile normal saline solution into each tube, and stand tubes upright in the sand in the capsule, or in the plasticine block.

3. Make a scratch with the glass cutting knife on the blood pipette above the upper level of the clear serum, and snap off and discard the empty portion of the tube.

4. Remove 0.1 c.c. of the serum from the blood pipette tube, and mix it thoroughly with the fluid in tube No. 1; and label **s.s.**, (specific serum), 10 per cent.

5. Remove 0.1 c.c. of the solution from tube No. 1 by means of a fresh pipette, and mix it with the contents of tube No. 2; and label **s.s.**, 1 per cent.

6. Remove 0.1 c.c. of the solution from tube No. 2 by means of a fresh pipette, and mix it with the contents of tube No. 3; and label **s.s.**, 0.1 per cent.

When the yield of serum from the specimen of blood which has been collected, or is available, is small, the above method of diluting is not practicable, and the dilution should be carried out by Wright's method in a capillary teat pipette.

Dilution of Serum by Means of a Teat Pipette.

Materials required:

Blood pipette containing sample of specific serum after centrifugalisation. Capsule of diluting fluid—normal saline solution. Supply of Pasteur pipettes (Fig. 13a). India-rubber teats. Small test-tubes. A block of plasticine to act as a test-tube stand. Grease pencil.

Method:

1. Mark three small test-tubes 10 per cent., 1 per cent. and 0.1 per cent. respectively, and stand them upright in the plasticine block.

2. Take a Pasteur pipette, nick the capillary stem just above the sealed end with a glass cutting knife, and snap off the sealed end with a quick movement so that the fracture is clean cut and at right angles to the long axis of the capillary stem—cut "square", in fact. Prepare several, say a dozen, in this manner.

3. Fit a rubber teat to the barrel of each of the pipettes.

4. Make a mark with the grease pencil on the stem of one of the pipettes about 2 or 3 cm. from the open extremity.

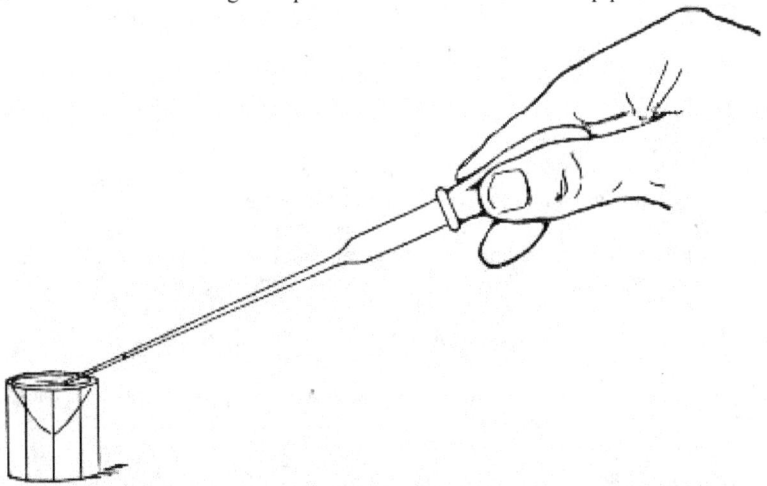

Fig. 193.—Filling the capillary teat pipette.

5. Compress the teat between the finger and thumb (Fig. 193) to such an extent as to drive out the greater part of the contained air.

6. Maintaining the pressure on the teat pass the stem of the pipette into the capsule holding the saline solution, until the open end of the pipette is below the level of the fluid.

7. Now cautiously relax the pressure on the teat and let the fluid enter the pipette and rise in the stem until it reaches the level of the grease pencil mark. As soon as this point is reached, check the movement of the column of fluid by maintaining the pressure on the teat, neither relaxing nor increasing it.

8. Withdraw the point of the pipette clear of the fluid, and again relax the pressure on the teat very slightly. The column of saline solution rises higher in the stem, and a column of air will now enter the pipette and serve as an index to separate the first volume of fluid drawn into the stem from the next succeeding one.

9. Again introduce the end of the pipette into the fluid and draw up a second volume of saline to the level of the grease pencil mark, and follow this with a second air index.

10. In like manner take up seven more equal volumes of saline solution and their following air bubbles. There are now nine equal volumes of normal saline in the pipette.

11. Now pass the point of the pipette into the blood tube and dip the open end below the surface of the serum. Proceeding as before, aspirate a volume of serum into the capillary stem up to the level of the pencil mark.

12. Eject the contents of the pipette into the small tube marked 10 per cent. by compressing the rubber teat between thumb and finger.

13. Mix the one volume of serum with the nine volumes of saline solution very thoroughly by repeatedly drawing up the whole of the fluid into the pipette and driving it out again into the test-tube.

14. Now take a clean pipette and proceed precisely as before, 4 to 10.

15. Having aspirated nine equal volumes of saline into this second pipette, now take up one similar volume of the fluid in the "10 per cent. tube."

16. Eject the contents of this pipette into the second tube marked 1 per cent. and mix thoroughly as before.

17. In similar fashion make the 0.1 per cent. solution and transfer to the third tube.

18. Further dilutions in multiples of ten can be prepared in the same way, and by varying the number of volumes of diluting fluid or serum any required dilution can be made (see Appendix, Dilution Tables).

Note.—The saline diluting fluid *must always* be taken into the pipette first, otherwise if the serum contains a very large amount of agglutinin the traces of this serum added to the saline solution may be sufficient to entirely vitiate the subsequent observations—whilst if more than one sample of serum is diluted from the same saline solution serious errors may be introduced into the experiments.

The Microscopical Reaction:

Apparatus Required:

Five hanging-drop slides (or preferably two slide), with two cells mounted side by side on each (Fig. 62, *a*), and one slide with one cell only.

Vaseline.
Cover-slips.
Platinum loop.

Grease pencil.
Eighteen to twenty-four-hour-old bouillon cultivation of the organism to be tested (*e. g.*, Bacillus typhi abdominalis)
Pipette end with the remainder of the specific serum labelled **s.s.**
Tubes containing the three solutions of the specific serum, 10, 1, and 0.1 per cent. respectively.
Pipette end with pooled normal serum labelled **p.s.**
Method.—
1. Make five hanging-drop preparations, thus:
(*a*) One loopful of bouillon cultivation + one loopful pooled serum; label "Control."
(*b*) One loopful culture + one loopful undiluted specific serum; label 50 per cent.
Mount these two cover-slips on a double-celled slide.
(*c*) One loopful bouillon culture + one loopful 10 per cent. serum; label 5 per cent.
Mount this on single-cell slide.
(*d*) One loopful bouillon culture + one loopful 1 per cent. serum; label 0.5 per cent.
(*e*) One loopful bouillon culture + one loopful 0.1 per cent. serum; label 0.05 per cent.
Mount these two cover-slips on a double-celled slide.
2. Note the time: Examine the control to determine that the bacilli are motile and uniformly scattered over the field—not collected into masses.
3. Next examine the 50 per cent. serum preparation.
If agglutinin is present and the test is giving a positive reaction, the bacilli *will* be collected in large clumps.
If the test is giving a negative reaction, the bacilli *may* be collected in large clumps owing to the viscosity of the concentrated serum.
4. Observe the 5 per cent. preparation microscopically.
If the bacilli are aggregated into clumps, positive reaction.
If the bacilli are *not* aggregated into clumps, observe until thirty minutes from the time of preparation before recording a negative reaction.
5. Examine the 0.5 and 0.05 per cent. preparations.
These may or may not show agglutination when the result of the examination of the 5 per cent. preparation is positive, according to the potency of the specific serum; and by the examination of a series of dilutions a quantitative comparison of the valency of specific sera from different sources, or of serum from the same animal at different periods during the course of active immunisation may be obtained.
Note.—The graduated pipettes supplied with Thoma's hæmatocytometer (intended for the collection of the specimen of blood required for the enumeration of leucocytes), giving a dilution of 1 in 10—*i. e.*, 10 per cent.—may be substituted for the graduated capillary pipettes referred to above, if the vessel in which the serum has been separated is of sufficiently large diameter to permit of their use.

The Macroscopical Reaction:
Sterile graduated capillary pipettes to contain 90 c. mm.
Eighteen to twenty-four-hours-old bouillon cultivation of the organism to be tested.
Three test-tubes containing the 10, 1, and 0.1 per cent. solutions of specific serum (about 90 c. mm. remaining in each).
Tube containing 50 per cent. solution of pooled serum.
Sedimentation pipettes (*vide* page 17) or teat pipettes.
Method.
1. Pipette 90 c. mm. of the bouillon culture into each of the tubes containing the diluted serum; and the same quantity into the tube containing the pooled serum.
2. Fill a sedimentation tube (by aspirating) or a teat pipette from the contents of each tube. Seal off the lower ends of the sedimentation tubes in the Bunsen flame.
3. Label each tube with the dilution of serum that it contains—viz., 5, 0.5, and 0.05 per cent.
4. Place the pipettes in a vertical position, in a beaker, in the incubator at 37°C., for one or two hours.
5. Observe the granular precipitate which is thrown down when the reaction is positive, and the uniform turbidity of the negative reaction as compared with the appearances in the control pooled serum.

OPSONIN.

Opsonin is the term applied by Wright to a substance, present in the serum of an inoculated animal, which is able to act upon or sensitise bacteria of the species originally injected, so as to render them an easy prey to the phagocytic activity of polymorphonuclear leucocytes. In the method for demonstrating opsonin about to be described, a comparison is made between the opsonic "power" of the pooled serum and the specific serum.

Apparatus:
Small centrifuge and tubes for same (made from the barrels of broken capillary pipettes by sealing the conical ends in the bunsen flame).
Capillary Pasteur pipettes.
India-rubber teats.
Grease pencil.
Bunsen burner with peep flame.
Electrical signal clock (see page 39) stop watch, or watch.
Rectangular glass box or tray to hold pipettes.
Incubator regulated at 37°C.
3 × 1 slides.
Piece of light rubber tubing.
Rectangular block of plasticine.
Flask of normal saline solution.

Flask of sodium citrate (1.5 per cent.) in normal saline solution.

Materials required, and their preparation:

Small tube of "washed cells" (red blood discs and leucocytes); human cells are used in estimating the opsonising power of the serum of experimental animals.

Small tube of emulsion of bacteria of the species responsible for the infection of the experimental animal.

Blood pipette containing specific serum.

Blood pipette containing "pooled" serum.

Washed Cells.—

1. Take a small centrifuge tube and half fill it with sodium citrate solution. Mark with the grease pencil the upper limit of the fluid.

2. Cleanse the skin of the distal phalanx of the second finger of the left hand above the root of the nail with lint and ether. Wind the rubber tubing tightly round the second phalanx; puncture with a sterile Hagedorn needle through the cleansed area of skin.

3. Take up a sufficiency of the issuing blood (more or less according to the number of tests to be performed) with a teat pipette, transfer it to the tube of citrate solution and mix thoroughly. Make a second mark on the tube at the upper level of the mixed citrate solution and blood.

4. Place the tube in the centrifuge, counterpoise accurately and centrifugalise until the blood cells are thrown down in a compact mass occupying approximately the same volume as is included between the two pencil marks.

The column of fluid in the tube now shows clear supernatant fluid (citrate solution and blood plasma) separated from the sharp cut upper surface of the red deposit of corpuscles by a narrow greyish layer of leucocytes.

5. Remove the supernatant column of citrate solution by means of a teat pipette, fill normal saline solution into the tube up to the upper pencil mark, and distribute the blood cells throughout the saline by means of the teat pipette. Centrifugalise as before.

6. Again remove the supernatant fluid and fill in a fresh supply of saline solution and centrifugalise once more.

7. Remove the supernatant saline solution as nearly down to the level of the leucocytes as can be safely done without removing any of the leucocytes.

8. Next distribute the leucocytes evenly throughout the mass of red cells by rotating the tube between the palms of the hands—just as is done with a tube of liquefied medium prior to pouring a plate.

9. Set the tube upright in the plasticine block near to one end.

Bacterial Emulsion.—

1. Take an 18- to 24-hour culture of the required bacterium (*e. g.*, Diplococcus pneumoniæ) grown upon sloped blood agar at 37° C. Pour over the surface of the medium some 5 c.c. of normal saline solution.

2. With a platinum loop emulsify the growth from the surface of the medium as evenly as possible in the saline solution.

3. Allow the tube to stand for a few minutes so that the large masses of growth may settle down; transfer the upper portion of the saline suspension to a centrifuge tube and centrifugalise thoroughly.

4. Examine a drop of the supernatant opalescent emulsion microscopically to determine its freedom from clumps and masses. If unsatisfactory prepare another emulsion, this time scraping up the surface growth with a platinum spatula, transferring it to an agate mortar and grinding it up with successive small quantities of normal saline. If satisfactory insert the tube in the plasticine block next to that containing the washed cells.

Specific Serum.—

Pooled Serum.—

These sera are collected and treated as already described (see page 379), and the portions of the blood pipettes containing them are arranged in the remaining space in plasticine block.

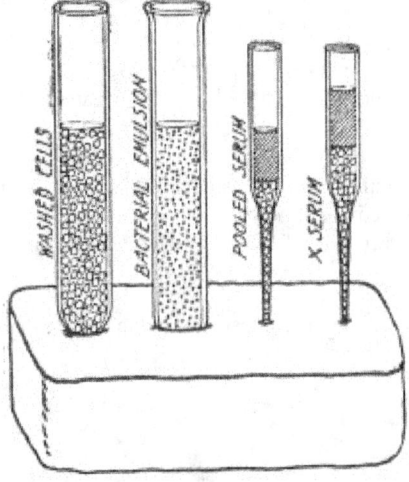

Fig. 194.—Plasticine block with materials arranged for opsonin estimations.

The plasticine block now presents the appearances shown in Fig. 194.

Method for Determining the Opsonic Index.—

1. Take a capillary pipette fitted with a teat, cut the distal end *square* and make a pencil mark about 2 cm. from the end.

2. Aspirate into the pipette one volume of washed cells, air index, one volume of bacterial emulsion, air index, and one volume of specific serum (see Fig. 195).

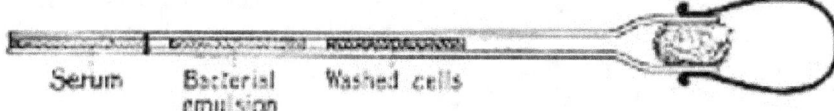

Fig. 195. Opsonin pipette.

3. Mix thoroughly on a 3 by 1 slide by compressing the teat and ejecting the contents of the pipette on to the surface of the slide, relaxing the pressure and so drawing the fluid up into the pipette again. These two processes should be repeated several times; finally take up the mixture in an unbroken column to the central portion of the capillary stem.

4. Seal the point of the pipette in the peep flame of the bunsen burner and remove teat.

5. Mark the pipette (with the grease pencil) with the distinctive number of the serum and place it in the glass box or tray.

6. Take another similarly prepared pipette and aspirate into it equal volumes of washed cells, bacterial emulsion and pooled serum. Treat precisely as in 3 and 4, label it "control" or "N.S." (normal serum) and place in the box by the side of the specific serum preparation.

7. Place the box with the pipettes in the incubator and set the signal clock to ring at 15 minutes (or start the stop watch).

8. At the expiration of the incubation time remove the pipettes from the incubator.

9. Cut off the sealed end of the specific serum preparation. Mix its contents thoroughly as in step 3, and then divide the mixture between two 3 by 1 slips and carefully spread a blood film (*vide* page 376) on each in such a way that only one-half of the surface of each slide is covered with blood—the free edge of the blood film approximating to the longitudinal axis of the slide.

Allow films to dry and label the slides with writing diamond.

10. Treat the contents of the control pipette in similar fashion.

11. Select the better film from each pair for fixing and staining.

12. Fixing and staining must be carried out under strictly comparable conditions, and to this end the slides are best handled by placing in a glass staining rack which can be lowered in turn into each of a series of glass troughs containing the various reagents (Fig. 196). Place the rack in the first trough which contains the alcoholic solution of Leishman's stain for two minutes to fix.

Transfer to the second trough containing the diluted stain for ten minutes.

Transfer to the third trough containing distilled water, and holding the trough over a sink, run in a stream of distilled water until washing is complete. Remove slides from the rack and dry.

Leishman's stain is the best for routine work for all bacteria other than B. tuberculosis. Films containing tubercle bacilli must of course be stained by the Ziehl Neelsen method.

Fig. 196. Glass staining trough for blood films.

13. Examine specific serum slide microscopically with 1/12 inch oil immersion. Find the edge of the blood film—along this the bulk of the leucocytes will be collected. Starting at one end of the film move the slide slowly across the microscope stage and as each leucocyte comes into view count and record the number of ingested bacteria. The sum of the contents of the first 50 consecutive polymorphonuclears that are encountered is marked down. (The *average* number of bacilli ingested per leucocyte = the "*phagocytic index*.")

14. In precisely similar manner enumerate the bacteria present in the first 50 cells of the control preparation. This number is recorded as the denominator of a vulgar fraction of which the numerator is the number recorded for the specific serum. This fraction, expressed as a percentage of unity = the *opsonic index*.

IMMUNE BODY.

Immune body or amboceptor is the name given to a substance present in the serum of an infected animal that has successfully resisted inoculation with some particular micro-organism, and which possesses the power of linking the complement normally present in the serum to bacteria of the species used as antigen in such a manner that the micro-organisms are rendered innocuous, and ultimately destroyed. The presence of the immune body in the serum can be demonstrated *in vitro* by the reaction elaborated by Bordet and Gengou, known as the complement fixation test, the existence or the absence of the phenomenon of complement fixation being rendered obvious macroscopically by the absence or presence of hæmolysis on the subsequent addition of "sensitised" red blood corpuscles, (e. g., a mixture of crythrocyte solution and the appropriate hæmolysin—two of the three essentials in the hæmolytic system, *vide* page 326).

Apparatus Required:
Sterile pipettes 1 c.c., (graduated in tenths).
16 × 2 cm. test-tubes.
9 × 1 cm. test-tubes.
Test-tube racks for each size of test-tube.

Reagents Required:
Normal saline solution.
Erythrocyte solution (human red cells, page 329) = E.
Hæmolytic serum (for human cells) = H.S.
Complement (fresh guinea-pig serum) = C.
Specific serum from inoculated animal, inactivated = S.S.
Control pooled serum from normal animals of same species, Inactivated = P.S.

Antigen (cultivation upon solid medium of the organism (*e. g.*, B. typhosus) which has already served as antigen in the inoculation of the experimental animal) = A.

To prepare the antigen for use, emulsify the whole of the bacterial growth in 5 c.c. normal saline solution.
Shake the emulsion in a test-tube with some sterilised glass beads to ensure a homogenous emulsion, and sterilise by heating to 60° C. in a water-bath for one hour.

Method.—
1. Take five small test-tubes, and number them 1 to 5 with a grease pencil.
2. Into tubes Nos. 1, 3, 4 and 5 pipette 0.1 c.c. of complement.
3. Into tubes Nos. 1 and 2 pipette 0.2 c.c. of the serum to be tested.
4. Into tube No. 4 pipette 0.2 c.c. of control serum.
5. Into tubes Nos. 1, 2, 3 and 4 pipette 1 c.c. of the bacterial emulsion which forms the antigen.
6. Place the whole set of tubes in the incubator at 37° C. for a period of one hour.
7. Remove the tubes from the incubator and pipette 1 c.c. erythrocyte solution and 4 minimal hæmolytic doses of the corresponding hæmolysin into each tube.
8. Mix thoroughly and return the tubes to the incubator at 37° C. for further period of one hour.
9. At the expiration of that time transfer the tubes to the ice chest, and allow them to stand for three hours.
10. Examine the tubes.

Tubes 3, 4 and 5 should show complete hæmolysis; tube 2 should give no evidence whatever of hæmolysis.
These tubes form the controls to the first tube, which contains the serum to be tested.

In tube No. 1 the absence of hæmolysis would indicate the presence in the serum of the inoculated animal of a specific antibody to the micro-organism used in the inoculations; since it shows that the complement has been bound by the immune body to the bacterial antigen, and none has been left free to enter into the hæmolytic system; on the other hand the presence of hæmolysis would show that no appreciable amount of antibody has yet been formed in response to the inoculations. In other words, there is an absence of infection, since the complement remained unfixed at the time of the addition of the erythrocyte solution and hæmolytic serum, and was ready to combine with those reagents to complete the hæmolytic system.

The method may be shown diagramatically as under using the symbols already indicated

Test-tubes.

①	②	③	④	⑤
0.1 c.c. C.	....	0.1 c.c. C.	0.1 c.c. C.	0.1 c.c. C.
0.2 c.c. S.S.A.	0.2 c.c. S.S.A.	.... A.	0.2 c.c. P.S.A.	....

Incubate at 37° C. for one hour.

1 c.c. E. H.S.[4]	1 c.c. E. H.S.[4]	1 c.c. E. H.S.[4]	1 c.c. E. H.S.[4]	1 c.c. E. H.S.[4]

Incubate at 37° C. for one hour.

(?)   No hæmolysis.   Hæmolysis.

Note.—It is sometimes more convenient to *sensitise* the erythrocytes just before they are needed. This is done forty-five minutes after the experiment has been started (page 394, step 6), that is to say, before the completion of the first period of incubation, thus:

1. Measure out into a sterile test-tube (or flask) five c.c. of erythrocyte solution.
2. Measure out twenty minimal hæmolytic doses of hæmolysin, add to the erythrocyte solution on the test-tube.
3. Allow the erythrocyte and hæmolysin to remain in contact for fifteen minutes at room temperature. The red cells are then sensitised and ready for use.
4. When the tubes are removed from the incubator at the end of the first hour (*i. e.*, step 7) add 1 c.c. sensitised red cells to each tube by means of a graduated pipette.
5. Mix thoroughly, return the tubes to the incubator at 37°C. and complete the experiment as previously described (steps 8 onward).

XIX. POST-MORTEM EXAMINATIONS OF EXPERIMENTAL ANIMALS.

The post-mortem examination should be carried out as soon as possible after the death of the animal, for it must be remembered that even in cold weather the tissues are rapidly invaded by numerous bacteria derived from the alimentary tract or the cavities of the body, and from external sources.

The following outlines refer to a complete and exhaustive necropsy, and in routine work the examination will rarely need to be carried out in its entirety.

Note.—Throughout the autopsy the searing irons must be freely employed, and it must be recollected that one instrument is only to be employed to seize or cut one structure. This done, it must be regarded as contaminated and a fresh instrument taken for the next step.

Apparatus Required:
Water steriliser.

Surgical instruments: { Scalpels.
{ Scissors.
{ Forceps.
{ Bone forceps.

Spear-headed platinum spatula (Fig. 199).
Searing irons (Fig. 198).
Tubes of media—bouillon and sloped agar.
Surface plates in petri dishes (of agar or one of its derivatives).

Platinum loop.
Aluminium "spreader."
Grease pencil.
Sterile capillary pipettes (Fig. 13, *a*).
Sterile glass capsules, large and small.
Cover-slips or slides.
Bottles of fixing fluid (*vide* page 114) for pieces of tissue intended for sectioning.

1. Place the various instruments, forceps, scissors, scalpels, etc., needed for the autopsy inside the steriliser and sterilise by boiling for ten minutes; then open the steriliser, raise the tray from the interior and rest it crosswise on the edges.

2. Heat the searing irons to redness in a separate gas stove.

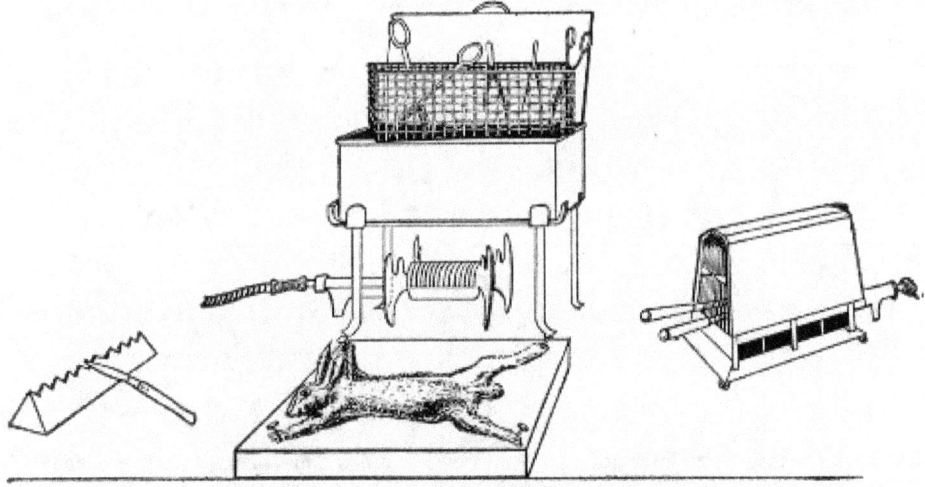

Fig. 197.—Apparatus for post-mortem examination, animal on board.

3. Drench the fur (or feathers) with lysol solution, 2 per cent. This serves the twofold purpose of preventing the hairs from flying about and entering the body cavities during the autopsy, and of rendering innocuous any vermin that may be present on the animal.

Fig. 198.—Searing iron.

4. Examine the cadaver carefully. Recollect that laboratory animals are not always hardy; death may be due to exposure to heat or cold, to starvation or over- or improper feeding or to the attack of rats—and not to the bacterial infection.

5. Fasten the body of the animal, ventral surface upward (unless there is some special reason for having the dorsum exposed), out on a board by means of copper nails driven through the extremities.

6. With sterile forceps and scalpel incise the skin in the middle line from the top of the sternum to the pubes. Make other incisions at right angles to the first out to the axillæ and groins, and reflect the skin in two lateral flaps. (Place the now infected instruments on the board by the side of the body or support them on a porcelain knife rest.)

Seat of Inoculation.—

7. Inspect the seat of inoculation. If any local lesion is visible, sear its exposed surface and with the platinum loop, remove material from the deeper parts to make tube and surface plate cultivations and cover-slip preparations.

Collect specimens of pus or other exudation in capillary pipettes for subsequent examination.

8. Inspect the neighbouring lymphatic glands and endeavour to trace the path of the virus.

9. Sear the whole of the exposed surface of the thorax with the searing irons.

Pleural Cavity.—

10. Divide the ribs on either side of the sternum and remove a rectangular portion of the anterior chest wall with sterile scissors and a fresh pair of forceps, exposing the heart. Place the infected instruments by the side of the first set.

11. Observe the condition of the anterior mediastinal glands, the thymus and the lungs. Collect a quantity of pleuritic effusion, if such is present, in a pipette for further examination later.

12. Raise the pericardial sac in a fresh pair of forceps and burn through this structure with a searing iron.

Collect a sample of pericardial fluid in a pipette for microscopical and cultural examination.

13. Grasp the apex of the heart in the forceps and sear the surface of the right ventricle.

14. Plunge the open point of a capillary pipette through the seared area into the ventricle and fill with blood.

Make cultivations and cover-slip preparations of the heart blood.

15. Collect a further sample of blood or serum for subsequent investigation as to the presence of antibodies.

Peritoneal Cavity.—

16. Sear a broad track in the middle line of the abdominal wall; open the peritoneal cavity by an incision in the centre of the seared line. Observe the condition of the omentum, the mesentery, the viscera and the peritoneal surface of the intestines.

17. Collect a specimen of the peritoneal fluid (or pus, if present) in a capillary pipette. Make cultivations, tube and surface plate, and cover-slip preparations from this situation.

18. Collect a specimen of the urine from the distended bladder in a large pipette (in the manner indicated for heart blood), for further examination, by cultivations, microscopical preparations, and chemical analysis.

19. Collect a specimen of bile from the gall bladder in similar manner.

20. Excise the spleen and place it in a sterile capsule. Later, sear the surface of this organ; plunge the spear-headed spatula through the centre of the seared area, twist it round between the finger and thumb, and remove it from the organ. Sufficient material will be brought away in the eye in its head to make cultivations. A repetition of the process will afford material for cover-slip preparations.

21. Seize one end of the spleen with sterile forceps. Sear a narrow band of tissue, right around the organ and divide the spleen in this situation with a pair of scissors. Holding the piece of spleen in the forceps, dab the cut surface on to a surface plate in a number of different spots.

22. In like manner examine the other organs—liver, lungs, kidneys, lymphatic glands (mesenteric, hepatic, lumbar, etc), etc. Prepare cultivations and cover-slip preparations.

23. Dissect out a long bone from one upper and one lower limb and one of the largest ribs. Prepare cultures from the bone marrow in each case. Set aside these bones for the subsequent preparation of marrow films.

24. Film preparations of bone marrow are best made by the Price-Jones method. Seize the bone in a pair of pliers and squeeze out some of the marrow; receive it in a platinum loop, and transfer to a watch glass of dissociating fluid and emulsify. The dissociating fluid is a neutral 10 per cent. solution of glycerine prepared as follows:—

Measure out 10 c.c. Price's best glycerine and 90 c.c. sterile ammonia-free distilled water. Mix. Titrate against n/10 sodic hydrate solution using phenolphthalein as the indicator. The initial reaction is usually + 0.1 to + 0.5; add the calculated amount of n/10 sodic hydrate solution to neutralise.

25. Place a loopful of fresh desiccating fluid on a 3 × 1 glass slide; add a similar loopful of the marrow emulsion, and spread very gently over the surface of the slip.

26. Allow film to dry in the air (protected from dust) without heating.

27. Stain with Jenner's polychrome stain (page 97) for two and a half minutes.

28. Wash with ammonia-free distilled water, dry thoroughly and mount in xylol balsam.

Cranial and Spinal Cavities.—

29. In some instances it may be necessary (e. g., experimental inoculation of rabies) to examine the cranial cavity or to remove the spinal cord. Return the viscera to the abdominal cavity; draw the flaps of skin together and secure with Michel's steel clips. Draw the copper nails securing the limbs to the board, reverse the animal and again nail the limbs down—the body now being dorsum uppermost.

30. Make a longitudinal incision in the mesial line from snout to root of tail, and four transverse incisions—one joining the roots of the two ears, one across the body at the level of the spinis of the scapulæ, another at the level of the costal margin and the last across the upper level of the pelvis. Reflect these flaps of skin.

31. With forceps and scalpel dissect out the muscles lying in the furrow on either side of the spinal processes.

32. Cut through the bases of the transverse processes with bone forceps. Cut away the vault of the skull, cut through the roots of the nerves and remove the brain and spinal cord, place in a large glass dish for examination. Prepare cultivations from the cerebro-spinal fluid. The removal of the brain and cord is a tedious process and during the dissection it is difficult to avoid injury to these structures.

The operation is, however, carried out very expeditiously and neatly with the aid of the surgical engine (*vide* page 361). A small circular saw is fitted to the hand piece. The bones of the skull are cut through and the whole of the vault removed, exposing the entire vertex of the brain. Similarly all the spinous processes can be removed in one string by running the saw down first one side of the spinal column and then the other. In this way ample space for the removal of the nervous tissues is obtained with a minimum of labour.

33. Having completed the preparation of cultures remove small portions of various organs at leisure and place each in separate bottles of fixing fluid for future sectioning. Affix to each bottle a label bearing all necessary details as to its contents.

34. If necessary, remove portions of the organs for preservation and display as museum specimens (*vide* page 404).

35. Gather up all the infected instruments, return them to the steriliser, and disinfect by boiling for ten minutes.

Fig. 199.—Spear-headed platinum spatula (actual size.)

36. Sprinkle dry sawdust into the exposed body cavities to absorb blood and fluid. Cover the body with blotting or filter paper, moistened with 2 per cent. lysol solution. Place in a galvanised iron pail, provided with a lid, ready for transport to the crematorium.

37. Cremate the cadaver together with the board upon which it is fixed.

38. Stain the cover-slip preparations by suitable methods and examine microscopically.

39. Incubate the cultivations and examine carefully from day to day.

40. Make full notes of the condition of the various body cavities and of the viscera immediately the autopsy is completed; and add the result of the microscopical and cultural investigation when available.

As part of the card index system in use in the author's laboratory already referred to (*vide* page 335) there is a special yellow card for P-M notes. On the face of the card are printed headings for various data—some of which are sometimes unintentionally omitted—and on the reverse is a schematic figure which can be utilised for indicating the position of the chief lesions in the cadaver of any of the laboratory animals.

AUTOPSY CARD. *Laboratory No.*

Date

Animal No. in Series ♂ ♀ Weight

Died (or killed)... ....o'clock.......m. Autopsy made .... .o'clock . . . m.

Notes on Post Mortem Examination.

General.

A. Seat of Inoculation.

B. Thoracic Cavity.

C. Abdominal Cavity.

D. Cranial Cavity.

Bacteriological Examination.	*Histological Examination.*	*Organs Preserved.*
A.		
B.		
C.		
D.		

Fig. 200.—Front of post-mortem card.

41. Finally, the results of the action of the organism or organisms isolated may be correlated with the symptoms observed during life and the observations summarised under the following headings:

Tissue changes:

1. Local—*i. e.*, produced in the neighbourhood of the bacteria.

Position:

(*a*) At primary lesion.

(*b*) At secondary foci.

Character:

(*a*) Vascular changes and tissue reactions.} Acute or chronic.

(*b*) Degeneration and necrosis.}

2. General (*i. e.*, produced at a distance from the bacteria, by absorption of toxins):

(*a*) In special tissues—*e. g.*, nerve cells and fibres, secreting cells, vessel walls, etc.

(*b*) General effects of malnutrition, etc.

Symptoms:

(*a*) Associated with known tissue changes.

(*b*) Without known tissue changes.

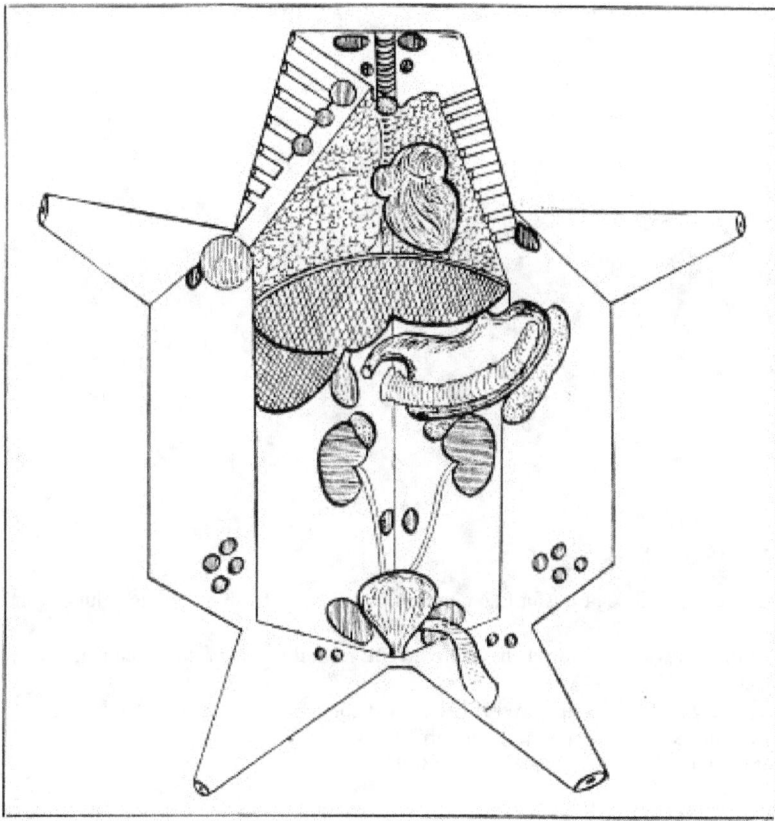

Fig. 201.—Back of post-mortem card.

Permanent Preparations—Museum Specimens.—

I. Tissues.—The naked-eye appearances of morbid tissues may be preserved by the following method:

1. Remove the tissue or organ from the cadaver as soon after death as possible, using great care to avoid distortion or injury.

2. Place it in a wide-mouthed stoppered jar, large enough to hold it conveniently, resting on a pad of cotton-wool, and arrange it in the position it is intended to occupy (but if it is intended to show a section of the tissue or organ, do not incise it yet).

3. Cover with the Kaiserling fixing solution, and stopper the jar; allow the tissues to remain in this solution for from forty-eight hours to seven days (according to size) to fix. Make any necessary sections.

Kaiserling modified solution is prepared as follows:

Weigh out

Potassium acetate   30 grammes.

Potassium nitrate   15 grammes.

and dissolve in

Distilled water   1000 c.c.

then add

Formalin   150 c.c.

Filter.

This fixing solution can be used repeatedly so long as it remains clear. Even when it has become turbid, if simple filtration is sufficient to render it clear, the filtrate may be used again.

4. Transfer the tissue to a bath of methylated spirit (95 per cent.) for thirty minutes to one hour.

5. Remove to a fresh bath of spirit and watch carefully. When the natural colours show in their original tints, average time three to six hours, remove the tissues from the spirit bath, dry off the spirit from the cut surfaces by mopping with a soft cloth, then transfer to the mounting solution.

Jore's mounting solution (modified) consists of

Glycerine   500 c.c.

Distilled water   750 c.c.

Formalin   2 c.c.

Equally good but much cheaper is Frost's mounting solution:

Potassium acetate   160 grammes.

Sodium fluoride	80 grammes.
Chloral hydrate	80 grammes.
Cane sugar (Tate's cubes)	3,500 grammes.
Saturated thymol water	8,000 c.c.

6. After twenty-four hours in this solution, or as soon as the tissue sinks, transfer to a museum jar, fill with fresh mounting solution, and seal.

6a. Or transfer to museum jar and fill with liquefied gelatine, to which has been added 1 per cent. formalin. Cover the jar and allow the gelatine to set. When solid, seal the cover of the jar in place.

7. To seal the museum preparation first warm the glass plate which forms the cover. This is most conveniently done by placing the cleaned and polished cover-plate upon a piece of asbestos millboard over a bunsen flame turned low.

8. Smear an even layer of hot cement over the flange of the jar. The cement is prepared as follows:

Weigh out and mix in an iron ladle

Gutta percha (pure)	4 parts.
Asphaltum	5 parts.

and melt together over a bunsen flame, stirring with an iron rod until solution is complete.

9. Invert the glass plate over the jar and press down firmly into the cement. Place a piece of asbestos board on the top and on that rest a suitable weight until the cement is cold and has thoroughly set.

10. Trim off any projecting pieces of cement with an old knife, burr over the joint between jar and cover-plate with a hot smooth piece of metal (*e. g.*, the searing iron).

11. Paint a narrow band of Japan black to finish off, round the joint, overlapping on to the cover-plate.

II. Tube Cultivations of Bacteria.—When showing typical appearances these may be preserved, if not permanently, at least for many years, as museum specimens, by the following method:

1. Take a large glass jar 25 cm. high by 18 cm. diameter, with a firm base and a broad flange, carefully ground, around the mouth. The jar must be fitted with a disc of plate glass ground on one side, to serve as a lid.

2. Smear a thick layer of resin ointment (B.P.) on the flange around the mouth of the jar.

3. Cover the bottom of the jar with a layer of cotton-wool and saturate it with formalin.

4. Remove the cotton-wool plug from the culture tubes and place them, mouth upward, inside the jar. (If water of condensation is present in any of the culture tubes, it should be removed by means of a capillary pipette before placing the tubes in the formalin chamber.)

5. Adjust the glass disc, ground side downward, over the mouth of the jar and secure it by pressing it firmly down into the ointment, with a rotary movement.

6. Remove the tubes from the formalin chamber after the lapse of a week, and dry the exterior of each.

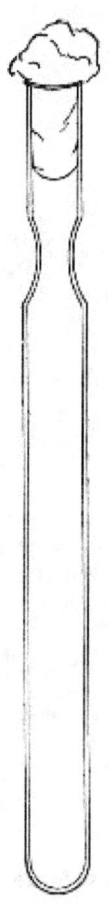
Fig. 202.—Bulloch's tubes.

7. Seal the open mouth of each tube in the blowpipe flame and label.

If the cultivations are intended for museum purposes when they are first planted, it is more convenient to employ Bulloch's tubes. These are slightly longer than the ordinary tubes, and are provided with a constriction some 2 cm. below the mouth (Fig. 202)—a feature which renders sealing in the blowpipe flame an easy matter.

XX. THE STUDY OF THE PATHOGENIC BACTERIA.

The student, who has conscientiously worked out the methods, etc., previously dealt with, is in a position to make accurate observations and to write precise descriptions of the results of such observations. He is, therefore, now entrusted with pure cultivations of the various pathogenic bacteria, in order that he may study the life-history of each and record the results of his own observations—to be subsequently corrected or amplified by the demonstrator. In this way he is rendered independent of text-book descriptions, the statements in which he is otherwise too liable to take for granted, without personally attempting to verify their accuracy.

During the course of this work attention must also be directed, as occasion arises, to such other bacteria, pathogenic or saprophytic, as are allied to the particular organisms under observation, or so resemble them as to become possible sources of error, by working them through on parallel lines—in other words the various bacteria should be studied in "groups." In the following pages the grouping in use in the author's elementary classes for medical and dental students and for candidates for the Public Health service is adopted, since a fairly long experience has completely vindicated the value and utility of this arrangement, and by its means a fund of information is obtained with regard to the resemblances and differences, morphological and cultural, of a large number of bacteria. The fact that some bacteria appear in more than one of these groups, so far from being a disadvantage, is a positive gain to the student, since with repetition alone will the necessary familiarity with the cultural characters of important bacteria be acquired. The study of the various groups will of course vary in detail with individual demonstrators, and with the student's requirements—the general line it should take is indicated briefly in connection with the first group only (pages 410-411). This section should be carefully worked through before the student proceeds to the study of bacterioscopical analysis.

It is customary to commence the study of the pathogenic bacteria with the Organisms of Suppuration. This is a large group, for all the pathogenic bacteria possess the power, under certain conditions, of initiating purely pyogenic processes in place of or in addition to their specific lesions, (*e. g.*, Bacillus tuberculosis, Streptococcus lanceolatus, Bacillus typhosus, etc.). There are, however, a certain few organisms which commonly express their pathogenicity in the formation of pus. These are usually grouped together under the title of "pyogenic bacteria," as distinct from those which only occasionally exercise a pyogenic rôle.

The organisms included in this group are:

1. Staphylococcus pyogenes albus. 2. Staphylococcus pyogenes aureus. 3. Staphylococcus pyogenes citreus. 4. Streptococcus pyogenes longus. 5. Micrococcus tetragenus. 6. Bacillus pyocyaneus. 7. Bacillus pneumoniæ.

and in certain special tissues

8. Micrococcus gonorrhœæ. 9. Micrococcus intracellularis meningitidis (Meningococcus). 10. Micrococcus catarrhalis. 11. Bacillus ægypticus (Koch-Weeks Bacillus).

The group may with advantage be subdivided as indicated in the following pages:

I. *Pyogenic cocci.*

Staphylococcus pyogenes albus. Staphylococcus pyogenes aureus. Staphylococcus pyogenes citreus. to contrast with Micrococcus candicans. Micrococcus agilis.

1. Prepare subcultivations from each:

Bouillon, } Agar streak, } Blood serum, } Litmus milk. } and incubate at 37°C. Agar streak, } Gelatine stab, } Potato. } and incubate at 20°C. Compare the naked-eye appearances of the cultures from day to day. Note M. agilis refuses to grow at 37°C.

2. Make hanging-drop preparations from the bouillon and agar cultivations after twenty-four hours' incubation. Examine microscopically and compare. Note the locomotive activity of M. agilis and the Brownian movement of the remaining micrococci.

3. Prepare cover-slip films from the agar cultures, after twenty-four hours' incubation. Stain for flagella by the modified Pitfield's method. Note M. agilis is the only micrococcus showing flagella.

4. Make microscopical preparations of each from all the various media after twenty-four and forty-eight hours and three days' incubation. Stain carbolic methylene-blue, carbolic fuchsin, and Gram's method. Examine the films microscopically and compare. Note in the Gram preparation, the Gram negative character of certain individual cocci in each film prepared from the three days' growth—such cocci are dead.

5. Stain section of kidney tissue provided (showing abscess formation by Staphylococcus aureus) by Gram's method, and counterstain with cosin.

6. Stain film preparation of pus from an abscess (containing Staphylococcus pyogenes aureus) with carbolic methylene-blue and also by Gram's method, counterstained with cosin.

7. Inoculate a white mouse subcutaneously with three loopfuls of a forty-eight-hour agar cultivation of the Staphylococcus aureus, emulsified with 0.2 c.c. sterile broth.

Observe carefully during life, and when death occurs make a careful post-mortem examination.

II. *Pyogenic cocci.*

Micrococcus gonorrhœæ. Micrococcus intracellularis meningitidis (meningococcus). Micrococcus catarrhalis. Micrococcus tetragenus. Micrococcus paratetragenus.

III. *Pyogenic cocci.*

Streptococcus pyogenes longus. Streptococcus of bovine mastitis. Streptococcus lanceolatus (Diplococcus pneumoniæ or pneumococcus). to contrast with Streptococcus brevis. Streptococcus lebensis.

IV. *Pyogenic bacilli.*

Bacillus pneumoniæ (Friedlaender). Bacillus rhinoscleromatis. Bacillus lactis aerogenes.

V. *Pyogenic bacilli.*

Bacillus pyocyaneus. to contrast with Bacillus fluorescens liquefaciens. Bacillus fluorescens non-liquefaciens.

VI. *Pneumonia group.*

Streptococcus lanceolatus (pneumococcus). Bacillus pneumoniæ (Friedlaender). Streptococcus pyogenes longus.

VII. *Diphtheroid group.*

Bacillus diphtheriæ (Klebs-Lœffler). Bacillus Hoffmanni. Bacillus xerosis. Bacillus septus.

VIII. *Coli-typhoid group.*

B. typhi abdominalis (B. typhosus). B. coli communis. B. enteritidis (Gaertner). to contrast with B. aquatilis sulcatus.

IX. *Escherich group.*

B. coli communis (Escherich). B. coli communior. B. lactis aerogenes. B. cloacæ.

X. *Gaertner group.*

Bacillus enteritidis (Gaertner). B. paratyphosus A. B. paratyphosus B. Bacillus choleræ suum (Hog Cholera). B. psittacosis.

XI. *Eberth group.*

B. typhosus (Eberth). B. dysenteriæ (Shiga). B. dysenteriæ (Flexner). B. fæcalis alcaligines.

XII. *Spirillum group.*

Vibrio choleræ. Vibrio metschnikovi. to contrast with Vibrio proteus (Finkler and Prior). Spirillum rubrum. Spirillum rugula.

XIII. *Anthrax group.*

Bacillus anthracis. to contrast with Bacillus subtilis. Bacillus mycoides. Bacillus mesentericus fuscus.

XIV. *Acid fast group.*

Bacillus tuberculosis (human). " " (bovine). " " (avian). " " (fish). to contrast with Bacillus phlei (Timothy grass bacillus). Butter bacillus of Rabinowitch.

XV. *Plague group.*

Bacillus pestis. B. septicæmiæ hæmorrhagicæ. B. suipestifer.

XVI. *Influenzæ group.*

B. influenzæ. Bacillus ægypticus (Koch-Weeks). Bacillus pertussis.
XVII. *Miscellaneous.*
Bacillus lepræ. Bacillus mallei. Micrococcus melitensis.
XVIII. *Streptothrix group.*
Streptothrix actinomycotica. Streptothrix maduræ. to contrast with Cladothrix nivea.
XIX. *Tetanus group.*
Bacillus tetani. Bacillus œdematis maligni. Bacillus chauvei (symptomatic anthrax).
XX. *Enteritidis sporogenes group.*
Bacillus enteritidis sporogenes. B. botulinus. B. butyricus. B. cadaveris.

FOOTNOTES:

See note on Vivisection License, page 334.

XXI. BACTERIOLOGICAL ANALYSES.

Each bacteriological or bacterioscopical analysis of air, earth, sewage, various food-stuffs, etc., includes, as a general rule, two distinct investigations yielding results of very unequal value:
1. Quantitative. 2. Qualitative.

The first is purely quantitative and as such is of minor importance as it aims simply at enumerating (approximately) the total number of bacteria present in any given unit of volume irrespective of the nature and character of individual organisms.

The second and more important is both qualitative and quantitative in character since it seeks to accurately identify such pathogenic bacteria as may be present while, incidentally, the methods advocated are calculated to indicate, with a fair degree of accuracy, the numerical frequency of such bacteria, in the sample under examination.

The general principles underlying the bacteriological analyses of water, sewage, air and dust, soil, milk, ice cream, meat, and other tinned stuffs, as exemplified by the methods used by the author, are indicated in the following pages, together with the methods of testing filters and chemical germicides; and the technique there set out will be found to be capable of expansion and adaptation to any circumstance or set of circumstances which may confront the student.

Controls.—The necessity for the existence of adequate controls in all experimental work cannot be too urgently insisted upon. Every batch of plates that is poured should include at least one of the presumably "sterile" medium; plate or tube cultures should be made from the various diluting fluids; every tube of carbohydrate medium that is inoculated should go into the incubator in company with a similar but uninoculated tube, and so on.

BACTERIOLOGICAL EXAMINATION OF WATER.

The bacteria present in the water may comprise not only varieties which have their normal habitat in the water and will consequently develop at 20° C., but also if the water has been contaminated with excremental matter, varieties which have been derived from, or are pathogenic for, the animal body, and which will only develop well at a temperature of 37° C. In order to demonstrate the presence of each of these classes it will be necessary to incubate the various cultivations at each of these temperatures.

Further, the sample of water may contain moulds, yeasts, or torulæ, and the development of these will be best secured by plating in wort gelatine and incubating at 20° C.

1. Quantitative.—

Collection of the Sample.—The most suitable vessels for the reception of the water sample are small glass bottles, 60 c.c. capacity, with narrow necks and overhanging glass stoppers (to prevent contamination of the bottle necks by falling dust). These must be carefully sterilised in the hot-air steriliser (*vide* page 31).

(*a*) If the sample is obtained from a **tap** or **pipe**, turn on the water and allow it to run for a few minutes. Remove the stopper from the bottle and retain it in the hand whilst the water is allowed to run into the bottle and three parts fill it. Replace the stopper and tie it down, but *do not seal it.*

(*b*) If the sample is obtained from a **stream**, **tank**, or **reservoir**, fasten a piece of stout wire around the neck of the bottle, remove the stopper, and retain it in the hand. Then, using the wire as a handle, plunge the bottle into the water, mouth downward, until it is well beneath the surface; then reverse it, allow it to fill, and withdraw it from the water. Pour out a few cubic centimetres of water from the bottle, replace the stopper, and tie it down.

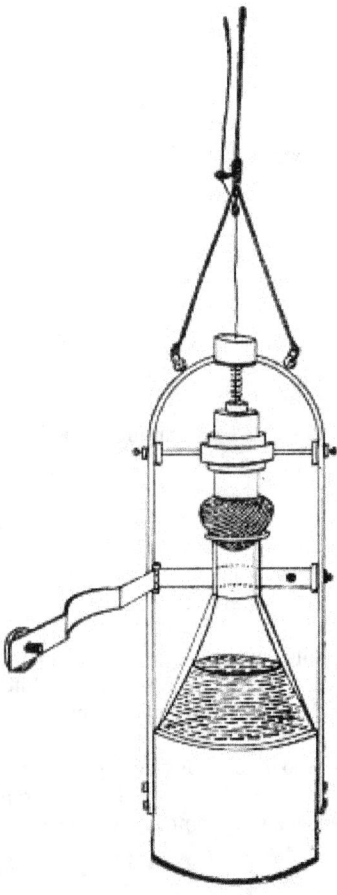

Fig. 203.—Esmarch's collecting bottle for water samples.

(c) If the sample is obtained from a **lake**, **river** or the **sea**; or when it is desired to compare samples taken at varying depths, the apparatus designed by v. Esmarch (Fig. 203) is employed. In this the sterilised bottle is enclosed in a weighted metal cage which can be lowered, by means of a graduated line, until the required depth is reached. At this point the bottle is opened by a thin wire cord attached to the stopper; when the bottle is full (as judged by the air bubbles ceasing to rise) the pull on the cord is released and the tension of the spiral spring above the stopper again forces it into the neck of the bottle. When the apparatus is taken out of the water, the small bottles are filled from it, and packed in the ice-box mentioned below.

An inexpensive substitute for Esmarch's bottle can be made in the laboratory thus:

Select a wide-mouthed glass stoppered bottle of about 500 c.c. capacity (about 20 cm. high and 8 cm. in diameter).

Remove the glass stopper and insert a rubber cork with two perforations in its place.

Through one perforation pass a piece of glass tubing about 5 cm. long and through the other a piece 22 cm. long, reaching to near the bottom of the bottle, each tube projecting about 2.5 cm. above the rubber stopper. Plug the open ends of the tubes with cotton wool. Secure the stopper in place with thin copper wire.

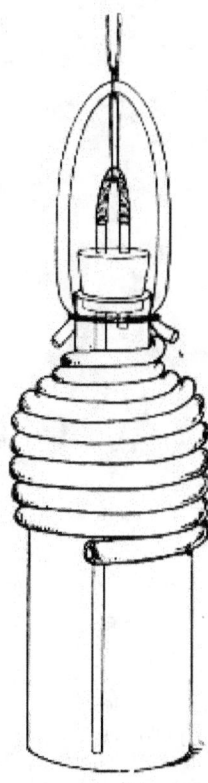

Fig. 204.—Thresh's deep water sampling bottle.

Sterilise the fitted bottle in the autoclave. Remove the cotton wool plugs and connect the projecting tubes by a piece of loosely fitting stout rubber pressure tubing about 5 cm. long, previously sterilised by boiling.

Take a piece of stout rubber cord about 33 cm. long, and of 10 mm. diameter (such as is used for door springs) thread a steel split ring upon it and secure the free ends tightly to the neck of the bottle by cord or catgut.

Attach the cord used for lowering the bottle into the water to the split ring on the rubber suspender. The best material for this purpose is cotton insulated electric wire knotted at every metre.

Connect the split ring also with the short piece of rubber tubing uniting the two glass tubes by a piece of catgut (or thin copper wire) of such length that when the bottle is suspended there is no pull upon the rubber tube, but which, however, will be easily jerked off when a sharp pull is given to the suspending cord.

Now wind heavy lead tubing about 1 cm. diameter around the upper part of the bottle, starting at the neck just above the shoulder. This ensures the sinking of the bottle in the vertical position (Fig. 204).

The apparatus being arranged is lowered to the required depth, a sharp jerk is then given to the suspending cord, which detaches the rubber tube and so opens the two glass tubes. Water enters through the longer tube and the air is expelled through the shorter tube. The bubbles of air can be seen or heard rising through the water, until the bottle is nearly full, a small volume of compressed air remaining in the neck of the bottle.

As the apparatus is raised, the air thus imprisoned expands, and prevents the entry of more water from nearer the surface.

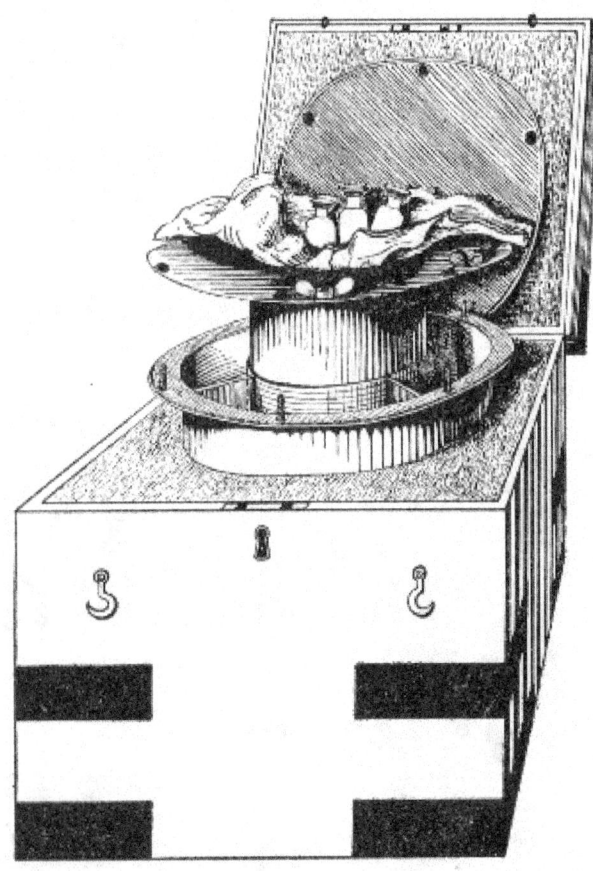

Fig. 205.—Ice-box for transmission of water samples, etc.

Transport of Sample.—If the examination of the sample cannot be commenced immediately, steps must be taken to prevent the multiplication of the bacteria contained in the water during the interval occupied in transit from the place of collection to the laboratory. To this end an ice-box such as that shown (in Fig. 205) is essential. It consists of a double-walled metal cylinder into which slides a cylindrical chamber of sufficient capacity to accommodate four of the 60 c.c. bottles; this in turn is covered by a metal disc—the three portions being bolted together by thumb screws through the overhanging flanges. When in use, place the bottles, rolled in cotton-wool, in the central chamber, pack the space between the walls with pounded ice, securely close the metal box by screwing down the fly nuts, and place it in a felt-lined wooden case. (It has been shown that whilst bacteria will survive exposure to the temperature of melting ice, practically none will multiply at this temperature.)

On reaching the laboratory, the method of examination consists in adding measured quantities of the water sample to several tubes of nutrient media previously liquefied by heat, pouring plate cultivations from each of these tubes, incubating at a suitable temperature, and finally counting the colonies which make their appearance on the plates.

Apparatus Required:

Plate-levelling stand. Case of sterile plates. Case of sterile pipettes, 1 c.c. (in tenths of a cubic centimetre). Case of sterile pipettes, 10 c.c. (in tenths of a cubic centimetre). Case of sterile capsules, 25 c.c. capacity. Tubes of nutrient gelatine. Tubes of nutrient agar. Tubes of wort gelatine. One 250 c.c. flask of sterile distilled water. Tall cylinder containing 2 per cent. lysol solution. Bunsen burner. Grease pencil. Water-bath regulated at 42° C.

Method.—

1. Arrange the plate-levelling platform with its water compartment filled with water, at 45° C.

2. Number the agar tubes, consecutively, 1 to 6; the gelatine tubes, consecutively, 1 to 6, and the wort tubes, 1, 2, and 3. Flame the plugs and see that they are not adherent to the lips of the tubes.

3. Place the agar tubes in boiling water until the medium is melted, then transfer them to the water-bath regulated at 42° C. Liquefy the nutrient gelatine and wort gelatine tubes by immersing them in the same water-bath.

4. Remove the bottle containing the water sample from the ice-box, distribute the bacterial contents evenly throughout the water by shaking, cut the string securing the stopper, and loosen the stopper, but do not take it out.

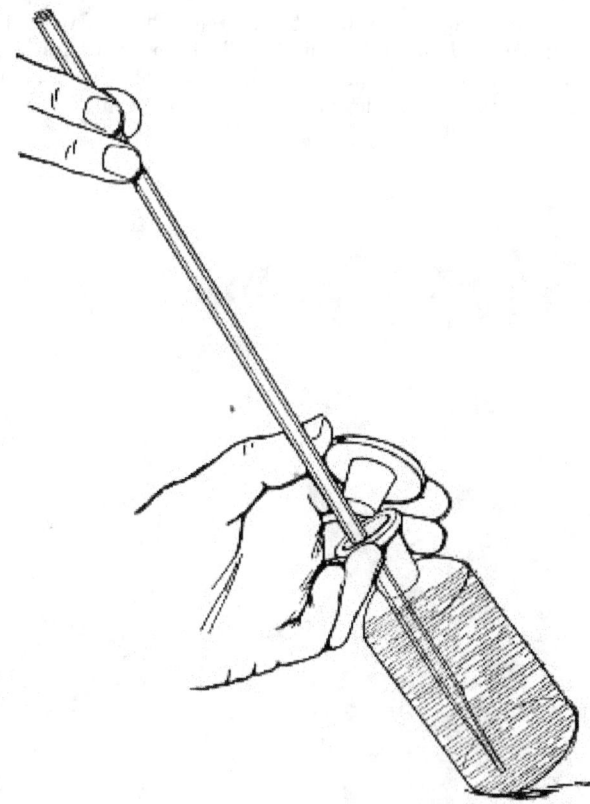

Fig. 206.—Withdrawing water from water sample bottle.

5. Remove one of the 1 c.c. pipettes from the case, holding it by the plain portion of the tube. Pass the graduated portion twice through the Bunsen flame. Tilt the bottle containing the water sample on the bench holding the neck between the middle and ring fingers of the left hand; grasp the head of the stopper between the forefinger and thumb, and remove it from the bottle.

6. Pass the pipette into the mouth of the bottle, holding its point well below the surface of the water (Fig. 206). Suck up rather more than 1 c.c. into the pipette and allow the pipette to empty; this moistens the interior of the pipette and renders accurate measurement possible. Now draw up exactly 1 c.c. into the pipette. Withdraw the pipette from the bottle, replace the stopper, and stand the bottle upright.

7. Take the first melted agar tube in the left hand, remove the cotton-wool plug, and add to its contents 0.5 c.c. of the water sample from the pipette; replug the tube and replace it in the water-bath. In a similar manner add 0.3 c.c. water to the contents of the second tube, and 0.2 c.c. to the contents of the third.

8. In a similar manner add 1 c.c. of the sample to the contents of the fourth tube.

9. Similarly, add 0.5 c.c. and 0.1 c.c. respectively to the contents of the fifth and sixth tubes.

10. Drop the pipette into the cylinder containing lysol solution.

11. Mix the water sample with the medium in each tube in the manner described under plate cultivations; pour a plate from each tube. Label each plate with (*a*) the distinctive number of the sample, (*b*) the quantity of water sample it contains, and (*c*) the date.

12. Pour the contents of a tube of liquefied agar—not inoculated—into a Petri dish to act as a control to demonstrate the sterility of the batch of agar employed.

13. Allow the plates to set, and incubate at 37° C.

14. Empty the water chamber of the levelling apparatus and refill it with ice-water.

15. By means of the sterile 10 c.c. pipette deliver 9.9 c.c. sterile distilled water into a sterile glass capsule.

16. Add 0.1 c.c. of the water sample to the 9.9 c.c. sterile water in the capsule. This will give a dilution of 1 in 100.

17. Plant the six tubes of nutrient gelatine in the following manner: To the first tube add 0.5 c.c. of the water sample direct from the bottle; to the second, 0.3 c.c.; and to the third, 0.2 c.c.; and pour a plate of each tube. To the fourth tube add 0.5 c.c. of the diluted water sample from the capsule; to the fifth, 0.3 c.c.; and to the sixth, 0.2 c.c.; and pour a plate from each.

18. Label each plate with the quantity of the water sample it contains—that is, 0.5 c.c., 0.3 c.c., 0.2 c.c., 0.005 c.c., 0.003 c.c., and 0.002 c.c.

19. Pour a control (uninoculated) gelatine plate.

20. Allow the plates to set, and incubate at 20°C.

21. To the first tube of liquefied wort gelatine add 0.5 c.c. water sample; to the second, 0.3 c.c.; and to the third, 0.2 c.c.

22. Label the plates, allow them to set, and incubate at 20° C.

23. Count and record the number of colonies that have developed upon the agar at 37° C. after forty-eight hours' incubation.

24. Note the number of colonies present on each of the gelatine and wort gelatine plates after forty-eight hours' incubation.

25. Replace the gelatine and wort plates in the incubator; observe again at three days, four days, and five days.

26. Calculate and record the number of organisms present per cubic centimetre of the original water from the average of the six gelatine plates at the latest date possible up to seven days—the presence of liquefying bacteria may render the calculation necessary at an earlier date, hence the importance of daily observations.

Method of Counting.—The most accurate method of counting the colonies on each of the plates is by means of either Jeffery's or Pakes' counting disc. Each of these discs consists of a piece of paper, upon which is printed a dead black disc, subdivided by concentric circles and radii, printed in white. In Jeffery's counter (Fig. 207), each subdivision has an area of 1 square centimetre; in Pakes' counter (Fig. 208), radii divide the circle into sixteen equal sectors, and counting is facilitated by concentric circles equidistant from the centre.

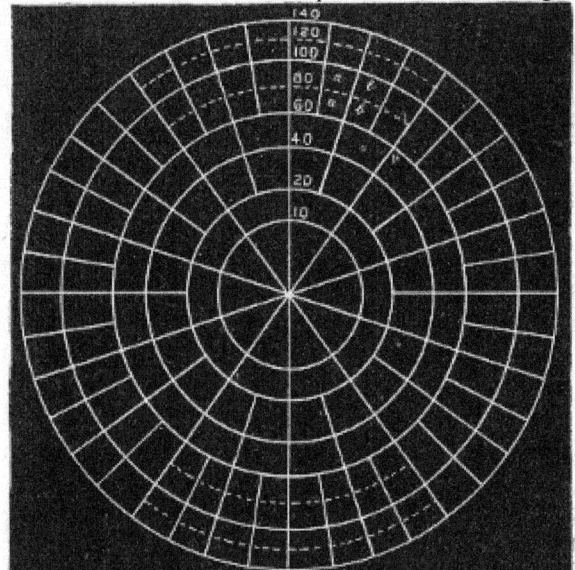

Fig. 207.—Jeffery's disc, reduced.

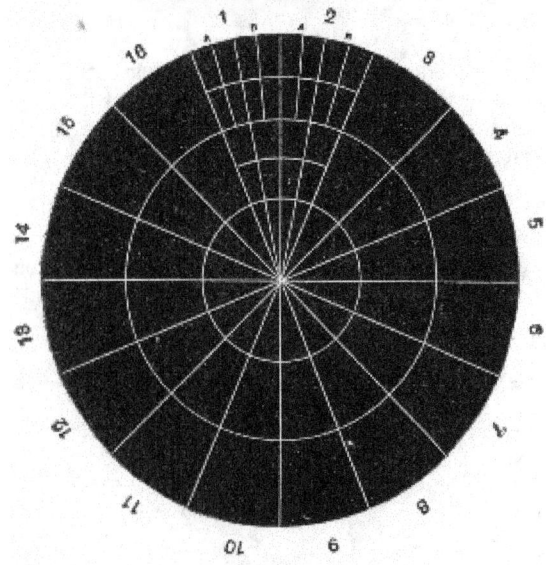

Fig. 208.—Pakes' disc, reduced.

(a) In the final counting of each plate, place the plate over the counting disc, and centre it, if possible, making its periphery coincide with one or other of the concentric circles.

(b) Remove the cover of the plate, and by means of a hand lens count the colonies appearing in each of the sectors in turn. Make a note of the number present in each.

(c) If the colonies present are fewer than 500, the entire plate should be counted. If, however, they exceed this number, enumerate one-half, or one-quarter of the plate, or count a sector here and there, and from these figures estimate the number of colonies present on the entire plate. In practice it will be found that Pakes' disc is more suitable for the former class of plate; Jeffery's disc for the latter. It should be recollected however that unless the plates have been carefully leveled and the medium is of equal thickness all over it is useless to try and average from small areas—since where the medium is thick all the bacteria will develop, where the layer is a thin one, only a few bacteria will find sufficient pabulum for the production of visible colonies.

It will be noted that the quantities of water selected for addition to each set of tubes of nutrient media have been carefully chosen in order to yield workable results even when dealing with widely differing samples. Plates prepared in agar with 0.1 c.c. and in gelatin with 0.02 c.c. can be counted even when large numbers of bacteria are present in the sample; whereas if micro-organisms are relatively few, agar plate 4 and gelatine plate 1 will give the most reliable counts. Again the counts of the plates in a measure control each other; for example, the second and third plates of each gelatine series should together contain as many colonies as the first, and the second should contain about half as many more than the third and so on.

2. Qualitative Examination.—

Collection of Sample.—The water sample required for the routine examination, which it will be convenient to consider first, amounts to about 110 c.c. It is collected in the manner previously described (*vide* page 416); similar bottles are used, and if four are filled the combined contents, amounting to about 240 c.c., will provide ample material for both the qualitative and quantitative examinations. Unless the examination is to be commenced at once, the ice-box must be employed, otherwise water bacteria and other saprophytes will probably multiply at the expense of the microbes indicative of pollution, and so increase the difficulties of the investigation.

In the routine examination of water supplies it is customary to limit the qualitative examination to a search for

A. B. coli and its near allies.

B. Streptococci,

organisms which are frequently spoken of as microbes of indication, as their presence is held to be evidence of pollution of the water by material derived from the mammalian alimentary canal, and so to constitute a danger signal.

C. Some observers still attach importance to the presence of B. enteritidis sporogenes, but as the search for this bacterium, (relatively scarce in water) necessitates the collection of a fairly large quantity of water it is not usually included in the routine examination.

In the case of water samples examined during the progress of an epidemic, of new supplies and of unknown waters the search is extended to embrace other members of the coli-typhoid group; and on occasion the question of the presence or absence of Vibrio choleræ or (more rarely) such bacteria as B. anthracis or B. tetani, may need investigation.

When pathogenic or excremental bacteria are present in water, their numbers are relatively few, owing to the dilution they have undergone, and it is usual in commencing the examination, to adopt one or other of the following methods:

A. *Enrichment*, in which the harmless non-pathogenic bacteria may be destroyed or their growth inhibited, whilst the growth of the parasitic bacteria is encouraged.

This is attained by so arranging the environment, (*i. e.*, Media, incubation temperature, and atmosphere) as to favor the growth of the pathogenic organisms at the expense of the harmless saprophytes.

B. *Concentration*, whereby all the bacteria present in the sample of water, pathogenic or otherwise, are concentrated in a small bulk of fluid.

This is usually effected by filtration of the water sample through a porcelain filter candle, and the subsequent emulsion of the bacterial residue remaining on the walls of the candle with a small measured quantity of sterile bouillon.

A. **Enrichment Method.**

(Dealing with the demonstration of bacteria of intestinal origin.)

Apparatus Required (*Preliminary Stage*):

Incubator running at 42° C. Case of sterile pipettes, 1 c.c. graduated in tenths. Case of sterile pipettes, 10 c.c. graduated in c.c. Case of sterile pipettes, graduated to deliver 25 c.c. Tubes of bile salt broth (*vide* page 180). Flask of double strength bile salt broth (*vide* page 199). Tubes of litmus silk. Sterile flasks, 250 c.c. capacity. Buchner's tubes. Tabloids pyrogallic acid. Tabloids sodium hydrate. Bunsen burner. Grease pencil.

(*Later stage*):

Incubator running at 37° C. Surface plates of nutrose agar (see page 232). Aluminium spreader. Tubes of various media, including carbohydrate media. Agglutinating sera, etc.

Method.—

1. Number a set of bile salt broth, tubes 1-5, and a duplicate set 1a-5a.
2. Number one flask 7 and another 8.
3. To Tubes No. 1 and 1a add 0.1 c.c. water sample.
To Tubes No. 2 and 2a add 1 c.c. water sample.
To Tubes No. 3 and 3a add 2 c.c. water sample.
To Tubes No. 4 and 4a add 5 c.c. water sample.
To Tubes No. 5 and 5a add 10 c.c. water sample.
4. Put up all the tubes in Buchner's tubes and incubate anaerobically at 42°C.

Note.—The bile salt medium is particularly suitable for the cultivation of bacteria of intestinal origin, and at the same time inhibits the growth of bacteria derived from other sources.

The anaerobic conditions likewise favor the multiplication of intestinal bacteria, and also their fermentative activity. The temperature 42° C. destroys ordinary water bacteria and inhibits the growth of many ordinary mesophilic bacteria.

5. Pipette 25 c.c. of double strength bile salt broth into flask 6, and 50 c.c. double strength bile salt broth into flask 7.
6. Pipette 25 c.c. water sample into flask 6, and 50 c.c. water sample into flask 7.
7. Incubate the two flasks aerobically at 42°C.
8. After twenty-four hours incubation note in each culture:

a. The presence or absence of visible growth.

b. The reaction of the medium as indicated by the colour change, if any, the litmus has undergone.

c. The presence or absence of gas formation, as indicated by a froth on the surface of the medium, and the collection of gas in the inner "gas" tube.

9. Replace those tubes which show no signs of growth in the incubator. Examine after another period of twenty-four hours (total forty-eight hours incubation) with reference to the same points.
10. Remove culture tubes which show visible growth from the Buchner's tubes, whether acid production and gas formation are present or not.
11. Examine all tubes which show growth by hanging-drop preparations. Note such as show the presence of chains of cocci.
12. Prepare surface plate cultivations upon nutrose agar from each tube that shows growth either macroscopically or microscopically, and incubate for twenty-four hours aerobically at 37° C.

13. Examine the growth on the plates either with the naked eye or with the help of a small hand lens. Practice will facilitate the recognition of colonies of the coli group, the typhoid group and the paratyphoid group; also those due to the growth of streptococci. The investigation from this stage proceeds along two divergent lines of enquiry—the first being concerned with the identity of the bacilli—typhoid bacilli, the second with that of the cocci.

A. B. Coli and its allies.

14. Pick off coliform or typhiform colonies; make streak or smear subcultivations upon nutrient agar; incubate aerobically for twenty-four hours at 37° C.

15. Examine the growth in each tube carefully both macroscopically and microscopically. If the growth is impure, replate on nutrose agar, pick off colonies and subcultivate again. When the growth in a tube is pure, add 5 c.c. sterile normal saline solution or sterile broth, and emulsify the entire surface growth with it.

16. Utilise the emulsion for the preparation of a series of subcultivations upon the media enumerated below, using the ordinary loop to make the subcultures upon solid media, but adding one-tenth of a cubic centimetre of the emulsion to each of the fluid media by means of a sterile pipette.

Gelatine streak. Agar streak. Potato. Nutrient broth. Litmus milk. Dextrose peptone solution. Lævulose peptone solution. Galactose peptone solution. Maltose peptone solution. Lactose peptone solution. Saccharose peptone solution. Raffinose peptone solution. Dulcite peptone solution. Mannite peptone solution. Glycerin peptone solution. Inulin peptone solution. Dextrin peptone solution.

17. Differentiate the bacilli after isolation by means of their cultural reactions and biological characters into members of:

I. The Escherich Group.

B. coli communis. B. coli communior. B. lactis aerogenes. B. cloacæ.

II. The Gærtner Group.

Bacillus enteritidis (of Gærtner). B. paratyphosus A. B. paratyphosus B. Bacillus choleræ suum.

III. The Eberth Group.

B. typhosus. B. dysenteriæ (Shiga). B. dysenteriæ (Flexner). B. fæcalis alcaligines.

18. Confirm these results by testing the organisms isolated against specific agglutinating sera obtained from experimentally inoculated animals.

If a positive result is obtained when using this method, it only needs a simple calculation to determine the smallest quantity (down to 0.1 c.c.) of the sample that contains at least one of the microbes of indication. For instance, if growth occurs in all the tubes from 4 to 10, and that growth is subsequently proved to be due to the multiplication of B. coli, then it follows that at least one colon bacillus is present in every 10 c.c. of the water sample, but not in every 5 c.c. If, on the other hand, the presence of the B. coli can only be proved in flask No. 7, then the average number of colon bacilli present in the sample is at least one in every 50 c.c. (*i. e.*, twenty per litre), but not one in every 25 c.c. and so on.

The general outline of the method of identifying the members of the coli-typhoid group is given in the form of an analytical schema—whilst the full differential details are set out in tabular form.

ANALYTICAL SCHEME FOR ISOLATION OF MEMBERS OF THE COLI AND TYPHOID GROUPS.

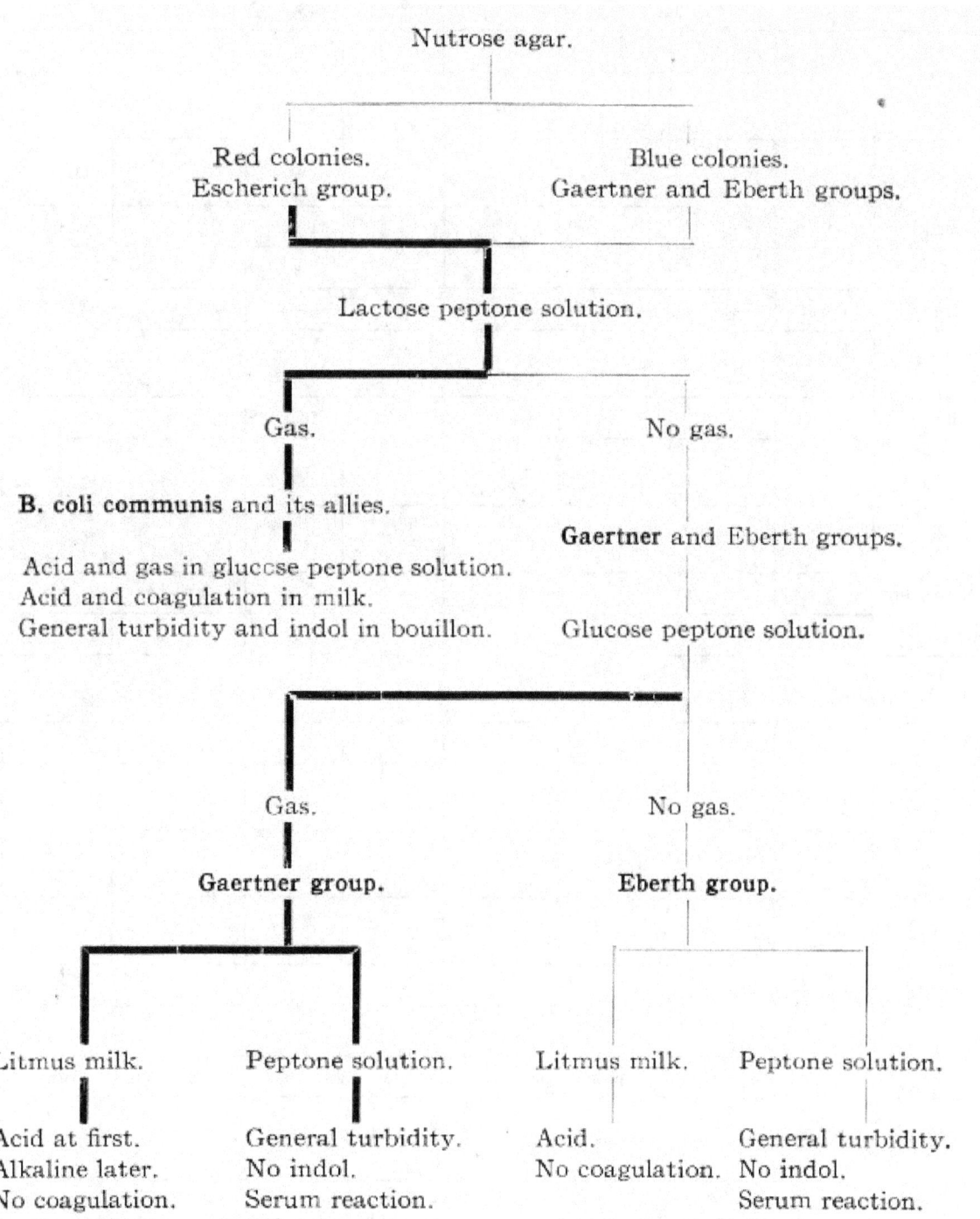

B. Streptococci.

19. Pick off streptococcus colonies and subcultivate upon nutrient agar exactly as directed in steps 14, 15 and 16.

20. Differentiate the streptococci isolated into members of the saprophytic group of short-chained cocci, or members of the parasitic (pathogenic) group of long-chained cocci, by means of their cultural characters, and record their numerical frequency in the manner indicated for the members of the coli-typhoid group.

DIFFERENTIAL TABLE OF COLI-TYPHOID GROUP

	Motility	Dextrose	Lævulose	Galactose	Maltose	Lactose	Sacchrarose	Raffinose	Dextrin
A = acid reaction G = gas formation		A G	A G	A G	A G	A G	A G	A G	A G
The Escherich									

Group.									
B. coli communis	+	+ +	+ +	+ +	+ +	+ +	O	+ +	+ +
B. coli communior	+	+ +	+ +	+ +	+ +	+ +	+ +	+ +	+ +
B. lactis aerogenes	-	+ +	+ +	+ +	+ +	+ +	O	O	+ +
B. acidi lactici	-	+ +	+ +	+ +	+ +	+ +	O	O	O
B. pneumoniæ	-	+ +	+ +	+ +	+ +	+ +	+ +	+ +	+ +
B cloaceæ(A)	+	+ +	+ +	+ +	+ +	+ +	+ +	+ +	+ +
The Gærtner Group.									
B. enteritidis	+	+ +	+ +	+ +	+ +	O	O	O	O
B. paratyphosus A	+	+ +	+ +	+ +	+ +	O	O	O	O
B. paratyphosus B	+	+ +	+ +	+ +	+ +	O	O	O	O
B. choleræ suum	+	+ +	+ +	+ +	+ +	O	O		O
B. suipestifer	+	+ +	+ +	+ +	+ +	O	O		O
The Eberth Group.									
B. typhosus	+	+	+	+	+	O	O	O	+
B. dysenteriæ (Shiga)	-	+	+	+	O	O	O	O	O
B. dysenteriæ (Flexner)	-	+	+	+	+	O	O	±	O
B. fæcalis alkaligines	+	O	O	O	O	O	O	O	O
Table Notes:	(B)	(C)							

	Inulin	Salicin	Glycerine	Dulcite	Mannite	Sorbite	Indol	Litmus Milk	
A=acid reaction G=gas formation	A G	A G	A G	A G	A G	A G		Early	Late
The Escherich Group									
B. coli communis	O	O	+ +	+ +	+ +	+ +	+	+	+ C
B. coli communior	O	O	+ +	+ +	+ +	+ +	+	+	+ C
B. lactis aerogenes	O	O	O	O	+ +	+ +	-	+	+ C
B. acidi lactici	O	O	O	+ +	+ +	+ +	+	+	+ C
B. pneumoniæ	O	O	+ +	+ +	+ +	+ +	-	+	+ C
B cloaceæ	O	O	+ +	O	+ +	- +	+	+	+ C
The Gærtner Group.									
B. enteritidis	O	O	O	+ +	+ +	+ +	-	±	-

B. paratyphosus A	O	±	O	+ +	+ +	+ +	-	+	O
B. paratyphosus B	O	O	O	+ +	+ +	+ +	-	+	-
B. choleræ suum	O	O	O	O	O	+ +	±	+	-
B. suipestifer	O	O	O	+ +	+ +	+ +	-	+	-
The Eberth Group.									
B. typhosus	O	O	O	O	+	+	-	+	+
B. dysenteriæ (Shiga)	O	O	O	O	O	O	-	+	-
B. dysenteriæ (Flexner)	O	O	O	O	+	O	±	+	-
B. fæcalis alkaligines	O	O	O	O	O	O	-	-	-
Table Notes:								(D)	(E)

Table Notes:
(A) * Liquefies gelatine.
(B) + = motile. - = non-motile.
(C) + = acid or gas production. ± = slight acid production. O = no change.
(D) + = indol production. ± = slight indol production. - = no indol formed.
(E) + = acid production. - = alkali production. O = no change in reaction. C = clot.

21. Determine the pathogenicity for mice (subcutaneous inoculation) and rabbits (intravenous inoculation) of the streptococci isolated.

On the facing insert page is reproduced a blank from the author's Laboratory Water Analysis Book, by means of which an exact record can be kept, with a minimum of labour, of every sample examined.

B. **Concentration Method.**

The remaining organisms referred to on page 426 are more conveniently sought for by the concentration method.

Collection of the Sample.—The quantity of water required for this method of examination is about 2000 c.c., and the vessel usually chosen for its reception is an ordinary blue glass Winchester quart bottle, sterilised in the hot-air oven, and over this a paper or parchment cap fastened with string. The bottle may be packed in a wooden box or in an ordinary wicker case. The method of collecting the sample is identical with that described under the heading of Quantitative Examination; there is, however, not the same imperative necessity to pack the sample in ice for transmission to the laboratory.

Apparatus required:
Sterile Chamberland or Doulton "white" porcelain open mouth filter candle, fitted with rubber washer.
Rubber cork to fit mouth of the filter candle, perforated with one hole.
Kitasato serum flask, 2500 c.c. capacity.
Geryk air pump or water force pump.
Wulff's bottle, fitted as wash-bottle, and containing sulphuric acid (to act as a safety valve between filter and pump).
Pressure tubing, clamps, pinch-cock.
Retort stand, with ring and clamp.
Rubber cork for the neck of Winchester quart, perforated with two holes and fitted with one 6 cm. length of straight glass tubing, and one V-shaped piece of glass tubing, one arm 32 cm. in length, the other 52 cm., the shorter arm being plugged with cotton-wool. The rubber stopper must be sterilised by boiling and the glass tubing by hot air, before use.
Flask containing 250 c.c. sterile broth.
Test-tube brush to fit the lumen of the candle, enclosed in a sterile test-tube (and previously sterilised by dry heat or by boiling).
Case of sterile pipettes, 10 c.c. in tenths.
Case of sterile pipettes, 1 c.c. in tenths.
Case of sterile pipettes, 1 c.c. in hundredths.
Tubes of various nutrient media (according to requirements).
Twelve Buchner's tubes with rubber stoppers.
Pyrogallic acid tablets.
Caustic soda tablets.

The Elements of Bacteriological Technique

No. _____   Sample of _____
                   From _____
                   Labelled _____

Date of Collection _____   Received _____   Reported _____

per _____

I. QUANTITATIVE

Gelatine Plates
- 0·2 cc.
- 0·3 cc.
- 0·5 cc.

Columns: 24 hrs. | 48 hrs. | 72 hrs. | 96 hrs. | 7 days. | | 24 hrs. | 48 hrs. | 72 hrs. | 96 hrs. | 7 days.

- 0·2 cc.
- 0·3 cc.
- 0·5 cc.

{ Living bacteria growing aërobically at 20° C. per cc. = ... }

Agar Plates
- 0·2 cc.
- 0·3 cc.
- 1·0 cc.

- 0·2 cc.
- 0·3 cc.
- 0·5 cc.

{ Living bacteria growing aërobically at 37° C. per cc. = ... }   Ratio =

Wort Gelatine Plates
- 0·2 cc.
- 0·3 cc.
- 0·5 cc.

- 0·5 cc.
- 1·0 cc.

{ Moulds growing aërobically at 20° C. per cc. = ... }

II. QUALITATIVE

A. In Bile Salt Broth (incubated anaërobically at 42° C.)   24 hrs. Acid Gas   48 hrs. Acid Gas

For B. coli communis.
- 1 = 0·1 cc. original water ...
- 2 = 1 cc. „
- 3 = 2 cc. „
- 4 = 5 cc. „
- 5 = 10 cc. original water, or 0·05 cc. Emulsion ...
- 6 = 20 cc. „ „ 0·10 cc. „
- 7 = 25 cc. „ „ 0·125 cc. „
- 8 = 50 cc. „ „ 0·25 cc. „
- 9 = 100 cc. „ „ 0·50 cc. „
- 10 = 250 cc. „ „ 1·25 cc. „
- 11 = 500 cc. „ „ 2·5 cc. „
- 12 = 1000 cc. „ „ 5·0 cc. „

Plated Tubes ...
Isolated Colonies
Shape ...
Motility ...
Gram ...
Indol ...
Gelatine Streak ...
Broth ...
Agar Streak ...
Litmus Milk ...
Potato ...
Dextrose Sol. ...
Laevulose „ ...
Galactose „ ...
Maltose „ ...
Lactose „ ...
Saccharose „ ...
Raffinose „ ...
Dulcite „ ...
Mannite „ ...
Glycerin „ ...
Inulin „ ...
Dextrin „ ...

{ B. coli communis present in ...., absent from .... cc., but not in .... cc. }

B. For Streptococci. Microscopical Examination in hanging drop of Bile Salt Broth Tubes (A.)   24 hrs.   48 hrs.

1, 2, 3, 4, 5, 6, 7, 8, 9, 10, 11, 12

ABSTRACT OF REPORT

Colonies on _____ 20° _____ per cc.
       „   „   37° _____ „
Ratio 20° to 37°
Moulds
B. Coli present, absent in _____ cc.
Streptococci present, absent in _____ cc.
B. Enteritidis present, absent in _____ cc.

{ Streptococci present in ...., absent from .... cc., but not in .... cc. }

C. In Litmus Milk (incubated anaërobically at 42° C. after partial sterilisation at 80° C. for 15 minutes).   24 hours acid clot gas.   48 hours acid clot gas.   Guinea pig. w.......inoc. with whey from tube

For B. enteritidis sporogenes.
- 1 = 1 cc. original water ...
- 2 = 2 cc. „
- 3 = 5 cc. „
- 4 = 10 cc. original water, or 0·05 cc. Emulsion
- 5 = 20 cc. „ „ 0·10 cc. „
- 6 = 25 cc. „ „ 0·125 cc. „
- 7 = 50 cc. „ „ 0·25 cc. „
- 8 = 100 cc. „ „ 0·50 cc. „
- 9 = 250 cc. „ „ 1·25 cc. „
- 10 = 500 cc. „ „ 2·50 cc. „

{ B. enteritidis sporogenes present in ...., absent from .... cc., but not in .... cc. }

Signed

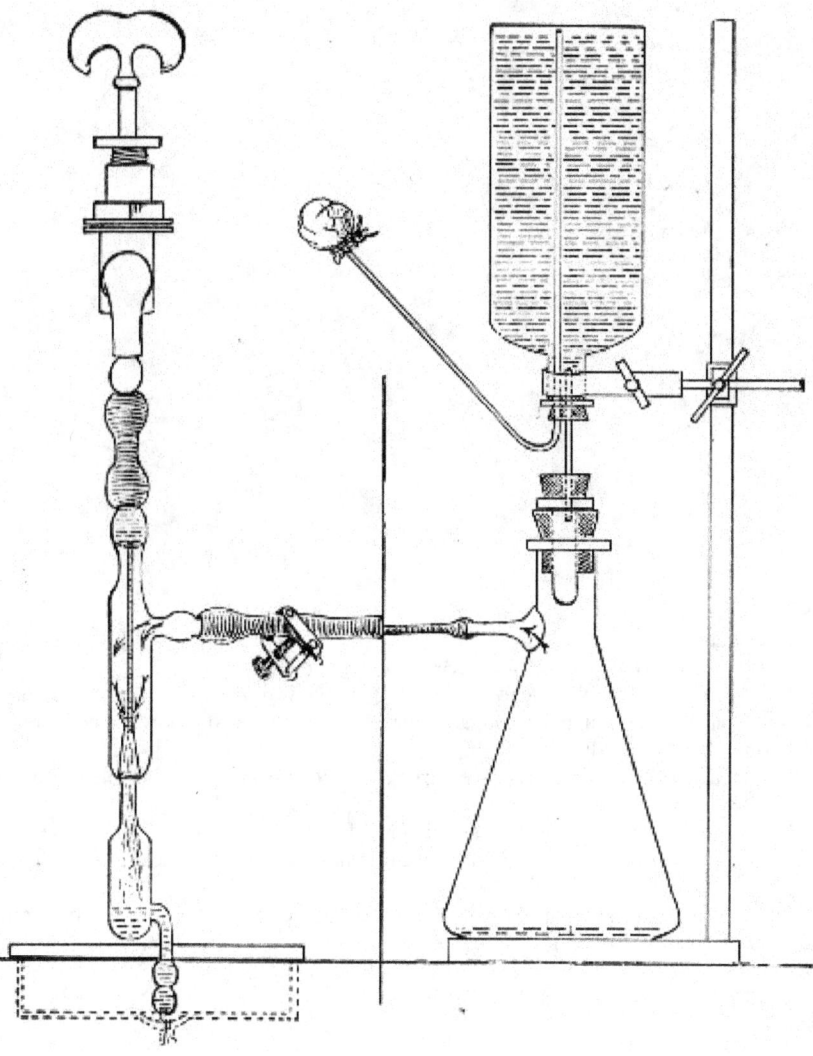

Fig. 209.—Water filtering apparatus. That portion of the figure to the left of the vertical line is drawn to a larger scale than that on the right, in order to show details of Sprengel's pump.

Method.—

1. Fit up the filtering apparatus as in the accompanying diagram (Fig. 209), interposing the wash-bottle with sulphuric acid between the filter flask and the force-pump (in the position occupied in the diagram by the central vertical line), and placing another screw clamp on the rubber tubing connecting the lateral arm of the filter flask with the wash-bottle.

Fig. 210. Sterile test-tube brush.

2. Filter the entire 2000 c.c. of water through the filter candle.

3. When the nitration is completed, screw up the clamps and so occlude the two pieces of pressure tubing.

4. Reverse the position of the glass tubes in the Wulff's bottle so that the one nearest the air pump now dips into the sulphuric acid.

5. Slowly open the metal clamps and allow air to gradually pass through the acid, and enter filter flask, and so restore the pressure.

6. Unship the apparatus, remove the cork from the mouth of the candle.

7. Pipette 10 c.c. of sterile broth into the interior of the candle, and by means of the sterile test-tube brush (Fig. 210) emulsify the slimy residue which lines the candle, with the broth.

Practically all the bacteria contained in the original 2000 c.c. of water are now suspended in 10 c.c. of broth, so that 1 c.c. of the suspension is equivalent, so far as the contained organisms are concerned, to 200 c.c. of the original water. (Some bacteria will of course be left behind on the walls of the filter and in its pores.)

Up to this point the method is identical, irrespective of the particular organism whose presence it is desired to demonstrate; but from this point onward the methods must be specially adapted to the isolation of definite groups of organisms or of individual bacteria.

The Coli-Typhoid Group.—

1. Number nine tubes of bile salt broth (*vide* page 180), consecutively from 1 to 9.

2. To No 1 add 1 c.c. } of the original water sample
2 add 2 c.c. } before the nitration is commenced.
3 add 5 c.c. }

3. To the remaining tubes of bile salt broth add varying quantities of the suspension by means of suitably graduated sterile pipettes, as follows:

No. 4 0.05 c.c. (equivalent to 10 c.c. of the original water sample).
No. 5 0.125 c.c. (equivalent to 25 c.c. of the original water sample).
No. 6 0.25 c.c. (equivalent to 50 c.c. of the original water sample).
No. 7 0.5 c.c. (equivalent to 100 c.c. of the original water sample).
No. 8 1.0 c.c. (equivalent to 200 c.c. of the original water sample).
No. 9 2.5 c.c. (equivalent to 500 c.c. of the original water sample).

4. Put up each tube anaerobically in a Buchner's tube and incubate at 42° C.

5. The subsequent steps are identical with those described under the Enrichment method (see page 428 to 431; Steps 8 to 18).

Alternative Methods.—

A few of the older methods for the isolation of the members of the coli-typhoid groups are referred to but they are distinctly inferior to those already described.

(A) The Carbolic Method:

1. Take ten tubes of carbolised bouillon (*vide* page 202) and number them consecutively from 1 to 10.
2. Inoculate each tube with a different amount of the water sample or suspension, as in the previous method.
3. Incubate aerobically at 37° C.
4. Examine the culture tubes after twenty-four hours' incubation.
5. From those tubes which shows signs of growth, pour plates in the usual manner, using carbolised gelatine (*vide* page 202) in place of the ordinary gelatine, and incubate at 20° C. for three, four, or five days as may be necessary.
6. Subcultivate from any colonies that make their appearance, and determine their identity on the lines laid down in the previous method.

(B) Parietti's Method:

1. Take nine tubes of Parietti's bouillon (*vide* page 202)—*i. e.*, three each of those containing 0.1 c.c., 0.2 c.c., and 0.5 c.c. of Parietti's solution respectively. Mark plainly on the outside of each tube the quantity of Parietti's solution it contains.
2. To each tube add a different amount of the original water, or of the suspension, and incubate at 37° C.
3. Examine the culture tubes after twenty-four and forty-eight hours' incubation, and plate in nutrient carbolised or potato gelatine from such as have grown.
4. Pick off suspicious colonies, if any such appear on the plates, subcultivate them upon the various media, and identify them.

(C) Elsner's Method: This method simply consists in substituting Elsner's potato gelatine (*vide* page 204) for ordinary nutrient gelatine in any of the previously mentioned methods.

(D) Cambier's Candle Method:

Treat a large volume of the water sample by the concentration method (*vide* page 434).

1. Remove the rubber stopper from the mouth of the filter candle, introduce 10 c.c. sterile bouillon into its interior, and emulsify the bacterial sediment; replug the mouth of the candle with a wad of sterile cotton-wool.
2. Remove the filter candle from the filter flask and insert it into the mouth of a flask or a glass cylinder containing sterile bouillon sufficient to reach nearly up to the rubber washer on the candle.
3. Incubate for twenty-four to thirty-six hours at 37° C.
4. From the now turbid bouillon in the glass cylinder pour gelatine plates and incubate at 20° C.
5. Subcultivate and identify any suspicious colonies that appear.

(The method depends upon the assumption that members of the typhoid and coli groups find their way through the porcelain filter from the interior to the surrounding bouillon at a quicker rate than the associated bacteria.)

B. Enteritidis Sporogenes.—

1. Transfer 5 c.c. of the emulsion from the filter candle to a sterile test-tube and plug carefully.
2. Place the test-tube in the interior of the benzole bath employed in separating out spore-bearing organisms (*vide* page 257), and expose to a temperature of 80° C. for twenty minutes.
3. Number ten tubes of litmus milk consecutively from 1 to 10.
4. Remove the test-tube from the benzole bath and shake well to distribute the spores evenly through the fluid.
5. To each tube of litmus milk add a measured quantity of the suspension corresponding to the amounts employed in isolating the coli group (*vide* page 437).
6. Incubate each tube anaerobically at 37° C. Anaerobic conditions can be obtained by putting the cultures up in Buchner's tubes or in Bulloch's apparatus. If, however, whole milk has been used in making the litmus milk the layer of cream that rises to the surface will be sufficient to ensure anaerobiosis; whilst if separated milk has been employed it will be sufficient to pour a layer of sterile vaseline or liquid paraffin on the surface of the fluid.
7. Examine after twenty-four hours' incubation. Note (if B. enteritidis sporogenes is present)—

(*a*) Acid reaction of the medium as indicated by the colour of the litmus or its complete decolourisation.

(*b*) Presence of clotting, and the separation of clear whey.

(*c*) Presence of gas, as indicated by fissures and bubbles in the coagulum, and possibly masses of coagulum driven up the tube almost to the plug.

8. Replace the tubes which show no signs of growth in the incubator for a further period of twenty-four hours and again examine with reference to the same points.
9. Remove those tubes which give evidence of growth from the Buchner's tubes and carefully pipette off the whey; examine the whey microscopically.
10. Inoculate two guinea-pigs each subcutaneously with 0.5 c.c. of the whey and observe the result.

Vibrio Choleræ.—

1. Number ten tubes of peptone water consecutively from 1 to 10.
2. To each of the tubes of peptone water add a measured quantity of the suspension, corresponding to those amounts employed in isolating the members of the coli group (*vide* page 437).
3. Incubate aerobically at 37° C. for twenty-four hours. Examine the tubes carefully for visible growth, especially delicate pellicle formation, which if present should be examined microscopically for vibrios, both by stained preparations or by fresh specimens with dark ground illumination.
4. Inoculate fresh tubes of peptone water from such of the tubes as exhibit pellicle formation—from the pellicle itself—and incubate at 37° C. for twenty-four hours.
5. Test the peptone water itself for the presence of indol and nitrite by the addition of pure concentrated H_2SO_4.
5. Prepare gelatine and agar plates in the usual way from such of these tubes as show pellicle formation.
6. Pick off from the plates any colonies resembling those of the Vibrio choleræ and subcultivate upon all the ordinary laboratory media.
7. Test the vibrio isolated against the serum of an animal immunised to the Vibrio choleræ for agglutination.

B. Anthracis.—

1. Transfer 5 c.c. of the emulsion from the filter candle to a sterile test-tube and plug carefully.

2. Place the test-tube in the interior of the benzole bath employed in separating out spore-bearing organisms (*vide* page 257), and expose to a temperature of 80° C. for twenty minutes.

3. Inoculate a *young* white rat subcutaneously (on the inner aspect of one of the hind legs) with 1 c.c. of the emulsion. Observe during life, and, if the animal succumbs, make a complete post-mortem examination.

4. Melt three tubes of nutrient agar in boiling water and cool to 42° C.

5. Number the tubes 1, 2, and 3. To No. 1 add 0.2 c.c., to No. 2 add 0.3 c.c., and to No. 3 add 0.5 c.c. of the suspension, and pour plates therefrom.

6. Incubate at 37° C. for twenty-four or forty-eight hours.

7. Pick off any colonies resembling those of anthrax and subcultivate on all the ordinary laboratory media.

8. Inoculate another young white rat as in 3, using two loopfuls of the agar subcultivation emulsified with 1 c.c. sterile bouillon. Observe during life, and if the animal succumbs, make a complete post-mortem examination.

B. Tetani.—

1. Proceed as detailed above in steps 1 and 2 for the isolation of the B. anthracis.

2. Add 1 c.c. of the suspension to each of three tubes of glucose formate broth, and incubate anaerobically in Buchner's tubes at 37° C.

3. From such of the tubes as show visible growth (with or without the production of gas) after twenty-four hours' incubation inoculate guinea-pigs, subcutaneously (under the skin of the abdomen), using 0.1 c.c. of the bouillon cultivation as a dose. Observe carefully during life, and, if death occurs, make a complete post-mortem examination.

4. From the same tubes pour agar plates and incubate anaerobically in Bulloch's apparatus, at 37° C.

5. Subcultivate suspicious colonies on the various media, incubate anaerobically, making control cultivations on glucose formate agar, stab and streak, to incubate aerobically and carry out further inoculation experiments with the resulting growths.

EXAMINATION OF MILK.

"One-cow" or "whole" milk, if taken from the apparently healthy animal (that is, an animal without any obvious lesion of the udder or teats) with ordinary precautions as to cleanliness, avoidance of dust, etc., contains but few organisms. In dealing with one-cow milk, from a suspected, or an obviously diseased animal, a complete analysis should include the examination (both qualitative and quantitative) of samples of (*a*) fore-milk, (*b*) mid-milk, (*c*) strippings, and, if possible, from each quarter of the udder. "Mixed" milk, on the other hand, by the time it leaves the retailer's hands, usually contains as many micro-organisms as an equal volume of sewage and indeed during the examination it is treated as such.

It is possible however to collect and store mixed milk in so cleanly a manner that its germ content does not exceed 5000 micro-organisms per cubic centimetre. Such comparative freedom from extraneous bacteria is usually secured by the purveyor only when he resorts to the process of pasteurisation (heating the milk to 65° C. for twenty minutes or to 77° C. for one minute) or the simpler plan of adding preservatives to the milk. Information regarding the employment of these methods for the destruction of bacteria should always be sought in the case of mixed milk samples, and in this connection the following tests will be found useful:

1. *Raw Milk* (Saul).

To 10 c.c. milk in a test tube, add 1 c.c. of a 1 per cent. aqueous solution of ortol (ortho-methyl-amino-phenol sulphate), recently prepared and mix. Next add 0.2 c.c. of a 3 per cent. peroxide of hydrogen solution. The appearance of a brick red color within 30 seconds indicates raw milk. Milk heated to 74° C. for thirty minutes undergoes no alteration in color; if heated to 75° C. for ten minutes only, the brick red color appears after standing for about two minutes.

2. *Boric Acid.*

Evaporate to dryness, 50 c.c. of the milk which has been rendered slightly alkaline to litmus, then incinerate.

Dissolve in distilled water, add slight excess of dilute hydrochloric acid and again evaporate to dryness.

Dissolve the residue in a small quantity of hot water and moisten a piece of turmeric paper with the solution. Dry the turmeric paper. *Rose* or *cherry-red* color = borax or boric acid.

3. *Formaldehyde* (Hehner).

To 10 c.c. milk in a test tube add 5 c.c. concentrated *commercial* sulphuric acid slowly, so that the two fluids do not mix. Hold the tube vertically and agitate very gently. *Violet zone* at the junction of the two liquids = formaldehyde.

4. *Hydrogen Peroxide.*

To 10 c.c. milk (diluted with equal quantities of water) in a test tube add 0.4 c.c. of a 4 per cent. alcoholic solution of benzidine and 0.2 c.c. acetic acid. *Blue coloration* of the mixture = hydrogen peroxide.

5. *Salicylic Acid.*

Precipitate the caseinogen by the addition of acetic acid and filter. To the filtrate add a few drops of 1 per cent. aqueous solution of ferric chloride. *Purple coloration* = salicylic acid.

6. *Sodium Carbonate or Bicarbonate.*

To 10 c.c. of the milk in a test tube add 10 c.c. of alcohol and 0.3 c.c. of a 1 per cent. alcoholic solution of rosolic acid. *Brownish* color = pure milk; *rose* color = preserved milk.

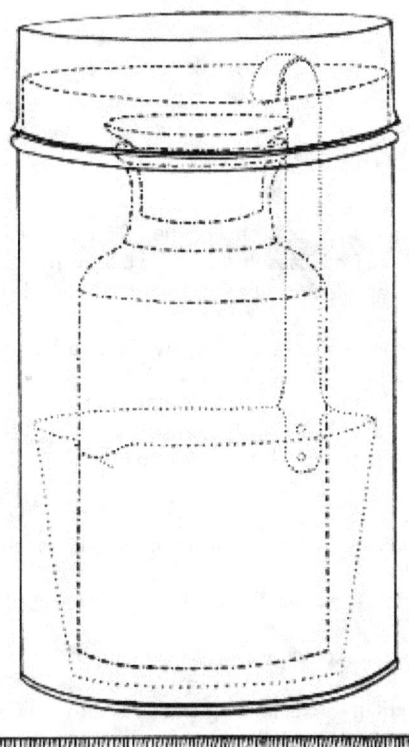

Fig. 211.—Milk-collecting bottle and dipper in case.

Quantitative.—

Collection of Sample.—

The apparatus used for the collection of a retail mixed milk sample consists of a cylindrical copper case, 16 cm. high and 9 cm. in diameter, provided with a "pull-off" lid, containing a milk dipper, also made of copper; and inside this, again, a wide-mouthed, stoppered glass bottle of about 250 c.c. capacity (about 14 cm. high by 7 cm. diameter), having a tablet for notes, sand-blasted on the side. The copper cylinder and its contents, secured from shaking by packing with cotton-wool, are sterilised in the hot-air oven (Fig. 26).

When collecting a sample,

1. Remove the cap from the cylinder.
2. Draw out the cotton-wool.
3. Lift out the bottle and dipper together.
4. Receive the milk in the sterile dipper, and pour it directly into the sterile bottle.
5. Enter the particulars necessary for the identification of the specimen, on the tablet, with a lead pencil, or pen and ink.
6. Pack the apparatus in the ice-box for transmission to the laboratory in precisely the same manner as an ordinary water sample.

"Whole" milk may with advantage be collected in the sterile bottle directly since the mouth is sufficiently wide for the milker to direct the stream of milk into it.

Condensed milk must be diluted with sterile distilled water in accordance with the directions printed upon the label, then treated as ordinary milk.

Apparatus Required:

Case of sterile capsules (25 c.c. capacity). Case of sterile graduated pipettes, 10 c.c. (in tenths of a cubic centimetre). Case of sterile graduated pipettes, 1 c.c. (in tenths of a cubic centimetre). Flask containing 250 c.c. sterile bouillon. Tall cylinder containing 2 per cent. lysol solution. Plate-levelling stand. Case of sterile plates. Tubes nutrient gelatine or gelatine agar. Tubes of wort gelatine. Tubes of nutrient agar. Water-bath regulated at 42° C. Bunsen burner. Grease pencil.

Method.—

1. Arrange four sterile capsules in a row; number them I, II, III, and IV.
2. Fill 9 c.c. sterile bouillon into the first, and 9.9 c.c. bouillon into each of the three remaining capsules.
3. Remove 1 c.c. milk from one of the bottles by means of a sterile pipette and add it to the bouillon in capsule I; mix thoroughly by repeatedly filling and emptying the pipette.
4. Remove 0.1 c.c. of the milky bouillon from capsule I, add it to the contents of capsule II, and mix as before.
5. In like manner add 0.1 c.c. of the contents of capsule II to capsule III; and then 0.1 c.c. of the contents of capsule III to capsule IV.

Then 1 c.c. of dilution I contains 0.1 c.c. milk sample. 1 c.c. of dilution II contains 0.001 c.c. milk sample. 1 c.c. of dilution III contains 0.00001 c.c. milk sample. 1 c.c. of dilution IV contains 0.0000001 c.c. milk sample.

6. Melt the gelatine and the agar tubes in boiling water; then transfer to the water-bath and cool them down to 42° C.
7. Number the gelatine tubes consecutively 1 to 12.
8. Inoculate the tubes with varying quantities of the material as follows:

To tube No. 1 add 1.0 c.c. of the milk sample. 2 add 0.1 c.c. of the milk sample. { 3 add 1.0 c.c. from capsule I. { 4 add 0.1 c.c. from capsule I. { 5 add 1.0 c.c. from capsule II. { 6 add 0.1 c.c. from capsule II. { 7 add 0.5 c.c. from capsule III. { 8 add 0.3 c.c. from capsule III. { 9 add 0.2 c.c. from capsule III. { 10 add 0.5 c.c. from capsule IV. { 11 add 0.3 c.c. from capsule IV. { 12 add 0.2 c.c. from capsule IV.

9. Pour plates from the gelatine tubes; label, and incubate at 20° C.

10. Liquefy five wort gelatine tubes and to them add 1.0 c.c. of the milk sample and a similar quantity of the diluted milk from capsules I, II, and III and IV respectively.

11. Pour plates from the wort gelatine; label, and incubate at 20° C.

12. Inoculate the liquefied agar tubes as follows:

To tube No. 1 add 0.1 c.c. of the milk sample. 2 add 0.1 c.c. from capsule I. 3 add 0.1 c.c. from capsule II. 4 add 0.1 c.c. from capsule III. 5 add 1.0 c.c. from capsule IV. } 6 add 0.1 c.c. from capsule IV. }

13. Pour plates from the agar tubes; label, and incubate at 37° C.

14. After twenty-four hours' incubation "inspect," and after forty-eight hours' incubation, "count" the agar plates and estimate the number of "organisms growing at 37° C." present per cubic centimetre of the sample of milk.

15. After three, four, or five days' incubation, "count" the gelatine plates and estimate therefrom the number of "organisms growing at 20° C." present per cubic centimetre of the sample of milk.

16. After a similar interval "count" the wort gelatine plates and estimate the number of moulds and yeasts present per cubic centimetre of the sample of milk.

Note.—Many observers prefer to employ gelatine agar (see page 193) for the quantitative examination. In this case gelatine-agar plates should be poured from tubes containing the quantities of material indicated in step 8, incubated at 28° C. to 30° C. and after five days the "total number of organisms developing at 28° C." recorded.

Qualitative.—The qualitative bacteriological examination of milk is chiefly directed to the detection of the presence of one or more of the following pathogenic bacteria and when present to the estimation of their numerical frequency.

Members of the Coli-typhoid group. Vibrio choleræ. Streptococcus pyogenes longus. Micrococcus melitensis. Staphylococcus pyogenes aureus. Bacillus enteritidis sporogenes. Bacillus diphtheriæ. Bacillus tuberculosis.

Some of these occur as accidental contaminations, either from the water supply to the cow farm, or from the farm employees, whilst others are derived directly from the cow.

In milk, as in water examinations, two methods are available, viz.: Enrichment and Concentration—the former is used for the demonstration of bacteria of intestinal origin, the latter for the isolation of the micro-organisms of diphtheria and tubercle. The first essential in the latter process is the concentration of the bacterial contents of a large volume of the sample into a small compass; but in the case of milk, thorough centrifugalisation is substituted for filtration.

Apparatus Required:

A large centrifugal machine. This machine, to be of real service in the bacteriological examination of milk, must conform to the following requirements:

1. The centrifugal machine must be of such size, and should carry tubes or bottles of such capacity, as to enable from 200 to 500 c.c. of milk to be manipulated at one time.

2. The rate of centrifugalisation should be from 2500 to 3000 revolutions per minute.

3. The portion of the machine destined to carry the tubes should be a metal disc, of sufficient weight to ensure good "flank" movement, continuing over a considerable period of time. In other words, the machine should run down very gradually and slowly after the motive power is removed, thus obviating any disturbance of the relative positions of particulate matter in the solution that is being centrifugalised.

4. The machine should preferably be driven by electricity, or by power, but in the case of hand-driven machines—

(a) The gearing should be so arranged that the requisite speed is obtained by not more than forty or fifty revolutions of the crank handle per minute, so that it may be maintained for periods of twenty or thirty minutes without undue exertion.

(b) The handle employed should be provided with a special fastening (*e. g.*, a clutch similar to that employed for the free wheel of a bicycle), or should be readily detachable so that, on ceasing to turn, the handle should not, by its weight and air resistance, act as a brake and stop the machine too suddenly.

One of the few satisfactory machines of this class is shown in figure 212.

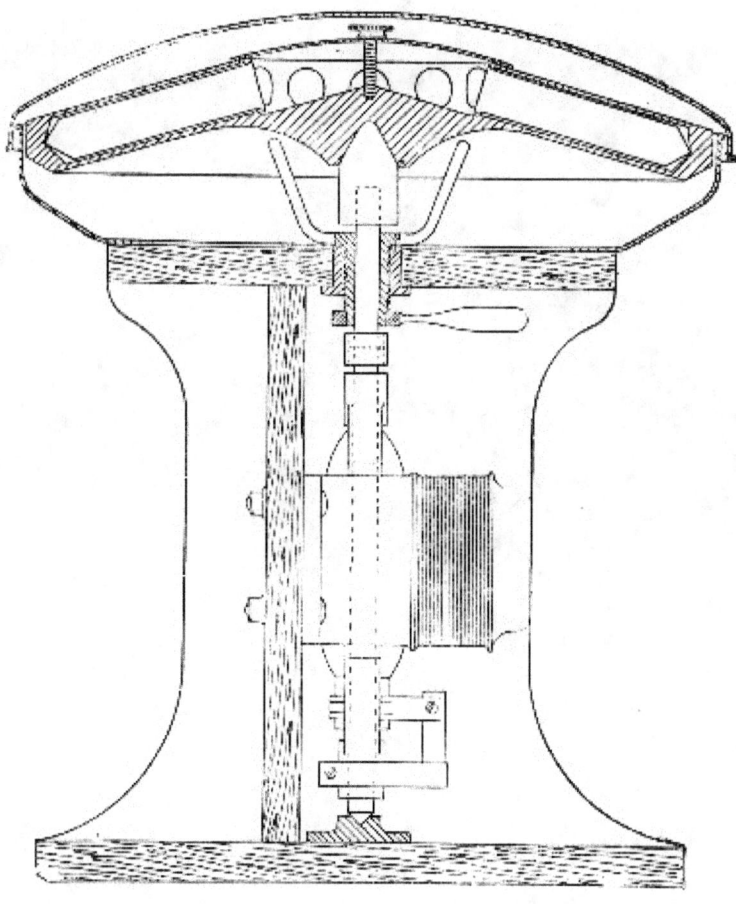

Fig. 212.—Electrically driven centrifugal machine, with flexible (broken) spindle encircled by the field magnets of the motor.

Sterile centrifugal tubes, of some 60-70 c.c. capacity, tapering to a point at the closed end, plugged with cotton-wool.

Small centrifugal machine to run two tubes of 10 c.c. capacity at 2500 to 3000 revolutions per minute preferably driven by electricity, of the type figured on page 327 (Fig. 162).

Sterile centrifugal tubes of 10 c.c. capacity with the distal extremity contracted to a narrow tube and graduated in hundredths of a cubic centimetre (Fig. 213).

Sterilised cork borer.
Case of sterile pipettes, 10 c.c. (in tenths of a cubic centimetre).
Case of sterile pipettes, 1 c.c. (in tenths of a cubic centimetre).
Sterile teat pipettes.
Flask of sterile normal saline solution.
Method.—

1. Fill 50 c.c. of the milk sample into each of four tubes, and replace the cotton-wool plugs by solid rubber stoppers (sterilised by boiling), and fit the tubes in the centrifugal machine.

Note.—One or two cubic centimetres of paraffinum liquidum introduced into the buckets of the centrifuge before the glass tubes are inserted will obviate any risk of breakage to the latter.

Fig. 213.—Milk sedimenting tubes.

Fig. 214.—Milk in centrifuge tube.

2. Centrifugalise the milk sample for thirty minutes at a speed of 2500 revolutions per minute.

3. Remove the motive power and allow the machine to slow down gradually.

4. Remove the tubes of milk from the centrifuge. Each tube will now show (Fig. 214):

(a) A superficial layer of cream (varying in thickness with different samples) condensed into a semi-solid mass, which can be shown to contain some organisms and a few leucocytes.

(b) A central layer of separated milk, thin, watery, and opalescent, and containing extremely few bacteria.

(c) A sediment or deposit consisting of the great majority of the contained bacteria and leucocytes, together with adventitious matter, such as dirt, hair, epithelial cells, fæcal débris, etc.

5. Withdraw the rubber stopper and remove a central plug of cream from each tube by means of a sterile cork borer; place these masses of cream in two sterile capsules. Label C^1 and C^2.

6. Remove all but the last one or two c.c. of separated milk from each tube, by means of sterile pipettes.

7. Mix the deposits thoroughly with the residual milk, pipette the mixture from each pair of tubes into one sterile 10 c.c. tube (graduated) by means of sterile teat pipettes, then fill to the 10 c.c. mark with sterile normal saline solution and mix together. Label D^1 and D^2.

8. Place the two tubes of mixed deposit in the centrifuge, adjust by the addition or subtraction of saline solution so that they counterpoise exactly, and centrifugalise for ten minutes.

Note.—Each tube now contains the deposit from 100 c.c. of the milk sample and the amount can be read off in hundredths of a centimetre. The multiplication of this figure by 100 will give the amount of "Apparent Filth," in "parts per million"—the usual method of recording this quality of milk.

9. Pipette off all the supernatant fluid and invert the tube to drain on to a pad of sterilised cotton-wool, contained in a beaker. (This wool is subsequently cremated.)

10. Examine both cream (C^1) and deposit (D^1) microscopically—

(a) In hanging-drop preparations.

(b) In film preparations stained carbolic methylene-blue, by Gram's method, by Neisser's method, and by Ziehl-Neelsen's method.

Note the presence or absence of altered and unaltered vegetable fibres; pus cells, blood discs; cocci in groups or chains, diphtheroid bacilli, Gram negative bacilli or cocci, spores and acid fast bacteria.

11. Adapt the final stages of the investigation to the special requirements of each individual sample, thus:

1. Members of the Coli-typhoid Group.—

1. Emulsify the deposit from the second centrifugal tube (D^2) with 10 c.c. sterile bouillon and inoculate three tubes of bile salt broth as follows:

To Tube No. 1 add 2.5 c.c. milk deposit emulsion (=25 c.c. original milk.) To Tube No. 2 add 1.0 c.c. milk deposit emulsion (=10 c.c. original milk.) To Tube No. 3 add 0.5 c.c. milk deposit emulsion (= 5 c.c. original milk.)

2. Inoculate tube of bile salt broth No. 4 with 1 c.c. of the original milk.

3. Inoculate further tubes of bile salt broth with previously prepared dilutions (see page 445) as follows:

To tube No. 5 add 1.0 c.c. from capsule I. To tube No. 6 add 0.1 c.c. from capsule I. To tube No. 7 add 1.0 c.c. from capsule II. To tube No. 8 add 0.1 c.c. from capsule II. To tube No. 9 add 1.0 c.c. from capsule III. To tube No. 10 add 0.1 c.c. from capsule III. To tube No. 11 add 1.0 c.c. from capsule IV. To tube No. 12 add 0.1 c.c. from capsule IV.

and incubate anaerobically (in Buchner's tubes) at 42° C. for a maximum period of forty-eight hours.

4. If growth occurs complete the investigation as detailed under the corresponding section of water examination (see pages 428 to 431).

Note.—The B. coli communis, derived from the alvine discharges of the cow, is almost universally present in large or small numbers, in retail milk. Its detection, therefore, unless in enormous numbers, (when it indicates want of cleanliness), is of little value.

2. Vibrio Choleræ.—Inoculate tubes of peptone water by using the same amounts as in the search for members of the Coli-typhoid groups (*vide ante* 1-3); incubate aerobically at 37° C. and complete the examination as detailed under the corresponding section of water examination (see page 439).

3. B. Enteritidis Sporogenes.—Inoculate tubes of litmus milk with similar amounts to those used in the previous searches, omitting tube No. 1 (*vide ante* 1-3) place in the differential steriliser at 80° C. for ten minutes and then incubate anaerobically at 37° C. for a maximum period of forty-eight hours. Complete the investigation as detailed under the corresponding section of water examination (see page 438).

4. B. Diphtheriæ.—

(A) 1. Plant three sets of serial cultivations, twelve tubes in each set, from (*a*) cream C^2, (*b*) deposit D^1 upon oblique inspissated blood-serum, and incubate at 37° C.

2. Pick off any suspicious colonies which may have made their appearance twelve hours after incubation, examine microscopically and subcultivate upon blood-serum and place in the incubator; return the original tubes to the incubator.

3. Repeat this after eighteen hours' incubation.

4. From the resulting growths make cover-slip preparations and stain carbolic methylene-blue, Neisser's method, Gram's method. Subcultivate such as appear to be composed of diphtheria bacilli in glucose peptone solution. Note those in which acid production takes place.

5. Inoculate guinea-pigs subcutaneously with one or two cubic centimetres forty-eight-hour-old glucose bouillon cultivation derived from the first subcultivation of each glucose fermenter, and observe the result.

6. If death, apparently from diphtheritic toxæmia, ensues, inoculate two more guinea pigs with a similar quantity of the lethal culture. Reserve one animal as a control and into the other inject 1000 units of antidiphtheritic serum. If the control dies and the treated animal survives, the proof of the identity of the organism isolated with the Klebs-Lœffler bacillus becomes absolute.

7. Inoculate guinea-pigs subcutaneously with filtered glucose bouillon cultivations (toxins?) and observe the result.

(B) 1. Emulsify the remainder of the deposit with 5 c.c. sterile bouillon and inoculate two guinea-pigs, thus: guinea-pig *a*, subcutaneously with 1 c.c. emulsion; guinea-pig *b*, subcutaneously with 2 c.c. emulsion; and observe the result.

2. If either or both of the inoculated animals succumb, make complete post-mortem examination and endeavour to isolate the pathogenic organisms from the local lesion. Confirm their identity as in A5 and 6 (*vide supra*).

5. Bacillus Tuberculosis.—

(A) 1. Inoculate each of three guinea-pigs (previously tested with tuberculin, to prove their freedom from spontaneous tuberculosis) subcutaneously at the inner aspect of the bend of the left knee, with 1 c.c. of the deposit emulsion remaining in one or other tube (D^1 or D^2).

2. Introduce a small quantity of the cream into a subcutaneous pocket prepared at the inner aspect of the bend of the right knee of each of these three animals. Place a sealed dressing on the wound.

3. Observe carefully, and weigh accurately each day.

4. Kill one guinea-pig at the end of the second week and make a complete post-mortem examination.

5. If the result of the examination is negative or inconclusive, kill a second guinea-pig at the end of the third week and examine carefully.

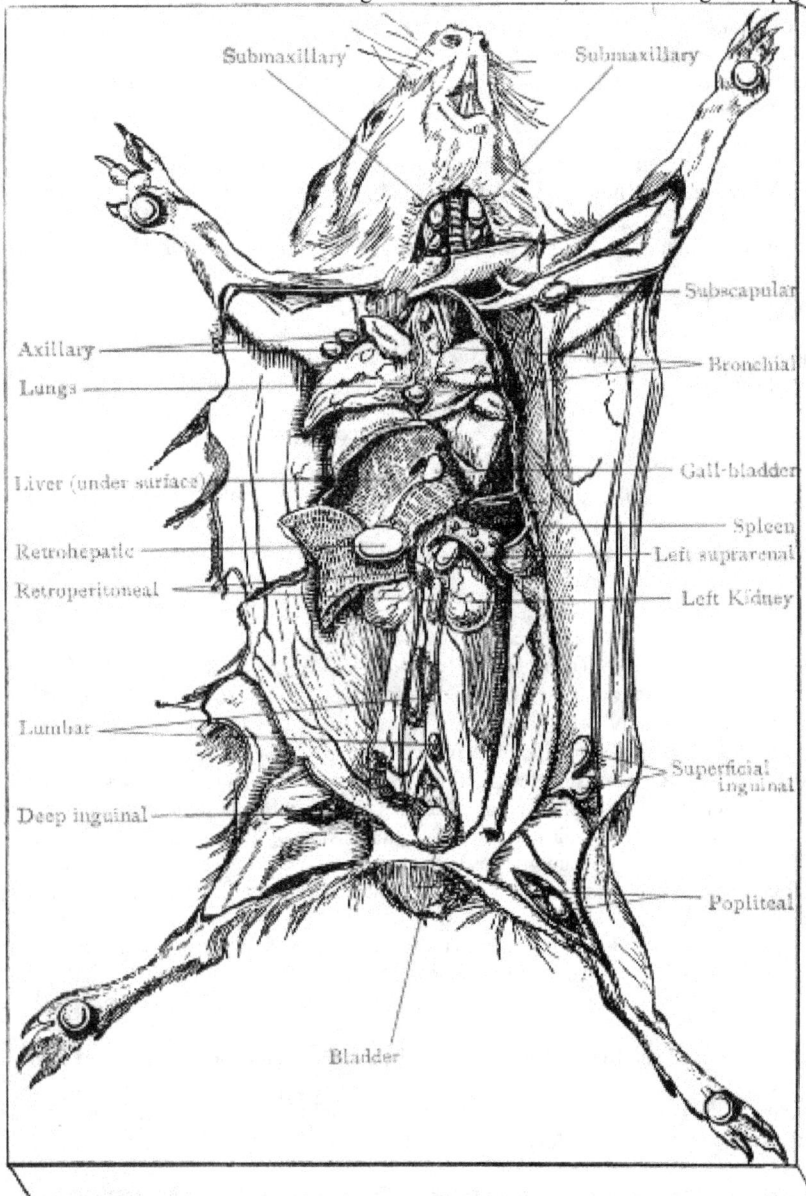

Fig. 215.—Cadaver of guinea-pig experimentally infected with B. tuberculosis.

6. If still negative or inconclusive, kill the third guinea-pig at the end of the *sixth* week. Make a careful post-mortem examination. Examine material from any caseous glands microscopically and inoculate freely on to Dorset's egg medium.

Note.—Every post-mortem examination of animals infected with tuberculous material should include the naked eye and microscopical examination of the popliteal, superficial and deep inguinal, iliac, lumbar and axillary glands on each side of the body, also the retrohepatic, bronchial and sternal glands, the spleen, liver and lungs (Fig. 215).

(B) 1. Intimately mix all the available cream and deposit from the milk sample, and transfer to a sterile Erlenmeyer flask.

2. Treat the mixture by the antiformin method (*vide* Appendix, page 502).

3. Inoculate each of two guinea-pigs, intraperitoneally, with half of the emulsion thus obtained.

4. Kill one of the guinea-pigs at the end of the first week and examine carefully.

5. Kill the second guinea-pig at the end of the second week and examine carefully.

6. Utilise the remainder of the deposit for microscopical examination and cultivations upon Dorset's egg medium.

Note.—No value whatever attaches to the result of a microscopical examination for the presence of the B. tuberculosis unless confirmed by the result of inoculation experiments.

6. Streptococcus Pyogenes Longus.—

(A) 1. Spread serial surface plates upon nutrose agar. Also plant serial cultivations upon sloped nutrient agar (six tubes in series).

2. If the resulting growth shows colonies which resemble those of the streptococcus, make subcultivations upon agar and in bouillon, in the first instance, and study carefully.

(B) 1. Plant a large loopful of the deposit D^2 into each of three tubes of glucose formate bouillon, and incubate anaerobically (in Buchner's tubes) for twenty-four hours at 37° C.

2. If the resulting growth resembles that of the streptococcus, make subcultivations upon nutrient agar.

3. Prepare subcultivations of any suspicious colonies that appear, upon all the ordinary media, and study carefully.

If the streptococcus is successfully isolated, inoculate serum bouillon cultivations into the mouse, guinea-pig, and rabbit, to determine its pathogenicity and virulence.

7. Staphylococcus Pyogenes Aureus.—

1. Examine carefully the growth upon the serial blood serum cultivations prepared to isolate B. diphtheriæ and the serial agar cultivations to isolate streptococci after forty-eight hours' incubation.

2. Pick off any suspicious orange coloured colonies, plant on sloped agar, and incubate at 20° C. Observe pigment formation.

3. Prepare subcultivations from any suspicious growths upon all the ordinary media, study carefully and investigate their pathogenicity.

8. Micrococcus Melitensis.—The milk from an animal infected with M. melitensis usually contains the organisms in large numbers and but few other bacteria.

1. Spread several sets of surface plates upon nutrose agar, each from one loopful of the deposit in tube D^1 or D^2.

2. Spread several sets of surface plates upon nutrose agar, each from one drop of the original milk sample.

3. Incubate aerobically at 37° C. and examine daily up to the end of ten days.

4. Pick off suspicious colonies, examine them microscopically and subcultivate upon nutrose agar in tubes; upon glucose agar and in litmus milk.

5. Test the subsequent growth against the serum of an experimental animal inoculated against M. melitensis to determine its agglutinability.

6. If apparently M. melitensis, inoculate growth from a nutrose agar culture after three days incubation intracranially into the guinea-pig.

ICE CREAM.

Collection of the Sample.—

1. Remove the sample from the drum in the ladle or spoon with which the vendor retails the ice cream, and place it at once in a sterile copper capsule, similar to that employed for earth samples (*vide* page 471).

2. Pack for transmission in the ice-box.

3. On arrival at the laboratory place the copper capsules containing the ice cream in the incubator at 20° C. for fifteen minutes—that is, until at least some of the ice cream has become liquid.

Qualitative and Quantitative Examination.—Treat the fluid ice cream as milk and conduct the examination in precisely the same manner as described for milk (*vide* page 443).

EXAMINATION OF CREAM AND BUTTER.

Collection of the Sample.—Collect, store, and transmit samples to the laboratory, precisely as is done in the case of ice cream.

Quantitative.—

Apparatus Required:

Sterile test-tube. Sterilised spatula. Water-bath regulated at 42° C. Case of sterile plates. Case of sterile graduated pipettes, 1 c.c. (in hundredths). Tubes of gelatine-agar (+10 reaction). Plate-levelling stand, with its water chamber filled with water at 42° C.

Method.—

1. Transfer a few grammes of the sample to a sterile test-tube by means of the sterilised spatula.

2. Place the tube in the water-bath at 42° C. until the contents are liquid.

3. Liquefy eight tubes of gelatine-agar and place them in the water-bath at 42° C, and cool down to that temperature.

4. Inoculate the gelatine-agar tubes with the following quantities of the sample by the help of a sterile pipette graduated to hundredths of a cubic centimetre—viz.,

To tube No. 1 add 1 c.c. liquefied butter. 2 add 0.5 c.c. liquefied butter. 3 add 0.3 c.c. liquefied butter. 4 add 0.2 c.c. liquefied butter. 5 add 0.1 c.c. liquefied butter. 6 add 0.05 c.c. liquefied butter. 7 add 0.03 c.c. liquefied butter. 8 add 0.02 c.c. liquefied butter. 9 add 0.01 c.c. liquefied butter.

5. Pour a plate cultivation from each of the gelatine-agar tubes and incubate at 28° C.

6. "Count" the plates after three days' incubation, and from the figures thus obtained estimate the number of organisms present per cubic centimetre of the sample.

Qualitative.—

Apparatus Required:

Sterile beaker, its mouth plugged with sterile cotton-wool.

Counterpoise for beaker.

Scales and weights.

Sterilised spatula.

Water-bath regulated at 42° C.

Separatory funnel, 250 c.c. capacity, its delivery tube protected against contamination by passing it through a cotton-wool plug into the interior of a small Erlenmeyer flask which serves to support the funnel. This piece of apparatus is sterilised *en masse* in the hot-air oven.

Large centrifugal machine.

Sterile tubes (for the centrifuge) closed with solid rubber stoppers.

Case of sterile pipettes, 10 c.c.

Case of sterile graduated pipettes, 1 c.c. (in tenths of a cubic centimetre).

Method.—

1. Weigh out 100 grammes of the sample in a sterile beaker.

2. Plug the mouth of the beaker with sterile cotton-wool and immerse the beaker in a water-bath at 42° C. until the contents are completely liquefied.

3. Fill the liquefied butter into the sterile separatory funnel.

4. Transfer the funnel to the incubator at 37° C. and allow it to remain there for four days.

At the end of this time the contents of the funnel will have separated into two distinct strata.

(a) A superficial oily layer, practically free from bacteria.

(b) A deep watery layer, turbid and cloudy from the growth of bacteria.

5. Draw off the subnatant turbid layer into sterile centrifugal tubes, previously warned to about 42° C., and centrifugalise at once.

6. Pipette off the supernatant fluid and fill the tubes with sterile 1 per cent. sodium carbonate solution previously warmed slightly; stopper the tubes and shake vigourously for a few minutes.

7. Centrifugalise again.

8. Pipette off the supernatant fluid; filling the tubes with warm sterile bouillon, shake well, and again centrifugalise, to wash the deposit.

9. Pipette off the supernatant fluid.

10. Prepare cover-slip preparations, fix and clear as for milk preparations, stain carbolic methylene-blue, Gram's method, Ziehl-Neelsen's method, and examine microscopically with a 1/12 inch oil-immersion lens.

11. Proceed with the examination of the deposit as in the case of milk deposit (see pages 450 *et seq.*).

EXAMINATION OF UNSOUND MEATS.

(Including Tinned or Potted Meats, Fish, Etc.)

The bacterioscopic examination of unsound food is chiefly directed to the detection of those members of the Coli-typhoid group—B. enteritidis of Gaertner and its allies—which are usually associated with epidemic outbreaks of food poisoning, and such anaerobic bacteria as initiate putrefactive changes in the food which result in the formation of poisonous ptomaines, consequently the quantitative examination pure and simple is frequently omitted.

A. Cultural Examination.

Quantitative.—

Apparatus Required:

Sterilised tin opener, (if necessary.)

Erlenmeyer flask (500 c.c. capacity) containing 200 c.c. sterile bouillon and fitted with solid rubber stopper.

Counterpoise.

Scissors and forceps.

Scales and weights.

Water steriliser.

Hypodermic syringe.

Syringe with intragastric tube.

Rat forceps.

Case of sterile capsules.

Filtering apparatus as for water analysis.

Case of sterile plates.

Case of sterile graduated pipettes, 10 c.c. (in tenths of a cubic centimetre).

Case of sterile graduated pipettes, 1 c.c. (in tenths of a cubic centimetre).

Plate-levelling stand.

Tubes of nutrient gelatine.

Tubes of nutrient agar.

Water-bath regulated at 42° C.

Bulloch's apparatus.

Method.—

1. Place the flask containing 200 c.c. sterile broth on one pan of the scales and counterpoise accurately.

2. Mince a portion of the sample by the aid of sterile scissors and forceps, and add the minced sample to the bouillon in the flask to the extent of 20 grammes.

3. Make an extract by standing the flask in the incubator running at 42° C. (or in a water-bath regulated to that temperature) for half an hour, shaking its contents from time to time. Better results are obtained if an electrical shaker is fitted inside the incubator and the flask kept in motion throughout the entire thirty minutes.

Now every centimetre contains the bacteria washed out from 0.1 gramme of the original food.

4. Inoculate tubes of liquefied gelatine as follows:

To tube No. 1 add 1.0 c.c. of the extract. 2 add 0.5 c.c. of the extract. 3 add 0.3 c.c. of the extract. 4 add 0.2 c.c. of the extract. 5 add 0.1 c.c. of the extract.

Pour plates from these tubes and incubate at 20° C.

5. Prepare a precisely similar set of agar plates and incubate at 37° C.

6. Pipette 5 c.c. of the extract into a sterile tube, heat in the differential steriliser at 80° C. for ten minutes.

7. From the heated extract prepare duplicate sets of agar and gelatine plates and incubate anaerobically in Bulloch's apparatus at 37° C. and 20° C. respectively.

8. After three days' incubation examine the agar plates both aerobic and anaerobic and enumerate the colonies developed from spores (7), and from vegetative forms and spores (5), and calculate and record the numbers of each group per gramme of the original food.

9. After seven days' incubation (or earlier if compelled by the growth of liquefying colonies) enumerate the gelatine plates in the same way.

10. Subcultivate from the colonies that make their appearance and identify the various organisms.

11. Continue the investigations with reference to the detection of pathogenic organisms as described under water (page 429 *et seq.*).

Qualitative.—

I. *Cultural.*

The micro-organisms sought for during the examination of unsound foods comprise the following:

Members of the Coli-typhoid groups (chiefly those of the Gaertner class).

B. anthracis.

Streptococci

Anaerobic Bacteria:

B. enteritidis sporogenes. B. botulinus. B. cadaveris.

The methods by which these organisms if present may be identified and isolated have already been described under the corresponding section of water examination with the exception of those applicable to B. botulinus, and B. cadaveris. These can only be isolated satisfactorily from the bodies of experimentally inoculated animals.

II *Experimental.*

Tissue.—

1. Feed rats and mice on portions of the sample and observe the result.
2. If any of the animals die, make complete post-mortem examinations and endeavour to isolate the pathogenic organisms.

Extract.—

1. Introduce various quantities of the bouillon extract into the stomachs of several rats, mice and guinea-pigs repeatedly over a period of two or three days by the intragastric method of inoculation (see page 367) and observe the result. Guinea-pigs and mice are very susceptible to infection by B. botulinus by this method; rabbits less so.
2. Inoculate rats, mice, and guinea-pigs subcutaneously into deep pockets, and intraperitoneally with various quantities of the bouillon extract, and observe the result.
3. Filter some of the extract through a Chamberland candle and incubate the filtrate to determine the presence of soluble toxins.
4. If any of the animals succumb to either of these methods of inoculation, make careful post-mortem examinations and endeavour to isolate the pathogenic organisms.

THE EXAMINATION OF OYSTERS AND OTHER SHELLFISH.

On opening the shell of an oyster a certain amount of fluid termed "liquor" is found to be present. This varies in amount from a drop to many cubic centimetres (0.1 c.c. to 10 c.c.)—in the latter case the bulk of the fluid is probably the last quantum of water ingested by the bivalve before closing its shell. In order to obtain a working average of the bacteriological flora of a sample, ten oysters should be taken and the body, gastric juice and liquor should be thoroughly mixed before examination. The examination, as in dealing with other food stuffs, is directed to the search for members of the Coli-typhoid group, sewage streptococci and perhaps also B. enteritidis sporogenes.

Apparatus Required:

Two hard nail brushes.

Liquid soap.

Sterile water in aspirator jar with delivery nozzle controlled by a spring clip.

Sterile oyster knives.

Sterile glass dish, with cover, sufficiently large to accommodate ten oysters.

Sterile forceps.

Sterile scissors.

Sterile towels or large gauze pads.

Sterile graduated cylinders 1000 c.c. capacity, with either the lid or the bottom of a sterile Petri dish inverted over the open mouth as a cover.

Glass rods.

Corrosive sublimate solution, 1 per mille.

Bile salt broth tubes.

Litmus milk tubes.

Surface plates of nutrose agar.

Case of sterile pipettes, 1 c.c. (in tenths of a c.c.)

Case of sterile pipettes, 10 c.c. (in tenths of a c.c.)

Case of sterile glass capsules.

Erlenmeyer flasks, 250 c.c. capacity.

Double strength bile salt broth.

Method.—

1. Thoroughly clean the outside of the oyster shells by scrubbing each in turn with liquid soap and nail brush under a tap of running water. Then, holding an oyster shell in a pair of sterile forceps wash every part of the outside of the shell with a stream of sterile water running from an aspirator jar; deposit the oyster inside the sterile glass dish. Repeat the process with each of the remaining oysters.

2. Before proceeding further, cleanse the hands thoroughly with clean nail brush, soap and water, then plunge them in lysol 2 per cent. solution, and finally in sterile water.

3. Spread a sterile towel on the bench.

4. Remove one of the oysters from the sterile glass dish and place it, resting on its convex shell, on the towel. Turn a corner of the sterile towel over the upper flat shell to give a firmer grip to the left hand, which holds the shell in position.

5. With the sterile oyster knife (in the right hand) open the shell and separate the body of the oyster from the inner surface of the upper flat shell. Bend back and separate the flat shell, leaving the body of the oyster in and attached to the concave shell. Avoid spilling any of the liquor.

(Some dexterity in opening oysters should be acquired before undertaking these experiments).

6. Cut up the body of the oyster with sterile scissors into small pieces and allow the liquor freed from the body during the process to mix with the liquor previously in the shell.

7. Transfer the comminuted oyster and the liquor to the cylinder.

8. Treat each of the remaining oysters in similar fashion.

9. Mix the contents of the cylinder thoroughly by stirring with a sterile glass rod. The total volume will amount to about 100 c.c.

10. Use 0.1 c.c. of the mixed liquor to inseminate each of a series of three nutrose surface plates.

11. Inoculate 0.1 c.c. of the mixed liquor into each of three tubes of litmus milk.

12. Add sterile distilled water to the contents of the cylinder up to 1000 c.c. and stir thoroughly with a sterile glass rod and allow to settle. The bacterial content of each oyster may be regarded, for all practical purposes, as comprised in 100 c.c. of fluid.

13. Arrange four glass capsules in a row and number I, II, III, IV. Pipette 9 c.c. sterile distilled water into each.

14. To capsule No. I add 1 c.c. of the diluted liquor, etc. from the cylinder, and mix thoroughly. To capsule II add 1 c.c. of dilution in capsule I and mix thoroughly. Carry over 1 c.c. of fluid from capsule II to capsule III, afterwards adding 1 c.c. of fluid from capsule III to capsule IV.

15. Label tubes of bile salt broth and inoculate with the following amounts of diluted oysters:

No. 6 with 10 c.c. cylinder fluid = 0.1 oyster. No. 5 with 1 c.c. cylinder fluid = 0.01 oyster. No. 4 with 1 c.c. capsule I fluid = 0.001 oyster. No. 3 with 1 c.c. capsule II fluid = 0.0001 oyster. No. 2 with 1 c.c. capsule III fluid = 0.00001 oyster. No. 1 with 1 c.c. capsule IV fluid = 0.000001 oyster.

16. Transfer 100 c.c. cylinder fluid (= 1 oyster) to an Erlenmeyer flask and add 50 c.c. double strength bile salt broth, and label 7.

17. Duplicate all the above indicated cultures.

18. Put up the tube cultures in Buchner's tubes and incubate anaerobically at 42° C.

If growth occurs in tube 1 the organism finally isolated, *e. g.*, B. coli, must have been present to the extent of one million per oyster.

19. Complete the examination for members of the Coli-typhoid group and sewage streptococci, as directed under Water Examination, page 429 (steps 11-21).

20. Inoculate a series of 6 tubes of litmus milk with quantities of the material similar to those indicated in step 15; heat to 80° C. for ten minutes, and incubate under anaerobic conditions at 37° C. Examine for the presence of B. enteritidis sporogenes as directed under Water Examination, page 438 (steps 7-10).

EXAMINATION OF SEWAGE AND SEWAGE EFFLUENTS.

Quantitative.—

Collection of the Sample.—As only small quantities of material are needed, the samples should be collected in a manner similar to that described under water for quantitative examination and transmitted in the ice apparatus used in packing those samples.

Apparatus Required.—As for water (*vide* page 420).

Method.—

1. Arrange four sterile capsules in a row and number them I, II, III, IV.

2. Pipette 9 c.c. sterile bouillon into capsule No. I.

3. Pipette 9.9 c.c. sterile bouillon into capsules II, III, and IV.

4. Add 1 c.c. of the sewage to capsule No. I by means of a sterile pipette, and mix thoroughly.

5. Take a fresh sterile pipette and transfer 0.1 c.c. of the mixture from No. I to No. II and mix thoroughly.

6. In like manner transfer 0.1 c.c. from No. II to No. III, and then 0.1 c.c. from No. III to No. IV.

Now 1 c.c. of dilution No. I contains 0.1 c.c. of the original sewage.
1 c.c. of dilution No. II contains 0.001 c.c. of the original sewage.
1 c.c. of dilution No. III contains 0.00001 c.c. of the original sewage.
1 c.c. of dilution No. IV contains 0.0000001 c.c. of the original sewage.

7. Pour a set of gelatine plates from the contents of each capsule, three plates in a set, and containing respectively 0.2, 0.3, and 0.5 c.c. of the dilution. Label carefully; incubate at 20° C. for three, four, or five days.

8. Enumerate the organisms present in those sets of plates which have not liquefied, probably those from dilution III or IV, and calculate therefrom the number present per cubic centimetre of the original sample of sewage.

Qualitative.—The qualitative examination of sewage is concerned with the identification and enumeration of the same bacteria dealt with under the corresponding section of water examination; it is consequently conducted on precisely similar lines to those already indicated (*vide* pages 426 to 441).

EXAMINATION OF AIR.

Quantitative.—

Apparatus Required:

Aspirator bottle, 10 litres capacity, fitted with a delivery tube, and having its mouth closed by a perforated rubber stopper, through which passes a short length of glass tubing.

Erlenmeyer flask, 250 c.c. capacity (having a wide mouth properly plugged with wool), containing 50 c.c. sterile water.

Rubber stopper to fit the mouth of the flask, perforated with two holes, and fitted as follows:

Take a 9 cm. length of glass tubing and bend up 3 cm. at one end at right angles to the main length of tubing. Pass the long arm of the angle through one of the perforations in the stopper; plug the open end of the short arm with cotton-wool.

Take a glass funnel 5 or 6 cm. in diameter with a stem 12 cm. in length and bend the stem close up to the apex of the funnel, in a gentle curve through a quarter of a circle; pass the long stem through the other perforation in the rubber stopper.

A battery jar or a small water-bath to hold the Erlenmeyer flask when packed round with ice.

Supply of broken ice.

Rubber tubing.

Screw clamps and spring clips, for tubing.

Water steriliser.

Retort stand and clamps.

Apparatus for plating (as for enumeration of water organisms, *vide* page 420).

Method.—

1. Fill 10 litres of water into the aspirating bottle and attach a piece of rubber tubing with a screw clamp to the delivery tube. Open the taps fully and regulate the screw clamp, by actual experiment, so that the tube delivers 1 c.c. of water every second. The screw clamp is not touched again during the experiment.

At this rate the aspirator bottle will empty itself in just under three hours. Shut off the tap and make up the contents of the aspirator bottle to 10 litres again.

2. Sterilise the fitted rubber cork, with its funnel and tubing, by boiling in the water steriliser for ten minutes.

3. Remove the cotton-wool plug from the flask, and replace it by the rubber stopper with its fittings. Make sure that the end of the stem of the funnel is immersed in the bouillon.

4. Place the flask in a glass or metal vessel and pack it round with pounded ice. Arrange the flask with its ice casing just above the neck of the aspirator bottle.

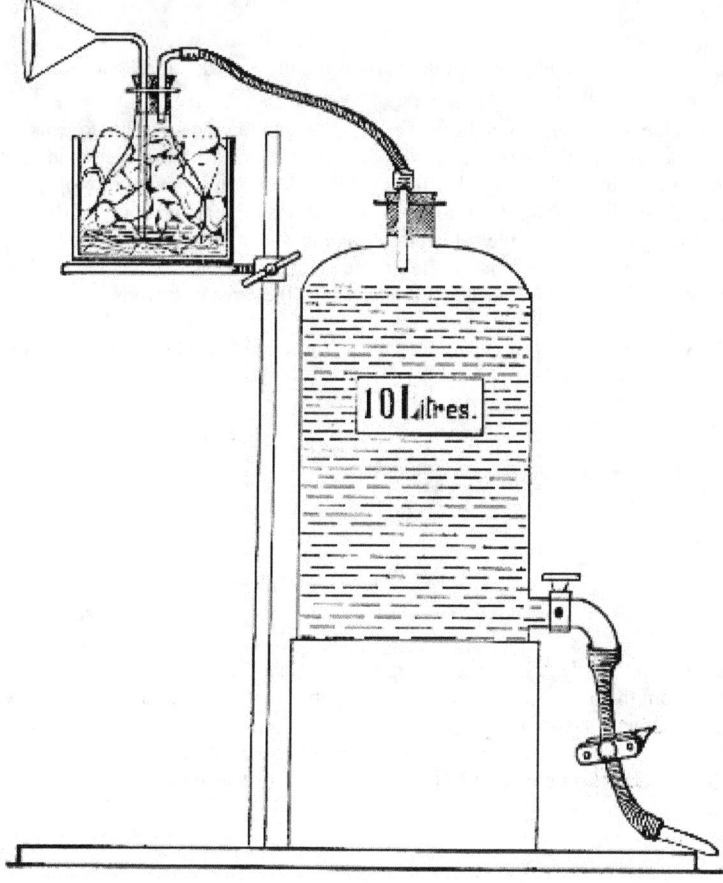

Fig. 216.—Arrangement of apparatus for air analysis.

5. Connect up the free end of the glass tube from the flask—after removing the cotton-wool plug—with the air-entry tube in the mouth of the aspirating bottle (Fig. 216).

6. Open the tap fully, and allow the water to run.

Replenish the ice from time to time if necessary.

(In emptying itself the aspirator bottle will aspirate 10 litres of air slowly through the water in the Erlenmeyer flask.)

7. When the aspiration is completed, disconnect the flask and remove it from its ice packing.

8. Liquefy three tubes of nutrient gelatine and add to them 0.5 c.c., 0.3 c.c., and 0.2 c.c., respectively, of the water from the flask, by means of a sterile graduated pipette, as in the quantitative examination of water. Pour plates.

9. Pour a second similar set of gelatine plates.

10. Incubate both sets of plates at 20° C.

11. Enumerate the colonies present in the two sets of gelatine plates after three, four, or five days and average the results from the numbers so obtained; estimate the number of micro-organisms present in 1 c.c., and then in the 50 c.c. of broth in the flask.

12. The result of air examination is usually expressed as the number of bacteria present per cubic metre (*i. e.*, kilolitre) of air; and as the number of organisms present in the 50 c.c. water only represent those contained in 10 litres of air, the resulting figure must be multiplied by 100.

Qualitative.—

1. Proceed exactly as in the quantitative examination of air (*vide supra*), steps 1 to 10.
2. Pour plates of wort agar with similar quantities of the air-infected water, and incubate at 37° C.
3. Pour plates of nutrient agar with similar quantities of the water and incubate at 37° C.
4. Pour similar plates of wort gelatine and incubate at 20° C.
5. Pick off the individual colonies that appear in the several plates, subcultivate them on the various media, and identify them.

EXAMINATION OF SOIL.

The bacteriological examination of soil yields information of value to the sanitarian during the progress of the process of homogenisation of "made soil" (*e. g.*, a dumping area for the refuse of town) and determines the period at which such an area may with propriety and safety be utilised for building purposes; or to the agriculturalist in informing him of the suitability of any given area for the growth of crops.

The surface of the ground, exposed as it is to the bactericidal influence of sunlight and to rapid alternations of heat and cold, rain and wind, contains but few micro-organisms. Again, owing to the density of the molecules of deep soil and lack of aeration on the one hand, and the filtering action of the upper layers of soil and bacterial antagonism on the other, bacterial life practically ceases at a depth of about 2 metres. The intermediate stratum of soil, situated from 25 to 50 cm. below the surface, invariably yields the most numerous and the most varied bacterial flora.

Collection of Sample.—A small copper capsule 6 cm. high by 6 cm. diameter, with "pull-off" cap secured by a bayonet catch, previously sterilised in the hot-air oven, is the most convenient receptacle for samples of soil.

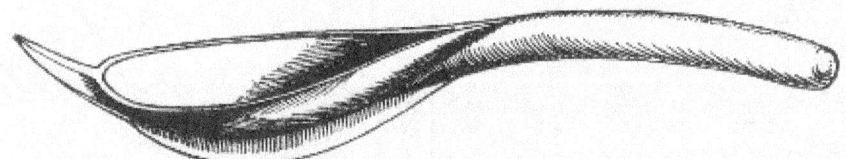

Fig. 217.—Soil scoop.

The instrument used for the actual removal of the soil from its natural position will vary according to whether we require surface samples or soil from varying depths.

(*a*) For **surface** samples, use an iron scoop, shaped like a shoe horn, but provided with a sharp spine (Fig. 217). This is wrapped in asbestos cloth and sterilised in the hot-air oven. When removed from the oven, wrap a piece of oiled paper, silk, or gutta-percha tissue over the asbestos cloth, and secure it with string, as a further protection against contamination.

On reaching the spot whence the samples are to be taken, the coverings of the scoop are removed, and the asbestos cloth employed to brush away loose stones and débris from the selected area. The surface soil is then broken up with the point of the scoop, scraped up and collected in the body of the scoop, and transferred to the sterile capsule for transmission.

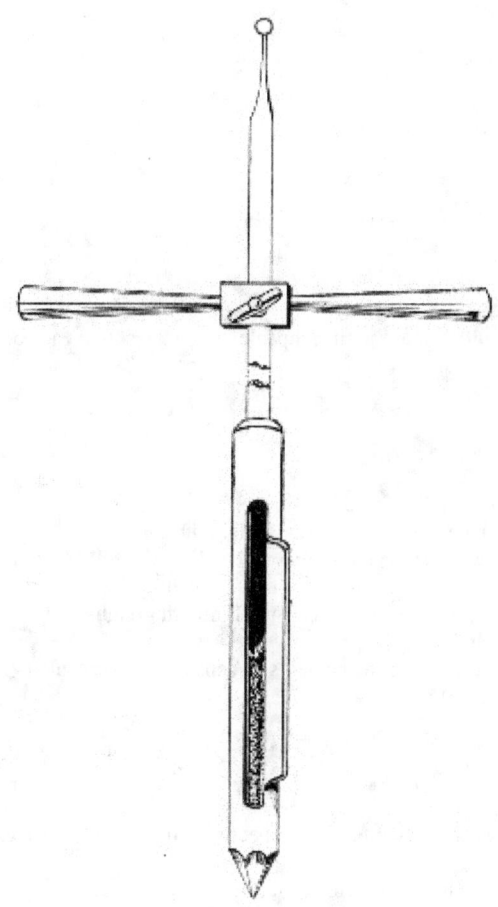

Fig. 218.—Fraenkel's borer.

(b) For **deep** samples collected at various distances from the surface, an experimental trench may be cut to the required depth and samples collected at the required points on the face of the section. It is, however, preferable to utilise some form of borer, such as that designed by Fraenkel (Fig. 218).

Fraenkel's Earth Borer.—This instrument consists of a stout hard-steel rod, 150 cm. long, marked in centimetres from the drill-pointed extremity. It is provided with a cross handle (adjustable at any point along the length of the rod by means of a screw nut). The terminal centimeters are thicker than the remainder of the rod, and on one side a vertical cavity about 0.5 cm. deep is cut. This is covered by a flanged sleeve so long as the borer is driven into the soil clockwise, and is opened for the reception of the sample of soil, when the required depth is reached, by reversing the screwing motion, and again closed before withdrawal of the borer from the earth by resuming the original direction of twist. It can be sterilised in a manner similar to that adopted for the scoop, or by repeatedly filling the cavity with ether and burning it off.

Quantitative.—Four distinct investigations are included in the complete quantitative bacteriological examination of the soil:
1. The enumeration of the aerobic organisms.
2. The enumeration of the spores of aerobes.
3. The enumeration of the anaerobic organisms (including the facultative anaerobes).
4. The enumeration of the spores of anaerobes.

Further, by a combination of the results of the first and second, and of the third and fourth of these, the ratio of spores to vegetative forms is obtained.

Apparatus Required:
Case of sterile capsules (25 c.c. capacity).
Case of sterile graduated pipettes, 10 c.c. (in tenths of a cubic centimetre).
Case of sterile graduated pipettes, 1 c.c. (in tenths of a cubic centimetre).
Flask containing 250 c.c. sterile bouillon.
Tall cylinder containing 2 per cent. lysol solution.
Plate-levelling stand.
12 sterile plates.
Tubes of nutrient gelatine.
Tubes of wort gelatine.
Tubes of nutrient agar.
Tubes of glucose formate gelatine.
Tubes of glucose formate agar.
Water-bath regulated at 42° C.

Bunsen burner.
Grease pencil.
Sterile mortar and pestle (agate).
Sterile wide-mouthed Erlenmeyer flask (500 c.c. capacity).
Sterile metal funnel with short wide bore delivery tube to just fit mouth of flask.
Solid rubber stopper to fit the flask (sterilised by boiling).
Pair of scales.
Counterpoise (Fig. 107).
Sterile metal (nickel) spoon or spatula.
Fractional steriliser (Fig. 140).
Method.—
1. Arrange four sterile capsules numbered I, II, III, and IV; pipette 9 c.c. sterile bouillon into the first capsule, and 9.9 c.c. into each of the remaining three.
2. Pipette 100 c.c. sterile bouillon into the Erlenmeyer flask.
3. Remove the cotton-wool plug from the flask and replace it by the sterile funnel.
4. Place flask and funnel on one pan of the scales, and counterpoise accurately.
5. Empty the sample of soil into the mortar and triturate thoroughly.
6. By means of the sterile spatula add 10 grammes of the earth sample to the bouillon in the flask.
The final results will be more reliable if steps 2, 3, 4, and 5 are performed under a hood—to protect from falling dust, etc.
7. Remove the funnel from the mouth of the flask; replace it by the rubber stopper and shake vigorously; then allow the solid particles to settle for about thirty minutes. One cubic centimetre of the turbid broth contains the washings from 0.1 gramme of soil.
8. Pipette off 1 c.c. of the supernatant bouillon, termed the "soil water," and add it to the contents of capsule I; mix thoroughly.
9. Remove 0.1 c.c. of the infected bouillon from capsule I and add it to capsule II, and mix.
10. In like manner add 0.1 c.c. of the contents of capsule II to capsule III, and then 0.1 c.c. of the contents of capsule III to capsule IV.
Then 1 c.c. fluid from capsule I contains soil water from .01 gm. earth.
Then 1 c.c. fluid from capsule II contains soil water from .0001 gm. earth.
Then 1 c.c. fluid from capsule III contains soil water from .000001 gm. earth.
Then 1 c.c. fluid from capsule IV contains soil water from .00000001 gm. earth.

(A) *Aerobes (Vegetative Forms and Spores).*—
11. Pour a set of gelatine plates from the contents of each capsule—two plates in a set, and containing respectively 0.1 c.c. and 0.4 c.c. of the diluted soil water. Label and incubate.
12. Pour similar sets of wort gelatine plates from the contents of capsules II and III, label, and incubate at 20° C.
13. Pour similar sets of agar plates from the contents of capsules II and III; label and incubate at 37° C.
14. Weigh out a second sample of soil—10 grammes—dry over a water-bath until of constant weight and calculate the ratio
wet soil weight ———— dry soil weight
15. "Count" the plates after incubation for three, four, or five days, and correcting the figures thus obtained by means of the "wet" to "dry" soil ratio estimate—
(a) The number of aerobic micro-organisms present per gramme of the soil.
(b) The number of yeasts and moulds present per gramme of the soil.
(c) The number of aerobic organisms "growing at 37° C." present per gramme of the soil.

(B) *Anaerobes (Vegetative Forms and Spores).*—
16. Pour similar sets of plates in glucose formate gelatine and agar and incubate in Bulloch's anaerobic apparatus.

(C) *Aerobes and Anaerobes (Spores Only).*—
17. Pipette 5 c.c. soil water into a sterile tube.
18. Place in the differential steriliser at 80° C. for ten minutes.
19. Pour two sets of four gelatine plates containing 0.1, 0.2, 0.5, and 1 c.c. respectively of the soil water; label and incubate at 20° C., one set aerobically, the other anaerobically in Bulloch's apparatus.
20. "Count" the plates (delay the enumeration as long as possible) and estimate the number of spores of aerobes and anaerobes respectively present per gramme of the soil.
21. Calculate the ratio existing between spores and spores + vegetative forms under each of the two groups, aerobic and anaerobic micro-organisms.

Qualitative Examination.—The qualitative examination of soil is usually directed to the detection of one or more of the following:
Members of the Coli-typhoid group.
Streptococci.
Bacillus anthracis.
Bacillus tetani.
Bacillus œdematis maligni.
The nitrous organisms.
The nitric organisms.
1. Transfer the remainder of the soil water (88 c.c.) to a sterile Erlenmeyer flask by means of a sterile syphon.
2. Fix up the filtering apparatus as for the qualitative examination of water, and filter the soil water.
3. Suspend the bacterial residue in 5 c.c. sterile bouillon (technique similar to that described for the water sample, *vide* pages 434-436).
Every cubic centimetre of suspension now contains the soil water from nearly 1 gramme of earth.

The methods up to this point are identical no matter which organism or group of organisms it is desired to isolate; but from this stage onward the process is varied slightly for each particular bacterium.

I. The Coli-typhoid Group.—

II. Streptococci.—

III. Bacillus Anthracis.—

IV. Bacillus Tetani.—

The methods adopted for the isolation of these organisms are identical with those already described under water (page 437 *et seq.*).

V. Bacillus Œdematis Maligni.—Method precisely similar to that employed for the B. tetani.

VI. The Nitrous Organisms.—

1. Take ten tubes of Winogradsky's solution No I (*vide* page 198) and number them consecutively from 1 to 10.

2. Inoculate each tube with varying quantities of the material as follows:

To tube No. 1 add 1.0 c.c. of the soil water. To tube No. 2 add 0.1 c.c. of the soil water. To tube No. 3 add 1.0 c.c. from Capsule I. To tube No. 4 add 0.1 c.c. from Capsule I. To tube No. 5 add 1.0 c.c. from Capsule II. To tube No. 6 add 0.1 c.c. from Capsule II. To tube No. 7 add 1.0 c.c. from Capsule III. To tube No. 8 add 0.1 c.c. from Capsule III. To tube No. 9 add 1.0 c.c. from Capsule IV. To tube No. 10 add 0.1 c.c. from Capsule IV.

Label and incubate at 30° C.

VII. The Nitric Organisms.—

3. Take ten tubes of Winogradsky's solution No II, number them consecutively from 1 to 10 and inoculate with quantities of soil water similar to those enumerated in section VI step 2. Label and incubate at 30° C.

4. Examine after twenty-four and forty-eight hours' incubation. From those tubes that show signs of growth make subcultivations in fresh tubes of the same medium and incubate at 30° C.

5. Make further subcultivations from such of those tubes as show growth, and again incubate.

6. If growth occurs in these subcultures, make surface smears on plates of Winogradsky's silicate jelly (*vide* page 198).

7. Pick off such colonies as make their appearance and subcultivate in each of these two media.

TESTING FILTERS.

Porcelain filter candles are examined with reference to their power of holding back *all* the micro-organisms suspended in the fluids which are filtered through them, and permitting only the passage of germ-free filtrates. In order to determine the freedom of the filter from flaws and cracks which would permit the passage of bacteria no matter how perfect the general structure of the candle might be, the candle must first be attached by means of a long piece of pressure tubing, to a powerful pump, such as a foot bicycle pump, fitted with a manometer. The candle is then immersed in a jar of water and held completely submerged whilst the internal pressure is gradually raised to two atmospheres by the action of the pump. Any crack or flaw will at once become obvious by reason of the stream of air bubbles issuing from it.

The examination for permeability is conducted as follows:

Apparatus Required:

Filtering apparatus: The actual filter candle that is used must be the one it is intended to test and must be previously carefully sterilised; the arrangement of the apparatus will naturally vary with each different form of filter, one or other of those already described (*vide* pages 42-48).

Plate-levelling stand.
Case of sterile plates.
Case of sterile pipettes, 10 c.c. (in tenths).
Case of sterile pipettes, 1 c.c. (in tenths).
Tubes of nutrient gelatine.
Flask containing sterile normal saline solution.
Sterile measuring flask, 1000 c.c. capacity.

Method.—

1. Prepare surface cultivations, on nutrient agar in a culture bottle, of the Bacillus mycoides, and incubate at 20° C., for forty-eight hours.

2. Pipette 5 c.c. sterile normal saline into the culture bottle and emulsify the entire surface growth in it.

3. Pipette the emulsion into the sterile measuring flask and dilute up to 1000 c.c. by the addition of sterile water.

4. Pour the emulsion into the filter reservoir and start the filtration.

5. When the filtration is completed, pour six agar plates each containing 1 c.c. of the filtrate.

6. Incubate at 37° C. until, if necessary, the completion of seven days.

7. If the filtrate is not sterile, subcultivate the organism passed and determine its identity with the test bacterium before rejecting the filter—since the filtrate may have been accidentally contaminated.

8. If the filtrate is sterile, resterilise the candle and repeat the test now substituting a cultivation of B. prodigiosus—a bacillus of smaller size.

9. If the second test is satisfactory, test the candle against a cultivation of a very small coccus, *e. g.*, Micrococcus melitensis, in a similar manner; in this instance continuing the incubation of cultivations from the filtrate for fourteen days.

TESTING OF DISINFECTANTS.

Methods have already been detailed (page 310) for the purpose of studying the vital resistance offered by micro-organisms to the lethal effect of germicides. But it frequently happens that the bacteriologist has to determine the relative efficiency of "disinfectants" from the standpoints of the sanitarian and commercial man rather than from the research worker's point of view. In pursuing this line of investigation, it is convenient to compare the efficiency, under laboratory conditions, of the proposed disinfectant with that of some standard germicide, such as pure phenol. In so doing, and in order that the work of different observers may be compared, conditions as nearly uniform as possible should be aimed at. The method described is one that has been in use by the writer for many years past, modified recently by the adoption of some of the recommendations of the Lancet Commission on the Standardisation of Disinfectants—particularly of the calculation for determining the phenol coefficient.

This method has many points in common with that modification of the "drop" method known as the Rideal-Walker test.

General Considerations.—

These may be grouped under three headings: Test Germ, Germicide, and Environment.

1. *Test Germ.*—**B. coli.**

As disinfectants are tested for sanitary purposes, it is obvious that a member of the coli-typhoid group should be selected as the test germ. B. coli is selected on account of its relative nonpathogenicity, the ease with which it can be isolated and identified by different observers in various parts of the world, the stability of its fundamental characters, and evenness of its resistance when utilised for these tests; finally since the colon bacillus is an organism which is slightly more resistant to the lethal action of germicides than the more pathogenic members of this group, a margin of safety is introduced into the test which certainly enhances its value.

B. coli should be recently isolated from a normal stool, and plated at least twice to ensure the purity of the strain; and a stock agar culture prepared which should be used throughout any particular test. For any particular experiment prepare a smear culture on agar and incubate at 37° C. for 24 hours anaerobically. Then emulsify the whole of the surface growth in 10 c.c. of sterile water. Transfer the emulsion to a sterile test-tube with some sterile glass beads and shake thoroughly to ensure homogenous emulsion. Transfer to a centrifuge tube and centrifugalise the emulsion to throw down any masses of bacteria which may have escaped the disintegrating action of the beads. Pipette off the supernatant emulsion for use in the test.

2. *Germicide.—*

a. *Disinfectant to be tested.—*

The first essential point is to test the unknown disinfectant, which may be referred to as germicide-x, on the lines set out on page 311 to determine its inhibition coefficient.

This constant having been fixed, prepare various solutions of germicide-x with sterilised distilled water by accurate volumetric methods, commencing with a solution somewhat stronger than that representing the inhibition coefficient. The solutions must be prepared in fairly large bulk, not less than 5 c.c. of the disinfectant being utilised for the preparation of any given percentage solution.

b. *Standard Control.*—**Phenol.**

The standard germicide used for comparison should be one which is not subject to variation in its chemical composition, and the one which has obtained almost universal use is Phenol.

The following table shows the effect of different percentages of carbolic acid upon B. coli for varying contact times, compiled from an experiment conducted under the standard conditions referred to under Environment. The results closely correspond to those recorded by the Lancet Commission on Disinfectants, 1909.

Percentage of phenol	Contact time in minutes.							
	2-1/2	5	10	15	20	25	30	35
1.20	-	-	-	-	-	-	-	-
1.10	-	-	-	-	-	-	-	-
1.0	+	-	-	-	-	-	-	-
0.9	+	-	-	-	-	-	-	-
0.85	+	+	-	-	-	-	-	-
0.80	+	+	+	-	-	-	-	-
0.75	+	+	+	+	-	-	-	-
0.7	+	+	+	+	+	-	-	-
0.65	+	+	+	+	+	+	+	-

- = No growth, *i. e.*, bacteria killed.
+ = Growth, *i. e.*, bacteria still living.

From this it will be seen that the following percentage solutions will need to be prepared, namely: 1.1 per cent., 1.0 per cent., 0.9 per cent., 0.75 per cent., 0.7 per cent., as controls for each experiment.

Prepare solutions of varying percentages by weighing out the quantity of carbolic acid required for each and dissolving in 100 c.c. of pure distilled water in an accurately standardised measuring flask. The solutions must be prepared freshly as required each day.

Environment.—

a. *General.—*

Close the windows and doors of the laboratory in which the investigation is carried out, to avoid draughts. Flush over the work bench and adjacent floor with 1:1000 solution of corrosive sublimate. Caution the assistant, if one is employed, to avoid unnecessary movement or speech.

b. *Contact Temperature,* **15-18° C.—**

This is the temperature at which contact between the germicide and the test germ takes place, and is of importance, since some germicides (*e. g.*, Phenol) appear to be more powerful at high temperatures. 18° C.—practically the ordinary room temperature—is a temperature at which the multiplication of B. coli is a comparatively slow process, but variation of a degree above this temperature or of two or three degrees below is of no moment. If the room temperature is below 15° C. when the experiments are in progress, arrange a water-bath regulated at 18° C. for the reception of the tubes containing the mixture of germ and germicide; if above 19° C. immerse the tubes in cold water, to which small pieces of ice are added from time to time to prevent the temperature rising above 18° C.

c. Relative Proportional Bulk of Test Germ and Germicide, **50:1.**—

Five cubic centimetres is a convenient amount of germicidal solution to employ, and to this 0.1 c.c. of the emulsion of test germ should be added.

d. Bulk of Sample Removed from Germ + Germicide Mixture at Each of the Time Periods, **0.1 c.c.**—

This is sufficient to afford a fair sample of the germ content of the mixture, and at the same time is insufficient to exert any inhibitory action when transferred to the subculture medium.

e. Subculture Medium. **Bile Salt Broth.**—

A *fluid* medium is essential in order to obtain immediate dilution of the germicide carried over; at the same time it is advantageous to employ a selective medium which favours the growth of the test germ to the exclusion of organisms likely to contaminate the preparation, and if possible one which affords characteristic cultural appearances.

Bile Salt Broth (page 180) combines these desiderata; it permits only the growth of intestinal bacteria, whilst the formation of an acid reaction and the production of gas in subcultures prepared from the germ-germicide mixture is fairly complete evidence of the presence of living B. coli.

The amount of medium present in each test-tube is a matter of importance, since the medium not only provides pabulum for the test germ, but also acts as a diluent to the germicide, to reduce its strength below its inhibition coefficient. For routine work each subculture tube contains 10 c.c. of medium, but it is obvious that if germicide-x possesses an inhibition coefficient of 0.1 per cent. the addition of 0.1 c.c. of a 10 per cent. solution to 10 c.c. of medium would effectually prevent the subsequent growth of the test germ after a contact period insufficient to destroy its vitality. Hence the preliminary tests may in some instances indicate the necessity for the presence of 12 c.c., 15 c.c. or more of the fluid medium in the culture tubes.

f. Incubation Temperature, **37° C.**—

g. Observation Period of the Subcultivations, **Seven Days.**—

In order to determine whether or no the test germs have been destroyed, observations must always be continued—when growth appears to be absent—up to the end of seven days before recording "no growth."

h. Identification of the Organisms Developing in the Subcultivations after Contact in the Germ + Germicide Solution.—

This is based on the naked eye characters of the growth in the bile salt broth, supplemented where necessary by plating methods, further subcultivations upon carbohydrate media and agglutination experiments. The sign (+) is used to indicate that growth of the test organism occurred in the subcultivations, and the sign (-) to indicate that the test germs have been destroyed and no subsequent growth has taken place.

Method.—

Apparatus Required:

Sterile test-tubes (narrow, not exceeding 1.3 cm. diameter).
Test-tube rack (Fig. 219).
Sterile graduated pipettes in case, 1 c.c. (in tenths).
Sterile graduated pipettes in case, 5 c.c. (in c.c.).
Circular rubber washers, 2.5 cm. diameter with central hole, sterilised by boiling immediately before use, then transferred to sterilised glass double dish.
Electric signal clock or stop watch.
Sterile forceps.
Sterilised glass beads.
Shaking machine.
Grease pencil.

Material Required:

Percentage solutions of germicide-x (*vide* page 481).
Percentage solutions of pure phenol (*vide* page 482).
Aqueous emulsion of B. coli (*vide* page 481).
Tubes of bile salt broth.

Preliminary Tests.—

a. Inhibition Coefficient.—

Determine the lowest percentage of germicide-x which inhibits growth of B. coli in the bile salt broth, and the highest percentage which fails to inhibit (page 311). On the result of this experiment determine the bulk of medium required in the subculture tubes and the percentage solutions to be employed in the trial trip. Assuming the inhibition coefficient to be 1:1000, it will be quite safe to employ the ordinary culture tubes containing 10 c.c. medium in the subsequent experiments.

b. Trial Trip.—

Determine the lethal effect of a series of five solutions of germicide-x (say 1:100, 1:250, 1:300, 1:500, 1:600) at contact times of 2-1/2, 5, 25 and 30 minutes in the following manner:

1. Arrange five test-tubes marked A to E in the lower tier of the test-tube rack.
2. Into tube A pipette 5 c.c. germicide-x 1:100 solution.
Into tube B pipette 5 c.c. germicide-x 1:200 solution.
Into tube C pipette 5 c.c. germicide-x 1:300 solution.
Into tube D pipette 5 c.c. germicide-x 1:500 solution.

Into tube E pipette 5 c.c. germicide-x 1:600 solution.

3. Arrange 20 tubes of bile salt broth in the upper tier of the test-tube rack in two rows, those in the front row numbered consecutively from left to right 1-10, those in the back row 11-20.

4. Place a square wire basket of about 50 tubes capacity close to the left of the test-tube rack, for the reception of the inoculated tubes.

5. Take a sterile 1 c.c. pipette from the case, pick up a sterile rubber washer with forceps and push the point of the pipette into the central hole.

6. Put down the forceps on the bench with the sterile points projecting over the edge. Without taking the tube from the rack remove the cotton-wool plug from tube A, and lower the pipette, with the rubber washer affixed, on to the open mouth of the tube; with the help of the forceps to steady the washer, push the pipette on through the hole until the point of the pipette has reached to within a few millimetres of the bottom of the tube (see fig. 219).

7. Adjust in the same way a pipette and a washer in the mouth of each of the other tubes, B, C, D and E.

8. Set the electric signal clock to ring for the commencement of the experiment and at subsequent intervals of 2-1/2, 5, 25 and 30 minutes.

9. Take up 0.5 c.c. of B. coli emulsion in sterile pipette graduated in tenths of a cubic centimetre and stand by.

10. As soon as the bell rings lift the pipette from tube A with the left hand and from the charged pipette held in the right hand deliver 0.1 c.c. of B. coli emulsion into the 1:100 solution. Then replace the pipette and washer.

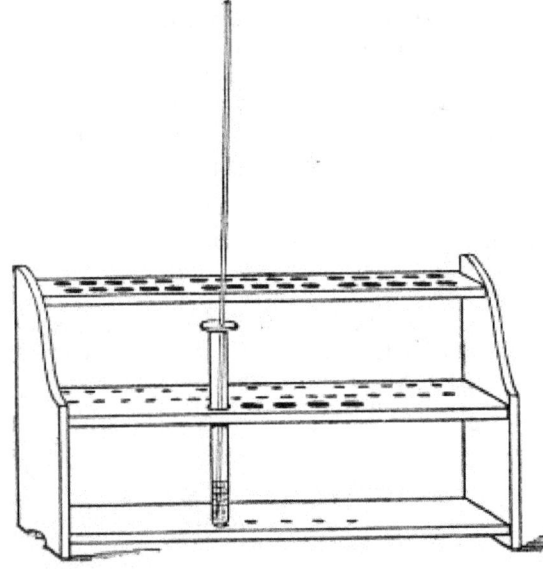

Fig. 219.—Test-tube rack.

11. Raise the tube with the left hand and shake it to mix germ and germicide, whilst returning the delivery pipette in the right hand.

12. Repeat the process with tubes B, C, D and E; then drop the infected delivery pipette in the lysol jar. The inoculation of the five tubes can be carried out very expeditiously, but a period of 10 seconds must be allowed for each tube.

13. When the bell rings at 2-1/2 minutes blow through the pipette in tube A (this agitates the germ + germicide mixture and ensures the collection of a fair sample); allow the mixture to enter the pipette, and as the column of fluid extends well above the terminal graduation, the right forefinger adjusted over the butt-end of the pipette before it is lifted will retain more than 0.1 c.c. of the mixture within the bore when the point of the pipette is clear of the fluid in the tube. Touch the point of the pipette on the inner wall of the tube, and allow any excess of fluid to escape, only retaining 0.1 c.c. in the pipette.

14. At the same time, with the left hand remove Bile Salt Tube No. 1 from the upper tier of the rack, take out the cotton-wool plug with the hand already holding the pipette (the relative positions of pipette, plug and culture tubes being practically the same as those of platinum loop, plug and culture tube shown in Fig. 68, page 74).

15. Insert the point of the pipette into the subculture tube, and blow out the mixture into the medium—replug the tube and drop it into the wire basket. Replace the washer-pipette in tube A.

As soon as the point of the pipette has entered the mouth of tube A it may be released, since it has already been so adjusted that it just clears the bottom of the test-tube, and the elastic washer will prevent any damage to the tube.

Steps 13, 14 and 15 occupy on an average 10 seconds.

16. Repeat steps 13, 14 and 15 with each of the other tubes B, C, D and E.

17. Repeat these various steps 13-16 when the bell rings at 5, 25 and 30 minutes.

18. Place all the inoculated tubes in the incubator at 37° C.

19. Examine the tubes at intervals of 24 hours, and record the results in tabular form as shown in Table page 491 (the figures in the squares indicate the number of hours at which the changes in the medium due to the growth of B. coli first appeared).

20. If a consideration of the tabulated results indicates strengths of Germicide-x lethal at 2-1/2 and 30 minutes the final test can be arranged, but if this result has not been attained, sufficient evidence will probably be available to enable a second trial test to be planned which will give the required information.

Final Test.—

c. Determination of Phenol Coefficient.—

X-Disinfectant.—This comprises two distinct tests, one of the Germicide-x, the other of the standard phenol.

1. Arrange five test-tubes clearly marked in the lower tier of the rack.
2. Pipette into each 5 c.c. respectively of the five percentage solutions of x-disinfectant which the trial run has already shown will include those affording lethal values at 2-1/2 and 30 minutes.
3. Arrange 20 tubes of bile salt broth in the upper tier of the test-tube rack in two rows, those in the front row numbered consecutively from left to right 1-10, those in the back row 11-20.
4. Arrange further 20 tubes of bile salt broth numbered 21-40 in two rows in a second smaller rack which can be stood on the upper tier of the rack as soon as the first 20 tubes have been inoculated.
5. Place a square wire basket of about 50 tube capacity close to the left of the test-tube rack, for the reception of the inoculated tubes.
6. Adjust a sterile 1 c.c. pipette in the mouth of each of the tubes, A, B, C, D and E, by means of a washer, as previously described.
7. Set the electric signal clock to ring for the commencement of the experiment and subsequently at 2-1/2, 5, 10, 15, 20, 25, 30 and 35 minutes.
8. Complete precisely as indicated in Trial Runs, steps 9-19.

Control Phenol.—

Immediately the subculture tube from the 30-minute contact period have been inoculated, carry out a precisely similar experiment, in which five percentage strengths of Phenol, (*e. g.*, 1.1, 1.0, 0.9, 0.75, 0.7) are arranged in the lower tier of the test-tube rack in place of the five strengths of Germicide-x.

Calculate the phenol coefficient by the following method:

(a) Divide the figure representing the percentage strength of the weakest lethal dilution of the carbolic acid control at the 2-1/2-minute contact period by the figure representing the percentage strength of the weakest lethal dilution of the x-disinfectant at the same period. The quotient = phenol coefficient at 2-1/2 minutes.

(b) Similarly obtain the phenol coefficient at 30 minutes contact period.

(c) Record the mean of the two coefficients obtained in (a) and (b) as the *mean phenol coefficient*, or simply as the **Phenol Coefficient**. The details of the Final Test of an actual determination are set out in the accompanying table.

TABLE 27

Organism employed, B. Coli Communis.
Culture Medium, Nutrient Agar (+10). Age, 24 hrs. Temp. of Incubation, 37°C.

Quantities used { Culture } { Emulsion } Emulsion 0.1 c.c. + 5 c.c. Germicide.

Room Temperature during Experiments, 17°C.											
Germicide	Strength	Time of exposure								Incubation	
		2-1/2	5	10	15	20	25	30	35	Time	Temp.
1 Germicide-x	4%	—	—	—	—	—	—	—	—	7 days.	37°C.
2 Germicide-x	3%	48	—	—	—	—	—	—	—	7 days.	37°C.
3 Germicide-x	2%	24	24	24	24	48	72	...	...	7 days.	37°C.
4 Germicide-x	1%	24	24	24	24	72	24	72	...	7 days.	37°C.
5 Germicide-x	0.5%	24	24	24	24	24	24	24	24	24 hours.	37°C.
1 Phenol	1.10%	—	—	—	—	—	—	—	—	7 days.	37°C.
2 Phenol	1.00%	24	...	...	...	...	...	...	...	7 days.	37°C.
3 Phenol	0.75%	24	24	24	24	48	...	...	...	7 days.	37°C.
4 Phenol	0.70%	24	24	24	24	24	72	...	...	7 days.	37°C.
5 Phenol	0.65%	24	24	24	24	24	48	24	24	2 days.	37°C.

$$\text{Phenol Coefficient} = \frac{((1.10/4.00) + (0.7/2.0))}{2} = \frac{0.27 + 0.35}{2} = \frac{.62}{2} = 0.31$$

APPENDIX.

METRIC AND IMPERIAL SYSTEMS OF WEIGHTS AND MEASURES.

The initial unit of the metric system is the Metre (*m.*) or unit of length, representing one-fourth-millionth part of the circumference of the earth round the poles.

The unit of mass is the Gramme (*g.*), and represents the weight of one cubic centimetre of water at its maximum density (viz. 4° C. and 760 mm. mercury pressure).

The unit of the measure of capacity is the Litre (*l.*), and represents the volume of a kilogramme of distilled water at its maximum density.

The decimal subdivisions of each of the units are designated by the Latin prefixes *milli* = 1/1000; *centi* = 1/100; *deci* = 1/10; the multiples of each unit by the Greek prefixes *deka* = 10; *hecto* = 100; *kilo* = 1000; *myria* = 10,000.

For a comparison of the values of some of the more frequently employed expressions of the Metric System and the Imperial System, the following may be found convenient for reference:

Length:
1 millimetre (= 1 mm.) = 1/25 of an inch.
1 centimetre (= 1 cm.) = 2/5 of an inch.
1 inch (1") = 25 millimetres or 2-1/2 centimetres.

Mass:
1 milligramme (= 1 mg.) = 0.01543 grain (or approximately 1/64 grain).
1 gramme (= 1 g.) = 15.4323 grains.
1 "kilo" or kilogramme (= 1 kgm.) = 2 pounds, 3-1/4 ounces avoirdupois.
1 pound avoirdupois (= 1 lb.) = 453.592 grammes.
1 ounce avoirdupois (= 1 oz.) = 28.35 grammes.
1 grain = 0.0648 gramme or 64.8 milligrammes.

Capacity:
1 cubic centimetre (= 1 c.c.) = 16.9 minims imperial measure.
1 litre (= 1 *l.*) = 35.196 fluid ounces imperial measure.
1 fluid ounce imperial measure (= 1 ℥) = 28.42 cubic centimetres.
1 pint imperial measure (= 1 O.) = 568.34 cubic centimetres.
1 gallon imperial measure (= 1 C.) = 4.546 litres, or 10 pounds avoirdupois, of pure water at 62° F. and under an atmospheric pressure of 30 inches of mercury.

Factors for Converting from one System to the Other.

To convert grammes into grains	× 15.432.
To convert grammes into ounces avoirdupois	× 0.03527.
To convert kilogrammes into pounds	× 2.2046.
To convert cubic centimetres into fluid ounces imperial	× 0.0352.
To convert litres into fluid ounces imperial	× 35.2.
To convert metres into inches	× 39.37.
To convert grains into grammes	× 0.0648.
To convert avoirdupois ounces into grammes	× 28.35.
To convert troy ounces into grammes	× 31.104.
To convert fluid ounces into cubic centimetres	× 28.42.
To convert pints into litres	× 0.568.
To convert inches into metres	× 0.0254.

TABLE FOR THE CONVERSION OF DEGREES CENTIGRADE INTO DEGREES FAHRENHEIT.

X.° C. = ((9x/5) + 32)° F.

Cent.	Faht.	Cent.	Faht.	Cent.	Faht.
0	32.0	34	93.2	68	154.4
1	33.8	35	95.0	69	156.2
2	35.6	36	96.8	70	158.0
3	37.4	37	98.6	71	159.8
4	39.2	38	100.4	72	161.6

5	41.0	39	102.2	73	163.4
6	42.8	40	104.0	74	165.2
7	44.6	41	105.8	75	167.0
8	46.4	42	107.6	76	168.8
9	48.2	43	109.4	77	170.6
10	50.0	44	111.2	78	172.4
11	51.8	45	113.0	79	174.2
12	53.6	46	114.8	80	176.0
13	55.4	47	116.6	81	177.8
14	57.2	48	118.4	82	179.6
15	59.0	49	120.2	83	181.4
16	60.8	50	122.0	84	183.2
17	62.6	51	123.8	85	185.0
18	64.4	52	125.6	86	186.8
19	66.2	53	127.4	87	188.6
20	68.0	54	129.2	88	190.4
21	69.8	55	131.0	89	192.2
22	71.6	56	132.8	90	194.0
23	73.4	57	134.6	91	195.8
24	75.2	58	136.4	92	197.6
25	77.0	59	138.2	93	199.4
26	78.8	60	140.0	94	201.2
27	80.6	61	141.8	95	203.0
28	82.4	62	143.6	96	204.8
29	84.2	63	145.4	97	206.6
30	86.0	64	147.2	98	208.4
31	87.8	65	149.0	99	210.2
32	89.6	66	150.8	100	212.0
33	91.4	67	152.6		

TABLE FOR THE CONVERSION OF DEGREES FAHRENHEIT INTO DEGREES CENTIGRADE.

$X° F. = (5(x - 32))/9° C.$

Faht.	Cent.	Faht.	Cent.	Faht.	Cent.	Faht.	Cent.	Faht.	Cent.
32	0.	68	20.0	104	40.0	140	60.0	176	80.0
33	0.6	69	20.6	105	40.6	141	60.6	177	80.6
34	1.1	70	21.1	106	41.1	142	61.1	178	81.1
35	1.7	71	21.7	107	41.7	143	61.7	179	81.7
36	2.2	72	22.2	108	42.2	144	62.2	180	82.2

37	2.8	73	22.8	109	42.8	145	62.8	181	82.8
38	3.3	74	23.3	110	43.3	146	63.3	182	83.3
39	3.9	75	23.9	111	43.9	147	63.9	183	83.9
40	4.4	76	24.4	112	44.4	148	64.4	184	84.4
41	5.0	77	25.0	113	45.0	149	65.0	185	85.0
42	5.6	78	25.6	114	45.6	150	65.6	186	85.6
43	6.1	79	26.1	115	46.1	151	66.1	187	86.1
44	6.7	80	26.7	116	46.7	152	66.7	188	86.7
45	7.2	81	27.2	117	47.2	153	67.2	189	87.2
46	7.8	82	27.8	118	47.8	154	67.8	190	87.8
47	8.3	83	28.3	119	48.3	155	68.3	191	88.3
48	8.9	84	28.9	120	48.9	156	68.9	192	88.9
49	9.4	85	29.4	121	49.4	157	69.4	193	89.4
50	10.0	86	30.0	122	50.0	158	70.0	194	90.0
51	10.6	87	30.6	123	50.6	159	70.6	195	90.6
52	11.1	88	31.1	124	51.1	160	71.1	196	91.1
53	11.7	89	31.7	125	51.7	161	71.7	197	91.7
54	12.2	90	32.2	126	52.2	162	72.2	198	92.2
55	12.8	91	32.8	127	52.8	163	72.8	199	92.8
56	13.3	92	33.3	128	53.3	164	73.3	200	93.3
57	13.9	93	33.9	129	53.9	165	73.9	201	93.9
58	14.4	94	34.4	130	54.4	166	74.4	202	94.4
59	15.0	95	35.0	131	55.0	167	75.0	203	95.0
60	15.6	96	35.6	132	55.6	168	75.6	204	95.6
61	16.1	97	36.1	133	56.1	169	76.1	205	96.1
62	16.7	98	36.7	134	56.7	170	76.7	206	96.7
63	17.2	99	37.2	135	57.2	171	77.2	207	97.2
64	17.8	100	37.8	136	57.8	172	77.8	208	97.8
65	18.3	101	38.3	137	58.3	173	78.3	209	98.3
66	18.9	102	38.9	138	58.9	174	78.9	210	98.9
67	19.4	103	39.4	139	59.4	175	79.4	211	99.4
								212	100.0

Percentage Formula for addition of salts, etc., to completed media.

Formula for preparing any desired percentage of a given salt, etc., in tubed media; *e. g.*, to make 4 per cent. solution of KNO_3 in a series of tubes of broth each containing 10 c.c. of medium, when there is already available a 25 per cent. stock aqueous solution of potassium nitrate.

$$\frac{(N + X) Y}{\text{------}} = \frac{A (X)}{\text{-----}}$$

N = number of cubic centimetres contained in each tube.
X = amount of stock solution to be added to each tube.
Y = percentage required in the medium.
A = percentage of stock solution.
Then

$$\frac{(10 + X)\,4}{100} = \frac{25X}{100}$$

Therefore, $40 + 4X = 25X$.
Therefore, $21X = 40$.
$X = 1.9$ c.c.
This allows for solution added to the original bulk of medium.
Therefore, 10 c.c. broth + 1.9 c.c. of a 25 per cent. aqueous solution KNO_3 makes 11.9 c.c. medium containing 4 per cent. KNO_3.

TABLES FOR PREPARING DILUTIONS

(of Serum, Disinfectants or other substances.)

In estimating the agglutinin content or *titre* of a serum, testing disinfectants and for many other purposes, it becomes necessary to prepare a series of dilutions of the material under examination, and in order to avoid unnecessary expenditure of labour it is convenient to adhere to some definite scale of increment, such for example as the following:

From dilutions of 1:10 to 1:80 rise by increments of 5.

From dilutions of 1:80 to 1:200 rise by increments of 10.

From dilutions of 1:200 to 1:400 rise by increments of 25.

From dilutions of 1:400 to 1:500 rise by increments of 50.

From dilutions of 1:500 to 1:1000 rise by increments of 100.

From dilutions of 1: 1000 to 1:5000 rise by increments of 250.

From dilutions of 1: 5000 to 1:10,000 rise by increments of 1000.

From dilutions of 1:10,000 to 1:100,000 rise by increments of 5000.

From dilutions of 1:100,000 to 1:1,000,000 rise by increments of 100,000.

When dealing with a substance of unknown powers—and this is especially true with regard to agglutinating sera—it is customary to run a preliminary test, using a few widely separated dilutions such as may be obtained in the following manner:

First Dilution—I.
1 c.c. serum + 9 c.c. normal saline solution = 10 per cent. solution or 1: 10 dilution (of which 1 c.c. contains 0.1 c.c. of the original serum).

When dealing with fluids other than serum the diluent is usually distilled water; whilst if the original substance is a solid the instructions would read:
1 gram o.s. + 10 c.c. distilled water = 10 per cent. solution, etc.

Second Dilution—II.
1 c.c. first dilution + 9 c.c. normal saline solution = 1 per cent. solution or 1: 100 dilution.

Third Dilution—III.
1 c.c. second dilution + 9 c.c. normal saline solution = 1 per mille solution or 1: 1000 dilution.

Fourth Dilution—IV.
1 c.c. second dilution + 9 c.c. normal saline solution = 0.1 per mille solution or 1: 10,000 dilution.

The following tables showing the secondary dilutions that can readily be prepared from each of these four primary dilutions for use in the subsequent determination of the exact *titre* will probably be found of service by those who are not ready mathematicians.

TABLES FOR PREPARING DILUTIONS.

TABLE I Using 10 % stock solution First dilution + Diluent	TABLE II Using 1% stock solution Second dilution + Diluent
1: 10 = 1 c.c. + 0 c.c.	1: 100 = 1 c.c. + 0 c.c.
1: 15 = 1 c.c. + 0.5 c.c.	1: 110 = 1 c.c. + 0.1 c.c.

1: 20 = 1 c.c. + 1.0 c.c.	1: 120 = 1 c.c. + 0.2 c.c.
1: 25 = 1 c.c. + 1.5 c.c.	
1: 30 = 1 c.c. + 2.0 c.c.	1: 130 = 1 c.c. + 0.3 c.c.
1: 35 = 1 c.c. + 2.5 c.c.	1: 140 = 1 c.c. + 0.4 c.c.
1: 40 = 1 c.c. + 3.0 c.c.	1: 150 = 1 c.c. + 0.5 c.c.
1: 45 = 1 c.c. + 3.5 c.c.	1: 160 = 1 c.c. + 0.6 c.c.
1: 50 = 1 c.c. + 4.0 c.c.	1: 170 = 1 c.c. + 0.7 c.c.
1: 55 = 1 c.c. + 4.5 c.c.	
1: 60 = 1 c.c. + 5.0 c.c.	1: 180 = 1 c.c. + 0.8 c.c.
1: 65 = 1 c.c. + 5.5 c.c.	1: 190 = 1 c.c. + 0.9 c.c.
1: 70 = 1 c.c. + 6.0 c.c.	1: 200 = 1 c.c. + 1.0 c.c.
1: 75 = 1 c.c. + 6.5 c.c.	————————————
1: 80 = 1 c.c. + 7.0 c.c.	1: 200 = 1 c.c. + 1.0 c.c.
————————————	1: 225 = 1 c.c. + 1.25 c.c.
1: 80 = 1 c.c. + 7.0 c.c.	1: 250 = 1 c.c. + 1.5 c.c.
1: 90 = 1 c.c. + 8.0 c.c.	1: 275 = 1 c.c. + 1.75 c.c.
1: 100 = 1 c.c. + 9.00 c.c.	1: 300 = 1 c.c. + 2.0 c.c.
1: 110 = 1 c.c. + 10.0 c.c.	1: 325 = 1 c.c. + 2.25 c.c.
1: 120 = 1 c.c. + 11.0 c.c.	1: 350 = 1 c.c. + 2.5 c.c.
	1: 375 = 1 c.c. + 2.75 c.c.
1: 130 = 1 c.c. + 12.0 c.c.	1: 400 = 1 c.c. + 3.0 c.c.
1: 140 = 1 c.c. + 13.0 c.c.	————————————
1: 150 = 1 c.c. + 14.0 c.c.	1: 400 = 1 c.c. + 3.0 c.c.
1: 160 = 1 c.c. + 15.0 c.c.	1: 450 = 1 c.c. + 3.5 c.c.
1: 170 = 1 c.c. + 16.0 c.c.	1: 500 = 1 c.c. + 4.0 c.c.
	————————————
1: 180 = 1 c.c. + 17.0 c.c.	1: 500 = 1 c.c. + 4.0 c.c.
1: 190 = 1 c.c. + 18.0 c.c.	1: 600 = 1 c.c. + 5.0 c.c.
1: 200 = 1 c.c. + 19.0 c.c.	1: 700 = 1 c.c. + 6.0 c.c.
————————————	
1: 200 = 1 c.c. + 19.0 c.c.	1: 800 = 1 c.c. + 7.0 c.c.
1: 225 = 1 c.c. + 21.5 c.c.	1: 900 = 1 c.c. + 8.0 c.c.
1: 250 = 1 c.c. + 24.0 c.c.	1: 1000 = 1 c.c. + 9.0 c.c.
1: 275 = 1 c.c. + 26.5 c.c.	————————————
1: 300 = 1 c.c. + 29.0 c.c.	1: 1000 = 1 c.c. + 9.0 c.c.
1: 325 = 1 c.c. +-31.5 c.c.	1: 2000 = 1 c.c. + 19.0 c.c.
1: 350 = 1 c.c. + 34.0 c.c.	1: 3000 = 1 c.c. + 29.0 c.c.

1: 375 = 1 c.c. + 36.5 c.c.	1: 4000 = 1 c.c. + 39.0 c.c.
1: 400 = 1 c.c. + 39.0 c.c.	1: 5000 = 1 c.c. + 49.0 c.c.
————————————	————————————
1: 400 = 1 c.c. + 39.0 c.c.	
1: 450 = 1 c.c. + 44.5 c.c.	
1: 500 = 1 c.c. + 49.0 c.c.	

TABLE III Using 0.1% stock solution Third dilution + Diluent	TABLE IV Using 0.01% stock solution Fourth Dilution + Diluent
1: 1000 = 1 c.c. + 0 c.c.	1: 10,000 = 1 c.c. + 0 c.c.
1: 1250 = 1 c.c. + 0.25 c.c.	1: 15,000 = 1 c.c. + 0.5 c.c.
1: 1500 = 1 c.c. + 0.5 c.c.	1: 20,000 = 1 c.c. + 1.0 c.c.
1: 1750 = 1 c.c. + 0.75 c.c.	1: 25,000 = 1 c.c. + 1.5 c.c.
1: 2000 = 1 c.c. + 1.0 c.c.	1: 30,000 = 1 c.c. + 2.0 c.c.
1: 2250 = 1 c.c. + 1.25 c.c.	1: 35,000 = 1 c.c. + 2.5 c.c.
1: 2500 = 1 c.c. + 1.5 c.c.	1: 40,000 = 1 c.c. + 3.0 c.c.
1: 2750 = 1 c.c. + 1.75 c.c.	1: 45,000 = 1 c.c. + 3.5 c.c.
1: 3000 = 1 c.c. + 2.0 c.c.	1: 50,000 = 1 c.c. + 4.0 c.c.
1: 3250 = 1 c.c. + 2.25 c.c.	1: 55,000 = 1 c.c. + 4.5 c.c.
1: 3500 = 1 c.c. + 2.5 c.c.	1: 60,000 = 1 c.c. + 5.0 c.c.
1: 3750 = 1 c.c. + 2.75 c.c.	1: 65,000 = 1 c.c. + 5.5 c.c.
1: 4000 = 1 c.c. + 3.0 c.c.	1: 70,000 = 1 c.c. + 6.0 c.c.
1: 4250 = 1 c.c. + 3.25 c.c.	1: 75,000 = 1 c.c. + 6.5 c.c.
1: 4500 = 1 c.c. + 3.5 c.c.	1: 80,000 = 1 c.c. + 7.0 c.c.
1: 4750 = 1 c.c. + 3.75 c.c.	1: 85,000 = 1 c.c. + 7.5 c.c.
1: 5000 = 1 c.c. + 4.0 c.c.	1: 90,000 = 1 c.c. + 8.0 c.c.
1: 5000 = 1 c.c. + 4.0 c.c.	1: 100,000 = 1 c.c. + 9.0 c.c.
1: 6000 = 1 c.c. + 5.0 c.c.	———————
1: 7000 = 1 c.c. + 6.0 c.c.	1: 100,000 = 0.1 c.c. + 0.9 c.c.
	1: 200,000 = 0.1 c.c. + 1.9 c.c.
1: 8000 = 1 c.c. + 7.0 c.c.	
1: 9000 = 1 c.c. + 8.0 c.c.	1: 300,000 = 0.1 c.c. + 2.9 c.c.
1: 10,000 = 1 c.c. + 9.0 c.c.	1: 400,000 = 0.1 c.c. + 3.9 c.c.
————————————	1: 500,000 = 0.1 c.c. + 4.9 c.c.
1: 10,000 = 1 c.c. + 9.0 c.c.	————————————
1: 15,000 = 1 c.c. + 14.0 c.c.	1: 500,000 = 0.1 c.c. + 4.9 c.c.
1: 20,000 = 1 c.c. + 19.0 c.c.	1: 600,000 = 0.1 c.c. + 5.9 c.c.

1: 25,000 = 1 c.c. + 24.0 c.c.	1: 700,000 = 0.1 c.c. + 6.9 c.c.
1: 30,000 = 1 c.c. + 29.0 c.c.	
————————————	1: 800,000 = 0.1 c.c. + 7.9 c.c.
	1: 900,000 = 0.1 c.c. + 8.9 c.c.
	1:1,000,000 = 0.1 c.c. + 9.9 c.c.

TEMPERATURE PRESSURE TABLE.

Temperature Centigrade	Mm. of Hg.	Pounds per sq. in. absolute pressure	Atmospheres
98°	707.1	13.7	0.93
99°	733.1	14.2	0.96
100°	760.0	14.7	1.00
101°	787.8	15.2	1.03
102°	816.0	15.8	1.07
103°	845.2	16.3	1.11
104°	875.4	16.9	1.15
105°	906.4	17.5	1.19
106°	938.3	18.1	1.23
107°	971.1	18.8	1.27
108°	1004.9	19.4	1.32
109°	1039.6	20.1	1.36
110°	1075.3	20.8	1.41
111°	1112.0	21.5	1.46
112°	1149.8	22.2	1.51
113°	1188.6	22.9	1.56
114°	1228.4	23.7	1.61
115°	1269.4	24.5	1.67
116°	1311.4	25.3	1.72
117°	1354.6	26.2	1.78
118°	1399.0	27.0	1.84
119°	1444.5	27.9	1.90
120°	1491.2	28.8	1.96
121°	1539.2	29.7	2.02

122°	1588.4	30.7	2.09
123°	1638.9	31.7	2.15
124°	1690.7	32.7	2.22
125°	1743.8	33.7	2.29

TABLE FOR DESICCATION AT LOW TEMPERATURES IN VACUO.

Temperature Centigrade	Mm. of Hg.
21°	18.4
22°	19.6
23°	20.8
24°	22.1
25°	23.5
26°	24.9
27°	26.4
28°	28.0
29°	29.7
30°	31.5
31°	33.3
32°	35.3
33°	37.3
34°	39.5
35°	41.7
36°	44.1
37°	46.6
38°	49.2
39°	51.9
40°	54.8
41°	57.8
42°	61.0
43°	64.3
44°	67.7
45°	71.3

46°	75.1
47°	79.0
48°	83.1
49°	87.4
50°	91.9

ANTIFORMIN METHOD

For the detection of B. Tuberculosis.

Antiformin was introduced into bacteriological technique by Uhlenhuth in 1908 for the purpose of demonstrating tubercle bacilli when present in small numbers, in sputum or other material. It is a powerful oxidising agent and rapidly destroys most bacteria, but tubercle and other acid-fast organisms resist its lethal action for considerable periods, and upon this fact the method is based.

To prepare Antiformin measure out and mix:—

Eau de Javelle (Liquor sodæ chlorinatæ—B.P.) 50 c.c.
Sodic hydrate 15 per cent. aqueous solution 50 c.c.

Method.

1. Introduce the sputum or other material (e. g. milk deposit and cream; pus; minced gland or other organ; caseous material; broken down foci, etc.) into a sterile tube and then add an equal volume of antiformin.

2. Close the tube with a rubber cork and shake vigorously (a sample of antiformin that does not "foam" at this stage is of little use). Disintegration of the material at once starts, associated bacteria are destroyed and the mixture rapidly becomes a homogenous but turbid fluid—a process which may be hastened by:—

3. Placing the tube in the incubator at 37° C. for 30 minutes—shaking from time to time.

4. Centrifugalise the fluid thoroughly, at high speed.

5. Pipette off the supernatant fluid, fill up with sterile distilled water, cork the tube and shake to distribute the deposit throughout the water. Again centrifugalise.

6. Repeat steps 4 and 5 twice more.

7. Employ one portion of the final deposit to inoculate guinea pigs.

8. Plant the remainder of the deposit freely on Dorset's Egg medium; cap and incubate at 37°C.

Note.—If only microscopical films are needed, fill up the centrifuge tube with Ligroin (a petroleum ether) in place of sterile distilled water in step 5 and prepare the films from the *surface* of the fluid, to stain by the Ziehl-Neelsen process.